The ARRL Operating Manual

AF804216

WILFRED S. JOBBINS

Edited by
Robert Halprin, K1XA

non commercial
citizen of world.
ITU DOC FCC

The American Radio Relay League
Newington, CT USA 06111

Mallorytown 2 May 1983

Copyright © 1980 by

The American Radio Relay League, Inc.

Copyright secured under the Pan-American Convention

International Copyright secured

This work is Publication No. 44 of the Radio Amateur's Library, published by the League. All rights reserved. No part of this work may be reproduced in any form except by written permission of the publisher. All rights of translation are reserved.

Printed in USA

Quedan reservados todos los derechos

Library of Congress Catalog Card Number: 80-66645

ISBN: 0-87259-654-0

$5.00 in USA
$5.50 elsewhere

Foreword

Been on the air lately?

If so, you'll appreciate this completely revised *Operating Manual*, which furnishes all the basics (and then some) of making the most effective use of your Amateur Radio station to achieve maximum satisfaction and enjoyment. The scope of earlier manuals has been maintained, as has the diversity of the recent *Operating Guide*. Along with all the up-to-date trends in contesting, DX, traffic/emergency matters and the other traditional aspects of operating, this book contains the latest information on using the OSCAR satellites, uhf-vhf and fm, interfacing home computers and beginner's concerns, plus a special section on shortwave listening. So this publication will be of interest to all active hams, regardless of tenure.

Editor Halprin has recruited a team of experts to explain operating practices in an easily understood manner. Kathy Adrian, WB1AAE, coordinated the production end of things, making sure the book was put together in a cohesive format. In short, whatever your favorite area of operating, now or in the future, the *ARRL Operating Manual* will be a worthy addition to your Amateur Radio library.

Richard L. Baldwin, W1RU
General Manager

Contents

Chapter 1

Strange Magic

How do you explain to nonhams, filled to their gills with daily news of international tensions, how 4X4 QSLs get on the same wall with an SU? An EI9 near a GI3? That KG4 beside some CO2s . . .

Which world is for real? Oh, both worlds are for real, all right. Perhaps the one shows there's real hope for the other. — W9BRD

The Amateur Radio experience is aptly described in this quote by former *QST* "How's DX?" conductor Rod Newkirk. Ham radio supplies that necessary ingredient for friendship — something in common. Radio amateurs are goodwill ambassadors, spreading a special form of camaraderie that knows no territorial boundaries. In effect, have signal will travel.

Communicating with another unseen person, without wires, is a magical if not mystical excursion. That must be the basic appeal of Amateur Radio; why we operate and continue to come back for more. But isn't making a few contacts enough? Take the guy who worked Australia his first day on the air and then purged all his equipment; after all, he had reached the farthest point he ever would.

We beg to differ. It is written that if you don't operate, you are depriving yourself of 90% of the enjoyment of Amateur Radio. In essence, operating is what it's all about; at least it's what we're all about.

This book is for all amateurs (and would-be amateurs) who believe communication is indeed a spirited adventure. It's a manual for those who visited the operating world as tourists and found it so engrossing that they decided to stay. Let this volume serve, too, as a rather extended letter of introduction into additional facets of operating practices that you may have been interested in but never quite got acquainted with. Organized activities sponsored by ARRL make everyone feel at home.

Getting on the air is fun. It's good therapy; a legal and non-toxic respite, an oasis perhaps, in a hectic society. It permits you to become immersed in a completely different world, made up of electronic impulses and the exchange of specialized information, in a unique language. Positive reinforcement and recognition come from accomplishments of various sorts.

The mystic traveler through this realm has much to choose from; vivid and soul-stirring encounters with the ether. The uninitiated find it strange and hard to understand. But listen. One can enjoy the code like good music, because cw is, after all, an art form. In the aftermath of an emergency, it's the sense of competent service rendered, "the only game in town." It's the thrill of the hunt, as an amateur stalks a new country like a distant quarry, in a race with time before the signal is cut from the airwaves by the sun's ionospheric scythe. And it's the pride and respect that traffic handlers draw from the smooth running of their net, of which each member is a part-owner. It's the tired-yet-satisfied feeling after an honest contest effort. It's the adrenalin charge and exquisite feeling of relief, after making it through that massive and never-ending pile-up, just when it seemed oh so hopeless. Or it's the basic realization that someone on the other side of town or the other side of the globe thinks a lot like you do. The diverse Amateur Radio subcultures abound with personal enrichment.

Then there was the story of two colleagues on a weekly traffic circuit, which was once chronicled in *QST*: It was a neat arrangement; a station reeling off messages one after another. The other station would just tap the dit side of his paddle signifying, "I got it man, give me the next one." When they finished their message handling, they'd chew the rag for a bit, time permitting. Being contemporaries, it was natural. As time wore on, they were able to develop the ability to sense each other's moods, just in the way the code left their fingertips; ESP may result when two hams conduct a regular sked for a long period of time. So they'd chat, about what's happening here, what's happening there, about the mind games being spun by various YLs, about the shape of things to come and of course, a debriefing of contest last. They were on the same wavelength, no doubt, to the degree that they were no longer conscious of actually sending. It was an almost instantaneous connection to thought processes.

Although ham radio is the armchair traveler's dream, those QSL cards, log entries and acquaintanceships come alive in a trip to a foreign locale, perhaps to a southern clime. Being on the DX-end of that immense radio pipeline is a study in metaphysics and in many respects, a look inward. Besides the culture shock of being in a different land, the operating aspect is such that each frequency band is a cornucopia of contacts. It doesn't stop until you pull the plug, which is done with reluctance

More importantly, the kindness and hospitality of the local amateurs tends to restore one's faith in humanity, auguring the day when all people can live together in peace. Heady stuff, eh? Well, anything's possible when the sun is brilliant, the tradewinds caress and the ships in the distance pass leisurely (it seems) enroute to exotic ports.

Like they say, "mucho gusto!"

On a serious note, a resident amateur knows that operating can change dramatically into something more than just a pleasant diversion. After a tropical storm, his shack will be in the forefront of relief efforts. Amateur Radio again serves as a veritable electromagnetic lifeline for a community in need. The next voice you hear will be that of the Prime Minister . . .

All this is a far cry from the initial novelty of talking to another station a hundred miles down the pike "for nothing." The idea is that ham radio provides an endless variety of operating categories. If you avoid restricting yourself to one specialized area, you will find enough to engage your interest indefinitely, from first blush to twilight. The use of the top-notch standard operating procedures, as described herein, will indicate to your fellow amateurs that you are, simply, a *good* operator. And your understanding of the major operating techniques will make your operating even more enjoyable for all concerned.

A benign case of the Lake Erie swing was classified as "Martian Music" by one of the aforementioned uninitiated. The musical-performance concept has already been cited so it is gratifying to see the idea appreciated by an impartial observer. While the sun plays a crucial role in radio communication, a concerto has not emanated from that star's fourth planet. Not yet it hasn't — at least as far as we know — since there's been no mention in "Calls Heard" or its latterday equivalent. Presumably, it's all a matter of time.

The world of 200 meters and beneath is an exploration into inner space. Amateur Radio has been and will continue to be a sanctuary for the introverted and a way to alleviate same; to become a naturalized citizen, so to speak, of this planet. Astonishingly enough, blasting a modest quantity of rf into the sky aids in the discovery and development of the self. The maturing process is a perk of holding a license. (Dig the sophistication of the younger amateurs and this will become evident.) A course will reveal itself, but then it's "your over." Don't bother to ask what awaits at journey's end, for that is undeterminable in each story. Hamming can bring about a chain of events that provides direction in a rudderless existence. Furthermore, radio is more interesting than real life. Is there any other explanation why generations will bridge their gap on the amateur bands, while barely allowing an exchange of grunts at the dinner table? Oh, but that's another story. Put it this way: Amateur Radio provides a galvanizing lift, just like a late-inning home run, making it a whole new ball game.

We don't want to get ahead of the narrative, but all things being equal, voice operation is not without its parallel to ballads from another tract. It goes like this: Fine business, *B*roken-*O*ld-*B*ottles, you're down in the mud, see you down the coax or the sheet, reading the mail, loading up the skyhook, catching you on the low end, sending you a greenstamp, kicking on the afterburner, having an eyeball (with an 807 chaser), giving you a pair of nickels, throwing the big switch, going to the sandbox and so on. Our jargon is so rich, but this is one dessert that can be devoured without concern for calories, because it's good for you. It all starts when the authorities stamp your *PASS*port and you're ready to fly. Then you can hear your own call sign being sent by someone else in response to your electronic exhortation, and you can hear your own signal as repeated by an orbiting space satellite.

Yes, hamming is all so intriguing, creative and energy-efficient, no less. It's been said that an amateur amplifier consumes less energy than a 300-watt light bulb. And ham radio lets you become the program director of your own entertainment. Think of it. You are safely beyond the clutches of those who seek to unduly influence via the broadcast media. No commercials, no time constraints, no ratings. The only limitation is the technicolor frontier of your own imagination. We thought you'd like to know.

By the way, the "we" used here is *not* the editorial we for a change. It's been a group effort. The chapter contributors, plus the many amateurs who provided information and advice, gave their all to bring you a comprehensive publication devoted entirely to Amateur Radio operating. It will all unfold, before your very eyes, in the ensuing pages.

But what's this? In our zeal to get this book completed and on the book racks, did we forget to include a picture in this photobox?

No, nothing's missing.

Although you will find many pictures, graphics and charts throughout this edition, we can't furnish you with a photograph of the subject of this chapter. We can show people who use it, devices that generate and capture it, numbers and diagrams that refer to it and confirmation cards that prove it. The truth is, though, that the space above is all we can show you of radio.

Radio remains inaudible unless heterodyned to the audio range and transduced to acoustic waves. Even now it passes through you and this piece of paper at every conceivable angle, but it can't be felt or seen.

Yet, our version, Amateur Radio, saves human lives and enlivens even more.

What's the point? An invitation. We invite you to step beyond the prosaic and solemn technical specifications of Amateur Radio and tune into the myriad of doings *on the air;* that magical privilege of sending and receiving signals that traverse neighborhoods and nations, that vault high mountain ranges and deep ocean seas.

And something more. When you're out there, we invite you to take along the most important tool you can ever use in your enjoyment of anything — a sense of wonder.

Robert Halprin, K1XA
Editor

Chapter 2

Basic Amateur Radio

Getting started on the road of good ham radio operating will present unique puzzles to each of you. Some of the doubt, uncertainty, decisions and, yes, fun, have been experienced by all of us in the Amateur Radio community. The purpose of this chapter is to acquaint you with the basics of owning and operating a station and to make your entry-process into Amateur Radio operating as easy as possible.

Setting Up Your Station

There are at least a million combinations of gear, space, antennas and junk that come together to make an amateur station. It would be very rare indeed to find two that were exactly alike. In fact, it would be just as rare to find two amateurs who *agree* on what one should acquire to have a well-appointed station. So, what follows are some of the things you may wish to consider before making those all-important purchases.

Setting up a station is perhaps the most confusing and difficult task for a Novice. If your Novice instructor showed you how to construct dipoles, solder coax connectors and connect feedlines to radios, among other things, then your task is a little easier. But if like most classes, the emphasis was on passing the exam and not the practical aspects, you may not know where to begin or what to do once you have. You are not alone!

Ask For Help

Everyone starts out as a beginner and everyone has to learn. Most hams are pleased if you ask for help. If you don't know any local amateurs, contact your nearest radio club (for information on a club in your area, write to the Club & Training Department at ARRL headquarters). When the time comes for you to go out and plunk down your hard-earned cash for a radio, new or used, your Elmer (ham advisor) can be an invaluable source of knowledge and experience.

If you have "lots" of money to spend, your task is a relatively simple one. Any of the modern, state-of-the-art transceivers will serve you well. If you plan to purchase used gear, however, take your Elmer along. You'll want an experienced person to check the rig for selectivity (rejection of unwanted signals), sensitivity (good reception of weak signals) and stability (no drift in frequency — this applies to both receivers and transmitters). Other things you'll want to consider are calibration accuracy, internal filters (or lack thereof), versatility and convenience of operation.

There are a few accessories that will make your life in the shack a little easier. A 24-hour clock (or 12-hour at least) set for UTC (Universal Coordinated Time) will make logging and QSLing accurate. A SWR/wattmeter is helpful for making antenna adjustments and for tuning up properly. A keyer is nice, but not a necessity. From there, the list is endless. But by the time you start wishing for them, you'll *know* what other accessories you want.

Just remember, when setting up your shack aim for the most efficient arrange-

The Hall of Science Radio Club in New York City regularly conducts classes. (*WA2VOS photo*)

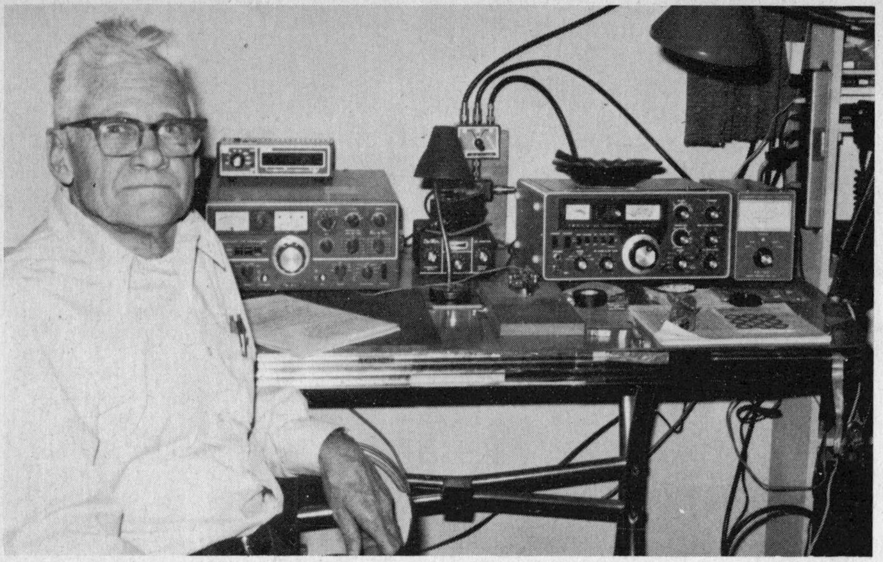

It's never too late to get your ticket says Walt, KA7CSK.

ment. Feedlines should be as short as possible, everything should be at your fingertips and there should still be room in front of you to write.

QRP vs. QRO

Power is an important aspect of operating that is appropriate to consider at the time of equipment purchases. There is a myth popular among amateurs, especially DXers, that maximum power (QRO) is necessary. It's nice sometimes — when the pile-ups are thick or propagation is poor, or in contests, where it might make the difference. Keeping control of an emergency frequency also requires a big signal. But for everyday operating, excessive power is hardly necessary and will do *nothing* to sharpen your operating skills. Power as low as 10 watts input (the ARRL definition of QRP) is sufficient to "work the world" — and have a good time doing it.

One of the nicest things about QRP transmitters is that they are relatively inexpensive to buy and likewise inexpensive and relatively easy to build. They make good Novice projects and are a good solution to budget problems. That way, you can use the receiver you acquired when you started studying or even put the bulk of your available funds into a good receiver now.

You do yourself a favor when you operate QRP, besides saving energy. It's not always easy to make the contacts you want with low power, but it teaches you skills you probably won't learn if you can "blast your way through" any pile-up at will. You'll learn to listen: what are the other station's operating habits; does he or she pick up the last call sign heard or the first? You'll also quickly learn the value of a good antenna. Ten watts to a beam with 10 dB of gain over a dipole will give you as good and probably better a signal than 100 watts into a random wire. The advantages are many. The big disadvantage is that if you're short on patience, QRP may cost you a few chewed fingernails or broken pencils!

CW Operation

Take some time and tune around the Novice bands. Hear that station whose fist is so poor that the letters run together and you can't copy anything? How about the one that sends CQ at 15 wpm, only to slow down to 3 wpm for the QSO? Or the one . . . but you get the idea.

To sound like a pro, take a bit of time and practice good cw procedures. Why not tape record your fist and then play it back to get an idea what you sound like to someone else? Work on the areas that need improvement. A good fist and skilled operating will make you *very* popular on the air.

Here are some items to keep in mind:

1) *Listen* before asking if a frequency is busy.

2) Send QRL? ("Is this frequency in use?") before calling CQ and then listen for an answer.

3) Keep your sending speed consistent with your copying ability.

4) Send short CQs and listen between each, rather than sending one l-o-n-g CQ. No one wants to talk to you badly enough that they will endure a 10-minute CQ call.

5) Don't over-identify during a QSO. Send your call a few times at the beginning — most Novices may need several chances to copy it accurately. But after that, you need only i-d every 10 minutes. Turn the transmission back to the other station by sending BK. That's much more conversational.

6) Learn the proper ending signals and Q-signals.

7) *Be courteous.* It may seem as if this

Correct CW Procedures

Establishing a contact — The best way to do this, especially at first, is to listen until you hear someone calling CQ and answer. If you hear a CQ, wait until the ham indicates he is listening, then call him, thus: K5KG K5KG DE K3YL K3YL $\overline{AR}$ ($\overline{AR}$ is equivalent to *over*).

In answer to your call, the called station will reply K3YL DE K5KG R . . .

That R (roger) means that he has received your call correctly. That's *all* it means — *received.* It *does not* mean (a) correct, (b) I agree (c) I will comply, etc. It is not sent unless everything from the previous transmission was received correctly. Perhaps K5KG heard someone calling him but didn't quite catch the call because of interference (QRM) or static (QRN). In this case, he might come back with QRZ? DE K5KG $\overline{AR}$ ("Who is calling me?").

Calling CQ — CQ means "I wish to contact *any* amateur station." Avoid the common operating discrepancy of calling CQ endlessly; it clutters up the air and drives off potential new friends. The typical CQ would go like this: CQ CQ CQ DE W1ASD W1ASD W1ASD K. The letter K is an invitation for any station to go ahead.

The QSO — During the contact, it is only necessary to identify your station once every 10 minutes. Keep the contact on a friendly and cordial level, remembering that the conversation is not private and many others, including nonamateurs, may be listening. Both on cw and phone, it is possible to be informal, friendly and conversational; this is what makes the Amateur Radio QSO enjoyable. During the contact, when you stand by, the recommended signal is $\overline{KN}$, meaning that you want only the contacted station to come back to you. If you don't mind someone else signing in, just K ("go") is sufficient.

Ending the QSO — When you decide to end the contact or the other ham expreses the desire to end it, don't keep talking. Briefly express your pleasure at having worked the other operator and then sign out: $\overline{SK}$ K3YL DE K5KG. If you are leaving the air, add CL to the end, right after your call sign.

These ending signals, which indicate to the casual listener the "status of the contact," establish Amateur Radio as a cordial and fraternal hobby. At the same time, they foster orderliness and denote organization. These signals have no legal standing; FCC regulations say little about our internal procedures. The procedures we ourselves adopt are even more important than that, because they indicate that we are not just a bunch of hobbyists playing around in random fashion, but that we are an established communications service (that we take pride in) with distinct and distinctive procedures tailored to our special needs.

Voice Equivalents to Code Procedure

Voice	*Code*	*Meaning*
over	$\overline{AR}$	after call to specific station
end of message	$\overline{AR}$	self-explanatory
wait, stand by	$\overline{AS}$	self-explanatory
roger	R	*all* received correctly
go	K	any station transmit
go only	$\overline{KN}$	addressed station only
clear	$\overline{SK}$	end of contact
closing station	CL	going off the air

A Hobby with No Barriers

One of the outstanding characteristics of hamming is that anyone can enjoy it. Amateur Radio has a basic appeal to the visually/physically handicapped: It provides a means of people-to-people contact on a basis of absolute equality. Ham radio has always been a natural way of making new friends, renewing old acquaintances and visiting — via radio — people and places all over the world. It is perfect for those with limited mobility. There are hundreds of handicapped individuals active in Amateur Radio. They have shown that with the proper training, patience and determination, almost any physical disability can be overcome.

Rex Kiser, WØGLU (foreground) and Dick Eichhorn, KBØAE (background) at the operating positions of club station WØZSW, located at the Courage Center, Golden Valley, Minnesota.

The HANDI-HAMS

There is an organized group which specializes in helping handicapped persons become a part of the hobby — the HANDI-HAMS. Bruce Humphrys, KØHR, is the director of the HANDI-HAMS; he described the program in January 1980 *QST*: "Courage Center is a comprehensive rehabilitation facility for persons with physical, speech, hearing and visual handicaps, based in Golden Valley, Minnesota. Besides the HANDI-HAM System, Courage Center offers a tremendously wide array of services for handicapped people . . . the HANDI-HAM System is one of the Center's services — serving more than 1000 people all over the world. The System provides three direct (and many indirect) services: (1) educational material, fraternity and close personal supervision; (2) Amateur Radio equipment on loan; and (3) specially designed devices for ease of station control. The System relies on a trained cadre of radio amateurs to help provide these services. We not only help new students to get their first license, but also help handicapped licensed hams to upgrade."

Every summer, under the guidance of KØHR, a radio camp is held at the nearby Camp Courage. FCC exams are conducted at the completion of each of the two sessions.

Further information can be obtained by contacting the Courage HANDI-HAMS, 3915 Golden Valley Rd., Golden Valley, MN 55422.

Getting the Ticket

All candidates for an amateur license must prove their ability to send and receive the international Morse code, and must pass a test on radio regulations and theory. Blind applicants may use a typewriter or a Braille-writer for the code test, or call out the words to be recorded by someone else. Similarly, answers to the written exam may be dictated to a sighted person, under the provisions of Section 97.28(d) of the U.S. amateur regulations. In addition, the FCC does not deny anyone the right to take the examination because of a disability that prevents travel to an FCC examination point. In such cases, a Commission examiner may conduct the exam (or appoint a volunteer examiner) at a location other than the regular examination point. The back of the applicant's Form 610 should be completed by a physician, certifying that the applicant is unable to appear at the normal test location because of a protracted disability preventing travel. Braille exams are also available to blind applicants by advance arrangement with the FCC District Engineer-In-Charge.

ARRL Involvement

The ARRL Blind and Handicapped Program provides sources of information that enable handicapped people to prepare for FCC amateur exams. ARRL hq. serves as an information center and clearinghouse for this purpose. The Blind and Handicapped Office maintains a current reference file containing lists of operating aids and clubs, and other pertinent information. *QST* publishes technical and general-interest articles of special relevance to the handicapped (reprints are available from Hq.).

Membership in the League is available to the visually handicapped at a reduced rate, without *QST*. This gives already-licensed sightless members an opportunity to take an active role in ARRL activities, vote in ARRL Director and SCM elections, run for League offices, and otherwise share in the privileges of full membership. Unlicensed blind persons, too, may show their support for the work of ARRL through an associate membership. For more information, write the Membership Services Department at ARRL headquarters.

QST is available on a flexible disc, directly from the Library of Congress, Division for the Blind and Handicapped, 1291 Taylor St. N.W., Washington, DC 20542. — *Rich Palm, K1CE*

old-fashioned virtue has gone out of style after a quick tune around the bands. But this is not necessarily so. You'll get a lot further in the long run if you are always courteous and setting a good example is an effective way of influencing your fellow hams. Courtesy pays — it makes the bands pleasant for all of us.

Your First Contact

Are your plams sweaty, knees weak, hands shaking and is there a queasy herd of butterflies (wearing spikes, perhaps) performing maneuvers in your stomach? If that description fits, chances are you're facing your first Amateur Radio QSO. If so, take heart! Although few may admit it, the vast majority of hams felt the same way before firing up the rig for that first contact.

Although nervousness is a natural reaction, there are some preparations you can make so that things go smoothly and a few of those butterflies are released. Practicing QSOs with a code practice oscillator is a good way to familiarize yourself with the typical QSO format. You can practice solo, with a friend or perhaps with several members of the local radio club.

Another thing that some people find helpful is to write down information in advance. While spelling in code may pose no problem when you are calm, nerves make it tough for some to remember where they were in a word, or for that matter, how to spell it.

One nervous Novice made notes of her own QSO data on index cards. One card contained her location, another her name (no kidding!); but after a half dozen contacts, she gradually forgot to use the cards and could spell her address in perfect code — asleep.

So will you.

If you are more concerned about copying than sending, a good security blanket is to have an experienced operator in the room copy along. That way you can be reasonably assured that you'll copy everything that the other station sends. You may also find that, much to your surprise, you do just fine on your own.

Perhaps the easiest way to slide into QSOs is to make a sked with a fellow

DATE TIME	STATION CALLED	CALLED BY	HIS SIGNAL RST	MY SIGNAL RST	FREQ.	EMISSION TYPE	POWER INPUT WATTS	TIME OF ENDING QSO	OTHER DATA NAME	QSLs S	QSLs R
JAN 8, 1980											
0200	KA7ZKO		479	579	21.110	CW	10	0220	Jim in Billings, MT – he's an electrical engineer working on communications systems – interesting QSO		
0230	KAØYOP		599	379	21.115	CW	10	0240	Adelle in St. Louis, Mo.		
JAN. 9											
2310		KA7VXA	449	469	21.120	CW	10	2330	John, Sun City, AZ. WANTS QSL running FT 101-E TO 4 el mono-band beam – lucky!		
JAN. 10											
2330		AA1K	599	599	28.110	CW	10	2340	Jon, Shelton, CT. An extra called me – took me a while to get his call – hi!		
JAN. 12											
2410	N5ZPZ		579	589	28.120	CW	10	2430	Jerry – Mobile, AL – he just got his new call		
2431	KA5VSP		589	599	28.120	CW	10	2440	Sarah – Houston, TX – I WAS her 50th state! She built her own gear.		

Fig. 1 — An example of a well-kept log.

Novice from your class, your teacher, your Elmer, etc. They will be most understanding of first-contact jitters. And you'll probably find that other stations will call *you* after you sign with your friend because you sound so smooth. There you are, in the middle of an unplanned QSO with a stranger and enjoying it!

Logging and QSLing

Although the FCC requires a minimal amount of information to be kept in an amateur log, you'll miss out on a lot of fun and some good memories if you don't keep more. A well-kept log will enable you to get a good return on your QSL cards, win awards, relive that great QSO two months ago, and protect you in the rare case of any questions about your operating practices. Fig. 1 is an example of a well-kept log. The basic information is there; you may want to add more or enter less. A first log may also serve to encourage another would-be ham.

Remember, when logging time, always use UTC (Universal Coordinated Time). Note the adjacent time-conversion chart for your information. This is "standard procedure" on the ham bands. UTC eliminates any problem when searching for a contact in your log for which you have received a QSL card or you would like to send one. For example, say you just received a card from 9Q5WH in the Republic of Zaire. The time listed is 2315Z (Z or "Zulu" is the same as UTC) on 24 October. Since you both used UTC, it will be easy to find 9Q5WH in your log, right there on October 24 at 2315Z. But suppose both of you used local time and suppose both of you had logged 10 pages worth of contacts that day. You would have had to search through pages before finding the QSO or else figure the number of hours difference between your local time and Zaire's. And wait a minute. Does that mean that 9Q5WH was in

Table 1

Some Facts About Time Conversion

The chart below has been arranged to show time zones used by most amateurs in the North American continent and Universal Coordinated Time, used universally as a standard. The advantage of UTC is that it is the universally understood reference throughout the world. ARRL recommends that all amateur logging be done in UTC.

Time zone letters may be used to identify the kind of time being used. For example, UTC is designated by the letter Z.

In converting from one time to another, be sure the day or date corresponds to the new time. That is, 2100 EST on January 1 would be 0200Z (UTC) on January 2, similarly, 0400Z on January 2 would be 2000 PST on January 1.

A good method is to use UTC (Z) for *all* amateur logging, schedule making, QSLing and other amateur work. Otherwise, confusion, with all the different time zones, is inevitable. Leave your clock on UTC.

UTC	EDT/AST	CDT/EST	MDT/CST	PDT/MST	PST
0000*	2000	1900	1800	1700	1600
0100	2100	2000	1900	1800	1700
0200	2200	2100	2000	1900	1800
0300	2300	2200	2100	2000	1900
0400	0000*	2300	2200	2100	2000
0500	0100	0000*	2300	2200	2100
0600	0200	0100	0000*	2300	2200
0700	0300	0200	0100	0000*	2300
0800	0400	0300	0200	0100	0000*
0900	0500	0400	0300	0200	0100
1000	0600	0500	0400	0300	0200
1100	0700	0600	0500	0400	0300
1200	0800	0700	0600	0500	0400
1300	0900	0800	0700	0600	0500
1400	1000	0900	0800	0700	0600
1500	1100	1000	0900	0800	0700
1600	1200	1100	1000	0900	0800
1700	1300	1200	1100	1000	0900
1800	1400	1300	1200	1100	1000
1900	1500	1400	1300	1200	1100
2000	1600	1500	1400	1300	1200
2100	1700	1600	1500	1400	1300
2200	1800	1700	1600	1500	1400
2300	1900	1800	1700	1600	1500
2400*	2000	1900	1800	1700	1600

Time changes one hour with each changes of 15° in longitude. The five time zones in the U.S. proper and Canada roughly follow these lines.

*0000 and 2400 are interchangeable. 2400 is associated with the date of the day ending, 0000 with the day just starting.

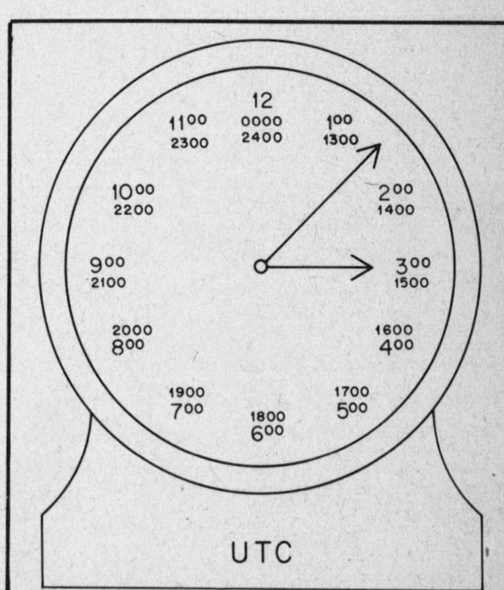

Fig. 2 — You can easily modify a common electric clock for universal time.

October 25th while you were still in October 24th? Start looking through another 10 pages of log sheets. Enough said? Use UTC.

If you don't own a 24-hour clock that can be set for UTC, you can use a conventional 12-hour job. Set it for UTC and for every hour after 1200 hours add 1200 to it. For example, your UTC-adjusted clock will read 0300Z in the evening and 1500Z in the daytime. Or better still, remove the glass on the face of the clock and write the afternoon times next to the 12-hour digits (see Fig. 2). And remember, the *date* changes at 0000 UTC.

That brings us to QSL cards. Your QSL card is like your ham radio signature. Eventually, your card will hang on shack walls all over the world. You will probably want to give some thought to your QSL. There are countless printers who have "stock" cards; they can be found in the ads in *QST* and other Amateur Radio magazines. Or you may want to have your own design custom printed.

If you don't want to go to the expense of a printer, you can make your own cards with linoleum block printing. Any good art store will have supplies and can tell you how to do it. Whatever you choose, have just a few (100 or 200) printed the first time around. You may discover that the card you chose is inconvenient for you to fill out or you may upgrade and get a new call sign. At any rate, consider your first cards an experiment — that way you won't have to pay for 900 cards you end up not using. Just be sure your card includes this basic information: your call, name and address, a place for the other station's call, a place for date, time, frequency, mode, signal report and your signature. See Fig. 3 for examples of serviceable QSL cards.

One of the most common disappointments to a beginning operator is not everyone will respond to a QSL card. Novices, though, are by and large the best

Hams from a local radio club aid victims of a simulated plane crash during a preparedness test.

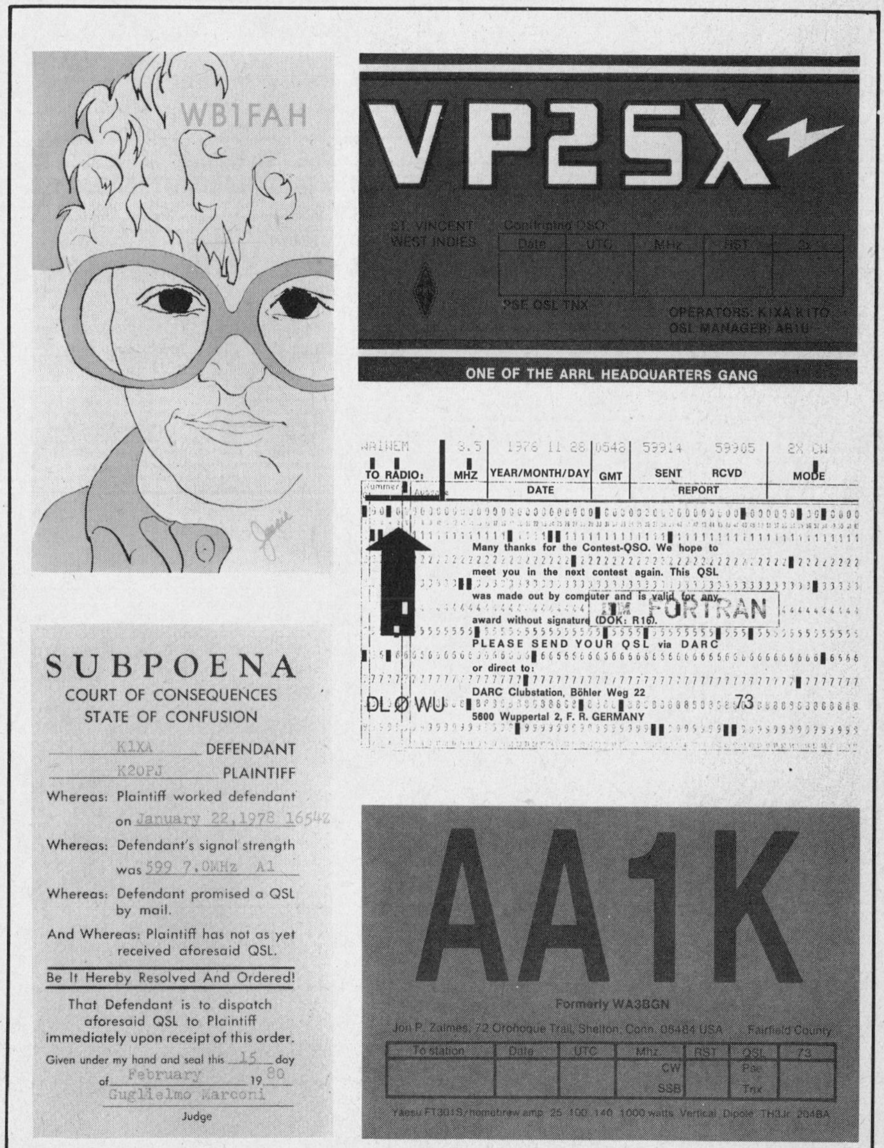

Fig. 3 — These typical QSL cards will guarantee you many happy returns.

Paul, WB3EMY, makes Washington, DC available on the air.

QSLers. There are some methods you can try to increase your returns:

1) Be sure to check your card after you've filled it out, to make sure you have entered all the information properly. Also, be sure that you have not written-over a mistake to correct it (that card won't be acceptable for awards purposes). Make out a new card if you make a mistake.

2) If you are sending a card to Alaska or Hawaii, enclose an s.a.s.e. These hams are flooded with requests for cards and it does get expensive (some hams in Vermont and Maine feel the same way). In fact, if you really want to get cards back, enclose a self-addresed-stamped-envelope *all the time*.

3) If one card doesn't get a return in a reasonable time, send another. If that doesn't work, and you want this particular card badly for one reason or another, make out a QSL on an index card, as if it were *his* card you were filling out (don't forget to put his call on it in big letters) with all the QSO information. Leave a place for him to sign it. Enclose the card with an s.a.s.e. and a note mentioning all that's required is his signature. You shouldn't need to resort to this, presumably, until you are down to those last one or two QSLs for Worked All States.

The subject of QSLing DX stations is covered comprehensively in chapter 4. Keep in mind that the average return-time on DX cards is 1 or 2 years, so be patient.

One final word regarding QSL cards. If a station should tell you QSL OM, it does not mean that he intends to send you a card. QSL literally means "I acknowledge receipt." What he is telling you is that he acknowledges your transmission. However, if he says I WILL QSL or QSL SURE VIA BURO or some such, then that means a card will be forthcoming. QSLL also means he will be sending you a card. This is a favorite Q-signal with European stations.

Contests

Contests abound on the ham bands — especially in the fall and winter. If you've

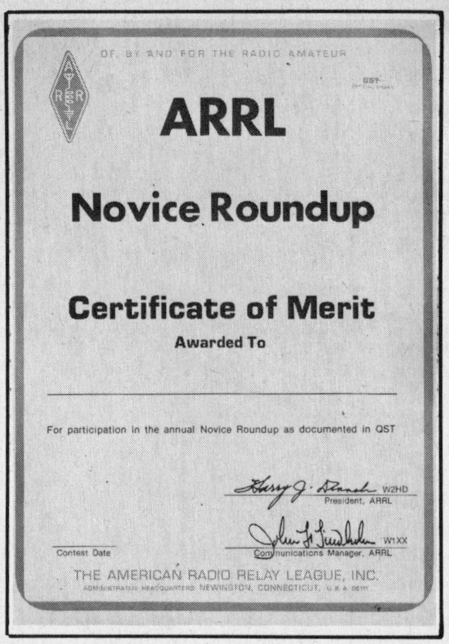

listened in on a few of them and become convinced that they aren't for you (30-wpm code and all that), hold the fort! There is a contest just for you. It's called the Novice Roundup. Held every February, the NR is a great opportunity to improve your operating skills. Besides making more contacts than you are likely to make in an ordinary week (or month for that matter), you'll learn to wade through pile-ups, from both ends, handle QRM effectively, improve your code speed and best of all, become a "savvy" operator.

Wait! Before you beg off because your code speed is barely 5 wpm and you just got on the air, remember, most Novice Roundup participants are exactly like you. Now is the best time to try your hand at contesting. You'll be surprised at how much fun you have, how fast you improve and, if you're not careful, you'll work all states!

Radio Clubs

If you attended a class to get your Novice ticket, then you are familiar with your local radio club. We hope it has led to your further involvement in club activities, since clubs are the backbone of the amateur fraternity. For all of you who studied on your own or attended a class that wasn't affiliated with a club, or whatever, clubs are where it's at. To find your local radio club, check with local hams or write the League's Club and Training Department.

There are many different kinds of clubs: repeater clubs, DX clubs, contest clubs, Amateur Radio Emergency Service groups, technical societies and the general-all-around-everything club. This last variety is the most common. What if your area lacks a club? Then start one.

The Club and Training Department will be pleased to supply a "club kit" to any group seriously contemplating forming a club. This package contains a lot of useful information on stimulating club activities, charters, etc., as well as affiliation papers to apply for ARRL's Affiliated Club status.

Most clubs provide a variety of activities for their members. Some are for fun, for public service, or for increasing your technical knowledge. Clubs frequently sponsor such things as auctions, hamfests, classes and transmitter hunts.

On the public service side of things, you'll find the club providing communications for the local government during emergencies as well as during parades, marathons and plenty of other events, such as a radiogram service at the area shopping mall and an effort in the ARRL Simulated Emergency Test. The list is as endless as your community's needs. All these activities train club members for work during emergencies, which is the primary purpose for the existence of Amateur Radio. *QST* documents monthly the service performed by amateurs during crisis situations.

And let's not forget Field Day. Even though its intent is emergency-preparedness, it is probably the high point of each club's year for fun, comaraderie and organized confusion.

Sound like fun? You bet. Well . . . get going and get involved. This book will show you how to do it up right.

Chapter 3

Rules and Regs — An Introduction

Amateur Radio is a more sophisticated pursuit than most. The operating skills, the electronics knowledge and the intellectual curiosity acquired by the Amateur Radio operator attests to a personal achievement that few other hobbies can match. The degree to which Amateur Radio operation is regulated by international, national and local government should not come as a surprise. The effects of an amateur's activity extend far beyond his radio shack.

Radio waves do not recognize national or state political boundaries; they recognize only the laws of physics. It wasn't long after the invention of radio that the United States and other countries realized that they must agree on frequency allocations, operating procedures, requisite skills for licensing operators and other matters if order were to be brought to the chaos of the radio bands.

By the year 1912, interference among Amateur, Navy and commercial radio stations was so severe that the U.S. Congress passed a law requiring U.S. radio operators to obtain a license from the Secretary of Commerce and Labor to legally operate transmitting equipment. The Radio Act of 1912 established federal authority over radio communications and created 19 regulations to govern it. The 15th regulation dealt with amateur stations:

"No private or commercial station not engaged in the transaction of *bona fide* commercial business by radio communication with the development and manufacture of radio apparatus for commercial purposes shall use a transmitting wavelength exceeding two hundred meters, or a transformer input exceeding one kilowatt; except by special authority

Note to Our Canadian Readers

For purely practical reasons, this chapter has been written to reflect the regulations of the Federal Communications Commission. A number of charts and tables that apply to Canadian regulations have been included, however.

Excerpts of the radio law in Canada and other information of value to Canadian amateurs may be obtained from the Department of Communications, Trebla Building, 473 Albert St., Ottawa, ON K1R 5B4, or from any of its regional offices:

Newfoundland: Department of Communications, P. O. Box 811, Corner Brook, NF A2H 6H6; Department of Communications, Sir Humphrey Gilbert Bldg., Room 612, Duckworth St., St. John's, NF A1C 5W1.

Prince Edward Island: Department of Communications, 180 Kent St., Charlottetown, PE C1A 1N7.

Nova Scotia: Department of Communications, 6009 Quinpool Rd., Halifax, NS B3K 5J7; Department of Communications, 500 King's Rd., Sydney, NS B1S 1B2.

New Brunswick: Department of Communications, 159 Main St., Bathurst, NB E2A 3Z1; Department of Communications, 77 Vaughan Harvey Blvd., Moncton, NB E1C 8P9; Department of Communications, P. O. Box 7285, Station A, Saint John, NB E2L 4G7.

Quebec: Department of Communications, 942 Chabanel St., Chicoutimi, PQ G7H 5W2; Department of Communications, 19th Floor, 2085 Union St., Montreal, PQ H3A 2C3; Department of Communications, 32 Frederic Herbert Ave., Noranda, PQ J9X 1V2; Department of Communications, Suite 436, 2 Place Quebec, Quebec, PQ G1R 2B5; Department of Communications, 701 Laure Blvd., Sept Iles, PQ G4R 1X8; Department of Communications, 1650 King St. West, Sherbrooke, PQ J1J 2C3; Department of Communications, 86 Edifice Publique, P. O. Box 67, Trois Rivieres, PQ G9A 5E3.

Ontario: Department of Communications, Alexandra Square, 135 James St. South, Hamilton, ON L8P 2Z6; Department of Communications, Federal Building, Room 154, Kenora, ON P9N 2X9; Department of Communications, Federal Building, Room 273, Clarence St., P. O. Box 633, Kingston, ON K7L 4X1; Department of Communications, 30 Duke St. West, Kitchener, ON N2H 3W5; Department of Communications, Government of Canada Bldg., 451 Talbot St., London, ON N6A 5C9; Department of Communications, Trebla Building, 473 Albert Street, Ottawa, ON K1R 5B4; Department of Communications, Suite 301, 222 McIntyre St. West, North Bay, ON P1B 8J5; Department of Communications, 421 Bay St., Sault Ste. Marie, ON P6A 5N3; Department of Communications, Dominion Public Bldg., 33 Court St. South, Thunder Bay, ON P7B 2W6; Department of Communications, 55 St. Claire Ave., East, Toronto, ON M4T 1M2; Department of Communications, 880 Ouellette St., Windsor, ON N9A 1C7.

Manitoba: Department of Communications, 436 Thompson Dr., Thompson, MB R8N 0C6; Department of Communications, 2300-One Lombard Pl., Winnipeg, MB R3B 2Z8.

Saskatchewan: Department of Communications, 2101 Scarth St., Regina, SK S4P 2H9; Department of Communications, 206 Circle Drive East, Saskatoon, SK S7K 0T5.

Alberta: Department of Communications, 205-8th Ave., S.E., Calgary, AB T2G 0K9; Department of Communications, Financial Building, Room 300, 10025-106 St., Edmonton, AB T5J 1G6; Department of Communications, 202-11117-100 St., Grande Prairie, AB T8V 2N2.

British Columbia: Department of Communications, 11-14th St. South, Cranbrook, BC V1C 2W9; Department of Communications, 471 Queensway, Kelowna, BC V1Y 6S5; Department of Communications, 3884-192-2nd St., P. O. Box 3396, Langley, BC V3A 3R7; Department of Communications, 1294-3rd Ave., Prince George, BC V2L 3E7; Department of Communications, Federal Building, Room 227, Prince Rupert, BC V8J 1G8; Department of Communications, Room 300, 325 Granville St., Vancouver, BC V6C 1S5.

Northwest Territories: Department of Communications, Post Office Building, P. O. Box 540, Fort Smith, NWT X0E 0P0; Department of Communications, Bellanca Bldg., P. O. Box 2700, Yellowknife, NWT H0E 1H0.

Yukon Territory: Department of Communications, 201-4133, 4th Ave., Whitehorse, YT Y1A 1H8.

Everyone has to fill out Form 610, the FCC's all-purpose application for an amateur license. There is no minimum age requirement for obtaining a license in the United States.

Table 1
Amateur Operator Licenses*

Class	*Code Test*	*Written Examination*	*Privileges*
Novice	5 wpm (Element 1A)	Elementary theory and regulations (Element 2)	Telegraphy in 3700-3750, 7100-7150, 21,100-21,200 and 28,100-28,200 kHz; 250 watts maximum input
Technician	5 wpm (Element 1A)	Elementary theory and regulations; general-level theory and regulations. (Elements 2 and 3)	All amateur privileges above 50.0 MHz plus Novice privileges.
General	13 wpm (Element 1B)	Elementary theory and regulations; general theory and regulations. (Elements 2 and 3)	All amateur privileges except those reserved for Advanced and Extra Class; See Section 97.7(a) and (b)
Advanced	13 wpm (Element 1B)	General theory and regulations, plus intermediate theory. (Elements 2, 3, and 4A)	All amateur privileges except those reserved to Extra Class; See Section 97.7(a).
Amateur Extra Class	20 wpm (Element 1C)	General theory and regulations, intermediate theory, plus special exam on advanced techniques. (Elements 2, 3, 4A, and 4B)	All amateur privileges.

*A licensed radio amateur will be required to pass only those elements that are not included in the examination for the amateur license currently held.

of the Secretary of Commerce contained in the license of the station . . ."

Today, thousands of regulations have replaced the 19 regulations of the Radio Act of 1912, and over 100 of them are applied specifically to Amateur Radio. The Communications Act of 1934, as amended, is the legislative authority of the U.S. Government for its continued regulation of wire and radio communication today. The Communications Act created a federal agency, the Federal Communications Commission, to execute and enforce the provisions of the Act. Part 97 of the Rules of the Federal Communications Commission regulates the Amateur Radio Service.

International law also affects Amateur Radio. The international agency that administers the International Radio Regulations is the International Telecommunication Union, an agency of the United Nations. Because the United States is a member-nation of the ITU, the international rules are already reflected in Part 97. However, one can especially appreciate the international ramifications of Amateur Radio by reading Article 41 of the ITU regulations, which is reproduced in the ARRL *License Manual*.

All Amateur Radio operators should have a good understanding of the regulations that apply to them. Though this chapter is no substitute for a *License Manual*, which contains all of Part 97, it will deal with most of the common situations involving the amateur rules.

Hams tend to complain that there are no specific rules to cover some circumstances. This should not be looked upon as a troublesome void that should be filled by more regulation. Rather, it should be regarded as an opportunity for radio amateurs to exercise good judgment and further hams' reputation as a self-regulating, disciplined radio service.

Every Amateur Radio operator should operate with one eye on the basis and purpose of the Amateur Radio Service, as set forth in §97.1 of the amateur rules. Adhere to these fundamental principles and you can't go wrong!

"*§97.1 Basis and purpose.*

The rules and regulations in this part are designed to provide an amateur radio service having a fundamental purpose as expressed in the following principles:

(a) Recognition and enhancement of the value of the amateur service to the public as a voluntary noncommercial communication service, particularly with respect to providing emergency communications.

(b) Continuation and extension of the amateur's proven ability to contribute to the advancement of the radio art.

(c) Encouragement and improvement of the amateur radio service through rules which provide for advancing skills in both the communication and technical phases of the art.

(d) Expansion of the existing reservoir within the amateur radio service of trained operators, technicians, and electronics experts.

(e) Continuation and extension of the amateur's unique ability to enhance international good will."

The License

You must have a license issued by the Federal Communications Commission to operate an Amateur Radio station in the United States, in any of its territories or possessions, or from any vessel or aircraft registered in the United States. The FCC sets no minimum age requirement for obtaining a license, nor does it require that the applicant be a U.S. citizen. All it requires is that the applicant pass an Amateur Radio examination. The procedure for applying for an amateur license is presented later in this chapter.

An Amateur Radio license incorporates two kinds of authorization. For the *operator*, the license authorizes operator privileges, issued after passing a code test and a written examination that assesses the applicant's knowledge of Amateur Radio regulations, radio theory and the proper operation of transmitting equipment. The license also authorizes an Amateur Radio *station* — operation of transmitting equipment at a specified location. It also authorizes portable and mobile operation as permitted under the regulations.

Alien Amateur Radio operators, licensed by a country with which the U.S. has signed a reciprocal-operating agreement, may apply to the FCC for permission to operate their stations without having to pass the FCC amateur examinations. See the separate section in this chapter dealing with reciprocal operating.

Classes of Operator License

There are five classes of amateur operator licenses in the United States: Novice, Technician, General, Advanced and Amateur Extra. The Novice class is the entry-level license requiring only a minimal amount of technical knowledge and Morse code proficiency. The Novice operator's license authorizes operation on four of the amateur frequency bands, but it limits the operator to A1 (cw) transmissions. Each successive class of license requires more technical knowledge and/or Morse code proficiency, but rewards the achiever with more privileges, including expanded frequency allocations and more modes of operation. Table 1 summarizes the five classes of operator licenses. Table 2 summarizes the U.S. amateur and mode allocations by class of license.

License Renewal/Modification

Once you have a license, it is a simple matter to keep it. Every operator license is renewable for a five-year term. To renew an Amateur Radio license, simply follow these steps:

(1) Fill out FCC Form 610 (available from any FCC office or from ARRL headquarters), and attach a photocopy, or the original, of your license to this form. (2) Mail the completed application to FCC, Box 1020, Gettysburg, PA 17325. (3) Retain copies of everything, if possible, as proof of filing before your license's expiration date. If you file your renewal application before the license-expiration date, you may continue to operate beyond the expiration date for the time it takes for your renewed license to arrive. Normally, application for renewal should be made approximately 90 days before the license's expiration date; however, renewal can be

Table 2
U.S. Amateur Frequency and Mode Allocations

Power Limits: All U.S. amateurs are limited to 250-watts dc input in the Novice segments. On all other segments, with certain exceptions in the 160-meter and 420-MHz bands, 1-kilowatt dc input is permitted. Also, there are erp limitations for stations in repeater operation. (See 97.67, FCC rules.) At all times the power level should be kept down to that necessary to maintain communications. (Revised as of March 1980.)

Bandwidth Limitations

FREQUENCY (OR PHASE) MODULATION: On frequencies below 29.0 MHz and between 50.1 and 52.5 MHz, the bandwidth of F3 emission shall not exceed that of an A3 emission having the same audio characteristics.

TELEVISION: On frequencies below 50 MHz, the bandwidth of A5 and F5 emissions shall not exceed that of an A3 single sideband emission. Between 50 and 225 MHz, single sideband or double sideband A5 may be used and the bandwidth shall not exceed that of an A3 single sideband or double sideband signal respectively. The bandwidth of F5 emission shall not exceed that of an A3 single sideband emission. Below 225 MHz, A3 and A5 emission may be used simultaneously on the same carrier frequency provided the total bandwidth does not exceed that of an A3 double sideband emission.

RTTY: When using frequency-shift keying the shift shall be less than 900 Hz. With audio frequency-shift keying, the highest fundamental modulating frequency shall not exceed 3000 Hz and the audio frequency shift shall be less than 900 Hz. Consult §97.61 of the amateur rules for special conditions to be met for ASCII transmissions (see chapter 13).

ALL MODES: The carrier frequency plus modulating frequencies must be contained within amateur allocations and within appropriate subbands.

NOTE: Some amateur bands are shared with other services. Some geographical limitations exist for the 420-MHz band. For details, and for information on specialized modes, see *ARRL License Manual*. For information on repeaters, see the *License Manual* and *Repeater Directory*.

160 METERS: Extra, Advanced and General may use some segments at 1.8-2.0 MHz. Limitations are on a geographical basis; see Table 5 elsewhere in this chapter.

Other — All modes, except as noted.

Extra, Advanced, General, Technician

MHz	*GHZ***
220-225*	10.0-10.5*
420-450*	24.0-24.25
1215-1300*	48.0-50.0
2300-2450	71.0-76.0
3300-3500	165.0-170.0
5650-5925	240.0-250.0
	All above 300

*Pulse not permitted.
**1 GHz = 1000 MHz.

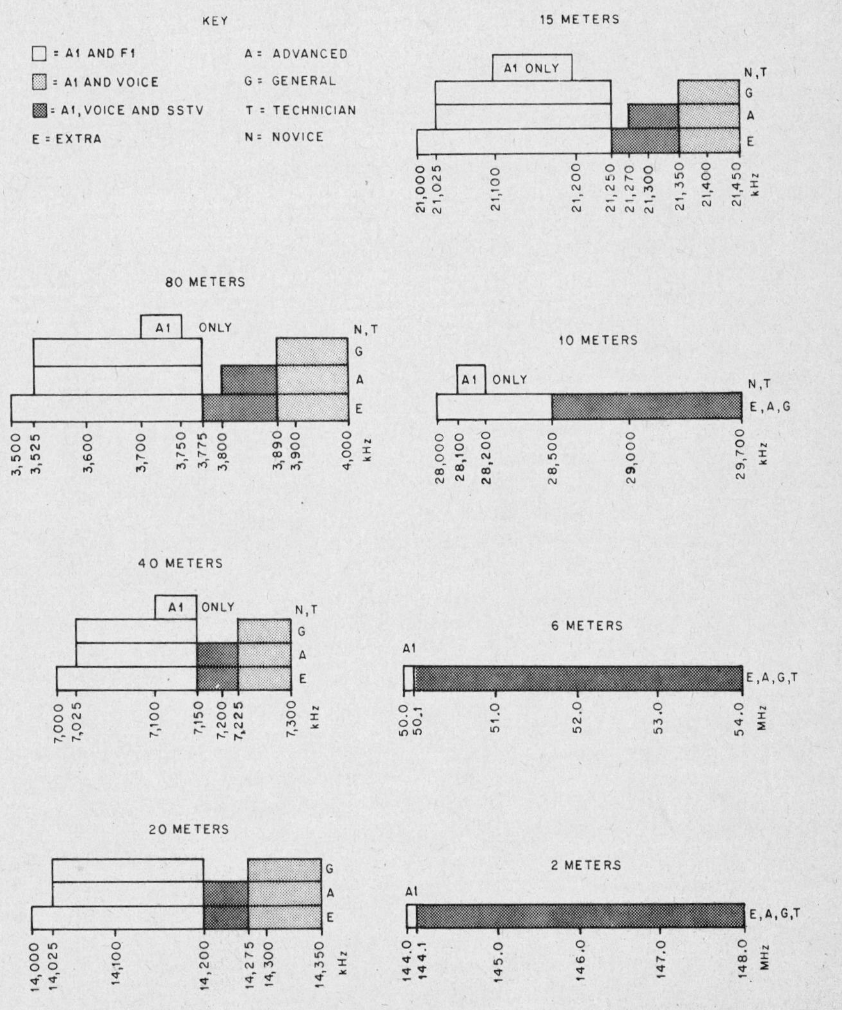

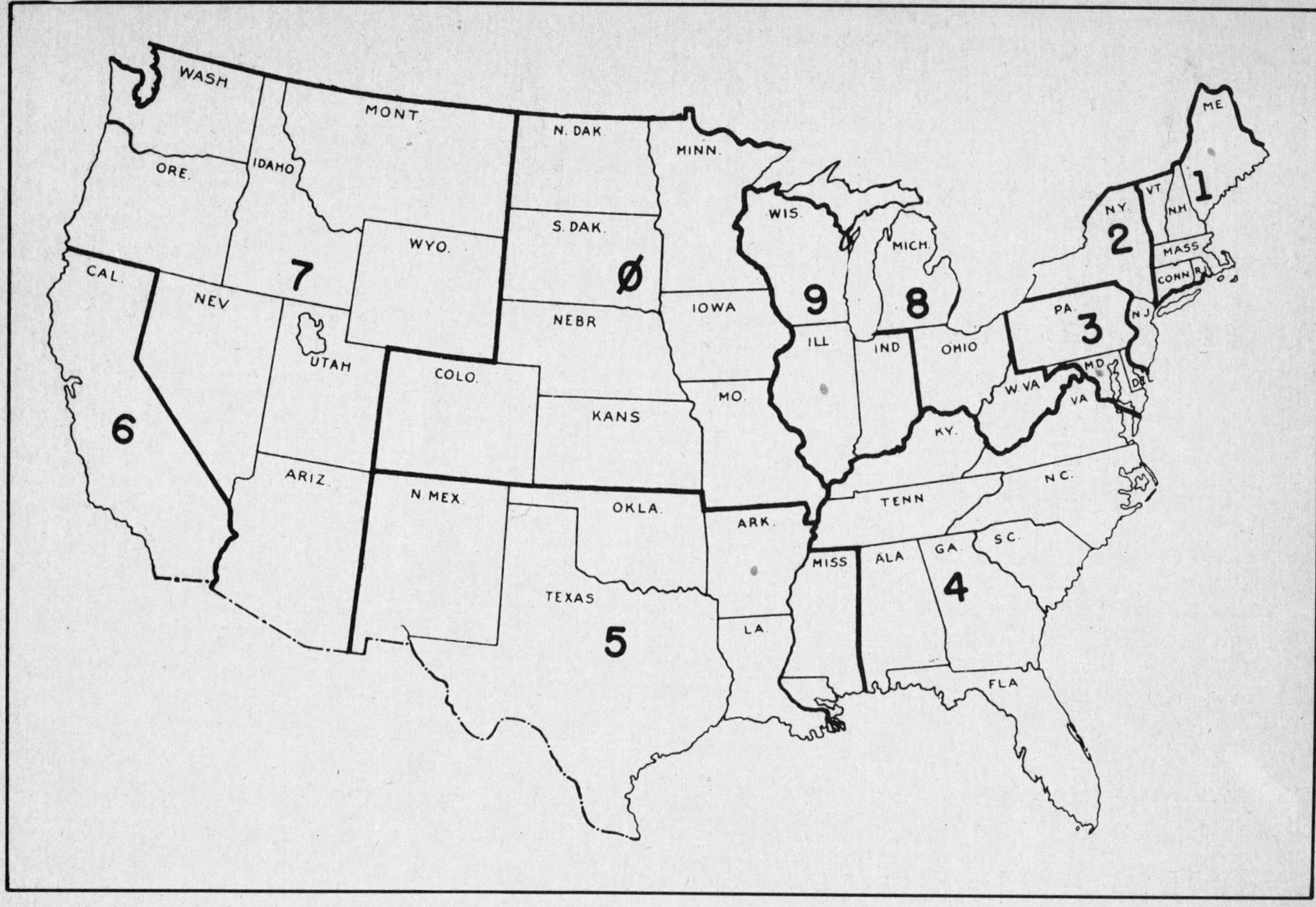

Fig. 1 — Continental U.S. Call Districts.

applied for at any time during the term of the license. (4) If your license has already expired, there is a one-year grace period during which you may renew and still keep your call sign without retesting. You must wait until the renewed license arrives before operating, however. There is also a five-year grace period during which you may renew your operator privileges without retesting; if the initial one-year grace period has passed, however, you will be assigned a new call sign. If the five-year grace period has passed, you must be re-examined for a new license. (5) If you are simply modifying your license (change of address, for example), you must fill out the Form 610; a letter is no longer sufficient for this purpose. Incidentally, your license will also be renewed automatically at this time.

If You Lose Your License

If your Amateur Radio license is lost, mutilated or destroyed, you must apply to the FCC for a duplicate. Send a letter to the FCC, Box 1020, Gettysburg, PA 17325, giving the circumstances under which the license was lost, mutilated or destroyed. After receiving your letter, FCC will issue a duplicate license bearing the same expiration date as the original license. If, subsequent to applying, you find the original license, either the original or the duplicate must be returned to the Commission.

Availability of License When Operating

If FCC personnel arrived while you were operating an Amateur Radio station, would you be able to produce the original or a photocopy of your license? The answer should be yes, because it is a violation of the amateur rules if you cannot. Section 97.82 of the rules requires an operator to have in his personal possession (or posted in a conspicuous place near the operating position) the original or a photocopy of his operator's license.

If you are guest-operating at a station, you must also have the original or a photocopy of the station license whose call letters you are using. This station's license should be posted in a conspicuous place at the operating position, or it could be in your personal possession. Conversely, if you give another ham permission to use your station while you are away, remember to leave behind a photocopy of your station license. When operating your station, your friend must be able to produce his own operator's license and your station license (or photocopies thereof) if he is visited by an FCC representative.

Even if operations are being conducted from a club station, the operator must be able to produce the club station license (or a photocopy) in addition to his own operator's license. The next time your radio club plans to operate from a remote location (during Field Day, for example), be sure everyone brings along his operator's license and the club trustee has the club station license.

U.S. Amateur Call Signs

The International Telecommunication Union (ITU) radio regulations outline the basic principles used in forming amateur call signs. According to these regulations, an amateur call sign must be comprised of one or two letters (the first or the second, in some instances, may be a numeral) as a *prefix*, followed by a numeral and then a *suffix* of not more than three letters. See chapter 15 — "Reference," for a complete listing of countries' call-sign prefixes.

Every U.S. Amateur Radio station call sign is a combination of a one- or two-letter prefix, a numeral and a one-, two-or three-letter suffix. The first letter of every U.S. Amateur Radio call sign is always an A, K, N or W. Take, for example, the call sign W1AW. W is the prefix and AW is the suffix.

For Amateur Radio stations located within the continental United States (the contiguous 48 states), the numeral in the

Table 3
FCC-Allocated Prefixes for Areas Outside Continental U.S.

Prefix	*Location*
AH1, KH1, NH1, WH1	Baker, Canton, Enderbury, Howland Is.
AH2, KH2, NH2, WH2	Guam
AH3, KH3, NH3, WH3	Johnston Is.
AH4, KH4, NH4, WH4	Midway Is.
AH5K, KH5K, NH5K, WH5K	Kingman Reef
AH5, KH5, NH5, WH5	(except K suffix) Palmyra, Jarvis Is.
AH6, KH6, NH6, WH6	Hawaii
AH7, KH7, NH7, WH7	Kure Is.
AH8, KH8, NH8, WH8	American Samoa
AH9, KH9, NH9, WH9	Wake, Wilkes, Peale Is.
AHØ, KHØ, NHØ, WHØ	Northern Mariana Islands
AL7, KL7, NL7, WL7	Alaska
KP1, NP1, WP1	Navassa Is.
KP2, NP2, WP2	Virgin Is.
KP3, NP3, WP3	Rancador Key, Quita Sueno Bank, Serrana Bank, Serranilla Bank
KP4, NP4, WP4	Puerto Rico

call sign designates the geographic area in which the station was *originally* licensed. The call-sign districts for the continental U.S. are illustrated in Fig. 1. For Amateur Radio stations under FCC jurisdiction but *not* located within the continental U.S., a prefix/numeral combination designates the geographic area in which the station was *originally* licensed. The prefix/numeral designators for U.S. Amateur Radio districts outside the continental U.S. are illustrated in Table 3.

The FCC used to require the numeral (or in the case of U.S. stations located outside the continental U.S., the prefix/numeral) to correspond to the call-sign district in which the station was licensed. In March 1978, however, FCC relaxed the rules and allowed any ham to retain his present call sign even if he moves his station location permanently to another call-sign district. Therefore, it is becoming more common to hear station call signs on the air which *do not* correspond to the call-sign district in which the station is located. For example, W9KDR is licensed to an address in Connecticut, a state in the first call district.

Some amateurs clarify their locations by including a second designator at the end of their call signs. For example, the voice identification, "This is W9KDR permanent W1" or "fixed W1" indicates that the station, though originally licensed in the ninth call district, is now licensed to an address in the first call district. When identifying using cw W9KDR/1 could be used to clarify the station's location.

It is important to make a distinction

United States Radio Districts

District No. 1, 1600 Customhouse, India & State Sts., Boston, MA 02109. The states of *Connecticut, Maine, Massachusetts, New Hampshire, Rhode Island and Vermont.*

District No. 2, 201 Varick St., corner of Houston St., New York, NY 10014. In the state of *New York*, the counties of Albany, Bronx, Columbia, Delaware, Dutchess, Greene, Kings, Nassau-New York, Orange, Putnam, Queens, Rensselaer, Richmond, Rockland, Schenectady, Suffolk, Sullivan, Ulster and Westchester; in the state of *New Jersey*, the counties of Bergen, Essex, Hudson, Hunterdon, Mercer, Middlesex, Monmouth, Morris, Passaic, Somerset, Sussex, Union and Warren.

District No. 3, 601 Market St., Philadelphia, PA 19106. In the state of *Pennsylvania*, all counties; in the state of *New Jersey*, the counties of Atlantic, Burlington, Camden, Cape May, Cumberland, Gloucester, Ocean and Salem; and the county of Newcastle in the state of *Delaware*.

District No. 4, 1017 Federal Building, 31 Hopkins Plaza, Baltimore, MD 21201. The state of *Maryland* (except that part lying in District 24); the counties of Kent and Sussex in the state of *Delaware*; in the state of *West Virginia*, all counties.

District No. 5, Military Circle, 870 North Military Highway, Norfolk, VA 23502. The state of *Virginia*, except District 24; and the state of *North Carolina*, all counties.

District No. 6, 1365 Peachtree St., N.E., Room 440, Atlanta, GA 30309. The states of *Georgia, South Carolina* and *Tennessee*, and the state of *Alabama* except that part lying in District 8.

District 6S, P. O. Box 8004, Savannah, GA 31402. Specialized office with no designated administrative area.

District No. 7, 51 S.W. First Ave., Miami, FL 33130. The state of *Florida* and in the state of *Alabama*, Baldwin and Mobile Counties.

District 7T, ADP Building, Suite 601, 1211 North Westshore Blvd., Tampa, FL 33607. Specialized office with no designated administrative area.

District No. 8, 829 F. Edward Hebert Federal Bldg., 600 South St., New Orleans, LA 70130. The states of *Arkansas, Louisiana* and *Mississippi*.

District No. 9, 5636 Federal Building, 515 Rusk Ave., Houston, TX 77002. In the state of Texas, the counties of Angelina, Aransas, Atascosa, Austin, Bandera, Bastrop, Bee, Bexar, Blanco, Brazoria, Brazos, Brooks, Burleson, Caldwell, Calhoun, Cameron, Chambers, Colorado, Comal, DeWitt, Dimmit, Duval, Edwards, Fayette, Fort Bend, Frio, Galveston, Gillespie, Goliad, Gonzales, Grimes, Guadalupe, Hardon, Harris, Hays, Hidalgo, Jackson, Jasper, Jefferson, Jim Hogg, Jim Wells, Karnes, Kendal, Kenedy, Kerr, Kinney, Kleberg, LaSalle, Lavaca, Lee, Liberty, Live Oak, Madison, Matagorda, Maverick, McMullen, Medina, Montgomery, Nacogdoches, Newton, Nueces, Orange, Polk, Real, Refugio, Sabine, San Augustine, San Jacinto, San Patricio, Starr, Travis, Trinity, Tyler, Uvalde, Val Verde, Victoria, Walker, Waller, Washington, Webb, Wharton, Willacy, Williamson, Wilson, Zapata and Zavala.

District 9B, Room 323, 300 Willow St., Beaumont, TX 77701. Specialized office with no designated administrative area.

District No. 10, Rm. 13E7 Fed. Bldg., 1100 Commerce St., Dallas, TX 75202. The state of *Texas* except that part lying in District 9 and the state of *Oklahoma*.

District No. 11, Suite 501, 3711 Long Beach, Blvd., Long Beach, CA 90807. The state of *Arizona*; in the state of *Nevada*, the county of Clarke; in the state of *California*, the counties of Imperial, Inyo, Kern, Los Angeles, Orange, Riverside, San Bernardino, San Diego, San Luis Obispo, Santa Barbara and Ventura.

District 11SD, 7840 El Cajon Blvd., La Mesa, CA 92041. Specialized office with no designated administrative area.

District No. 12, 323 A Customhouse, 555 Battery St., San Francisco, CA 94111. The state of *California* except that part lying in District 11; the state of *Nevada*, except the county of Clark, and the state of *Utah*.

District No. 13, 1782 Federal Office Bldg., 1220 S.W. 3d Ave., Portland, OR 97204. The state of *Oregon*, the state of *Idaho* except that part lying in District 14; in the state of *Washington*, the counties of Clark, Cowlitz, Klickitat, Skamania and Wahkiakum.

District No. 14, 3256 Federal Bldg., 915 2d Ave., Seattle, WA 98174. The state of *Montana*; the state of *Washington* except that part lying in District 13; in the state of *Idaho*, the counties of Benewah, Bonner, Boundary, Clearwater, Idaho, Kootenai, Latah, Lewis, Nez Perce and Shoshone.

District No. 15, Suite 2925, The Executive Tower, 1405 Curtis St., Denver, CO 80202. The states of *Colorado, New Mexico, North Dakota, South Dakota* and *Wyoming*.

District No. 16, 316 North Roberts St., St. Paul, MN 55101. The state of *Minnesota*, the state of *Michigan*, the counties of Alger, Baraga, Chippewa, Delta, Dickinson, Gogebie, Houghton, Iron, Keweenaw, Luce, Mackinac, Marquette, Menominee, Ontanagon and Schoolcraft, and *Wisconsin*, except for the part lying in District 18.

District No. 17, Room 320 Brywood Office Tower, 8800 E. 63rd Street, Kansas City, MO 64133. The states of *Iowa, Kansas, Missouri* and *Nebraska*.

District No. 18, 3935 Federal Bldg., 320 S. Dearborn, Chicago, IL 60604. The states of *Illinois* and *Indiana*; in the state of *Wisconsin*, the counties of Brown, Calumet, Columbia, Crawford, Dane, Dodge, Door, Fond du Lac, Grant, Green, Iowa, Jefferson, Kenosha, Keewanee, Lafayette, Manitowoc, Marinette, Milwaukee, Oconto, Outagamie, Ozaukee, Racine, Richland, Rock, Sauk, Sheboygan, Walworth, Washington, Waukesha and Winnebago; the state of *Kentucky* except that part lying in District 19.

District No. 19, 1054 New Federal Bldg., 231 W; Lafayette St, Detroit, MI 48226. The state of *Ohio*; the state of *Michigan* except that part lying in District 16; in the state of *Kentucky*, the counties of Bath, Bell, Boone, Bourbon, Boyd, Bracken, Breathitt, Campbell, Carter, Clare, Clay, Elliott, Estill, Fayette, Fleming, Floyd, Franklin, Gallatin, Garrard, Grant, Greenup, Harlan, Harrison, Jackson, Jessamine, Johnson, Kenton, Knott, Knox, Laurel, Lawrence, Lee, Leslie, Letcher, Lewis, Lincoln, Madison, Magoffin, Martin, Mason, McCreary, Menifee, Montgomery, Morgan, Nicholas, Owen, Owsley, Pendleton, Perry, Pike, Powell, Pulaski, Robertson, Rockcastle, Rowan, Scott, Wayne, Whitley, Wolfe, and Woodford.

District No. 20, 1307 Fed. Bldg., 111 W. Huron St., Buffalo, NY 14202. The state of *New York* except that part lying in District 2.

District No. 21, 7304 Prince Jonah Kuhio Bldg., 300 Ala Moana Boulevard, Honolulu, Hawaii 96850. The state of *Hawaii* and outlying Pacific possessions.

District No. 22, 747 Federal Bldg., Hato Rey, PR 00918, *Puerto Rico* and the *Virgin Islands*.

District No. 23, Federal Bldg., P. O. Box 2955 4th and F Streets, Anchorage, AK 99510. The state of *Alaska*.

District No. 24, Presidential Building, 6525 Belcrest Rd., Suite 830, Hyattsville, MD 20782. District of Columbia and in *Maryland*, the counties of Charles, Montgomery, and Prince Georges; and in *Virginia*, the counties of Arlington, Fairfax, Loudoun, and Prince Williams.

between clarifying one's *permanent* location and one's *portable* location, especially on voice. Using cw, the slant bar / (dahdididahdit) followed by a numeral, could designate either portable or permanent operation. If a ham decides to clarify his location status during voice identification, however, he should say "portable" when he means portable and "permanent" when he means permanent.

On-The-Air Identification Requirements

The amateur rules specifically prohibit unidentified transmissions, but what is required for proper station identification? An amateur station must be identified at the beginning of a transmission or series of transmissions, and at intervals not to exceed 10 minutes during a single transmission or series of transmissions of more than 10 minutes' duration. Also, at the end of a communication, a ham must give the call sign of one of the stations he has been in contact with (or the name of a network or group if applicable), followed by his own call sign. If the communications are being conducted using teleprinter, however, the operator is not required to give the other station's call sign. In all cases, the identification must be given using international Morse code or, when identifying on phone, in the English language.

Repeater Operation

If an amateur station is in repeater operation, the word "repeater" is transmitted at the end of the station call sign if the identification is by phone. If identification is by cw, the station call sign must be followed by the slant bar and the letter R or letters RPT.

Auxiliary Operation

If an amateur station is in auxiliary operation, the station call sign is followed by the word "auxiliary" if the identification is by phone. If the identification is by cw, the station call sign must be followed by the slant bar and the letter A or letters AUX.

Interim Designators

Interim designators must be used in conjunction with a call sign under some circumstances. These designators are used for "instant upgrading," an FCC rule that permits holders of an amateur license who successfully pass an examination for a higher class of operator license to use the new privileges immediately upon leaving the testing center. The FCC examiner issues the amateur an *interim amateur permit*, which authorizes the holder to use his new privileges for 90 days or until the issuance of the new, upgraded, permanent license (whichever comes first). The interim amateur permit also sets forth a two-letter designator for the FCC field office which administered the examination. For example, BS, BM, LB designate the Boston, Baltimore and Long Beach FCC field offices.

When operating under the authority of an interim amateur permit with privileges authorized by the permit, but which exceed the privileges of the licensee's permanent station license, the station must be identified in the following manner:

(1) On phone, by the transmission of the station call sign, followed by the word "interim," followed by the special identifier shown on the interim amateur permit. (2) On cw, by the transmission of the station call sign, followed by the slant bar, followed by the special identifier shown on the interim amateur permit, i.e., WB1ACZ/BS.

Additional Identification Requirements

To meet the identification requirements of the amateur rules, the call sign must be transmitted on each frequency being used. If identification is made by an automatic device used only for identification by cw, the code speed must not exceed 20 words per minute. While the Commission does not require use of a specific list of words to aid in sending one's call sign on phone, it does *encourage* the use of a nationally or internationally recognized standard phonetic alphabet as an aid for correct phone identification. The ITU phonetic alphabet appears in the DX Operating chapter.

Identification Requirements for Aliens

If a visiting amateur from another country is operating his station in the U.S. by virtue of a reciprocal operating agreement, special rules for station identification apply. These rules are as follows: (1) Cw operation: The amateur shall transmit the call sign issued to him by the licensing country followed by the slant bar, the U.S. amateur call sign prefix letter and number appropriate to the location of his station, i.e., G5BFU/W3. (2) Phone operation: The amateur shall transmit the call sign issued to him by the licensing country, followed by the words "fixed," "portable" or "mobile," as appropriate to the location of his station. The identification shall be made in English. (3) At least once during each contact with another amateur station, the alien shall indicate, in English, the geographical location of his station as nearly as possible by city and state, commonwealth or possession.

Permitted Communications

The U.S. amateur rules clearly set forth the kinds of stations with which amateurs may communicate. These stations are: (1) other amateur stations; (2) stations in other services licensed by the FCC and with U.S. Government stations for civil defense purposes, in accordance with the Radio Amateur Civil Emergency Service (RACES), in emergencies and, on a temporary basis, for test purposes; (3) any station that is authorized by the Commission to communicate with amateur stations.

Additionally, amateur stations may be used for transmitting signals, communications or energy to receiving equipment for the measurement of emissions, temporary observation of transmission phenomena, radio control of remote objects and similar experiments, and for the objectives set forth in the basis and purpose of the Amateur Radio Service.

The radio regulations of the ITU, to which the U.S. and Canada adhere, have this to say about amateur communications:

Article 41 — Amateur Stations

"*Section 1.* Radiocommunications between amateur stations of different countries shall be forbidden if the administration of one of the countries concerned has notified that it objects to such radiocommunications.

"*Section 2.* (1) When transmissions between amateur stations of different countries are permitted, they shall be made in plain language and shall be limited to messages of a technical nature relating to tests and to remarks of a personal character for which, by reason of their unimportance, recourse to the public telecommunications service is not justified. It is absolutely forbidden for amateur stations to be used for transmitting international communications on behalf of third parties. (2) The preceding provisions may be modified by special arrangements between the administrations of the countries concerned."

At this writing, there is no "banned-countries list." U.S. amateurs are free to communicate with amateur stations in any country. However, it is important to realize that in some countries there is no Amateur Radio Service; therefore, amateur communications with these countries are, *de facto*, not possible.

The U.S., Canada and some other countries have allowed amateurs to exchange messages on behalf of third parties. Such communications are on a very restricted basis, however.

Third-Party Communications

Third-party traffic is defined in the U.S. amateur rules as "Amateur Radio communications by or under the supervision of the control operator at an Amateur Radio station with another Amateur Radio station on behalf of anyone other than the control operator" (§97.3[v]); in other words, sending or receiving any communication on behalf of anyone other than yourself or the control operator of the other station. If you allow anyone to say "hello" into the microphone at your station, that is third-party traffic. Sending or receiving formal messages is third-party traffic (see the chapter dealing with this subject). Autopatching and phone patching are third-party traffic. You, the control operator of the station, are the first party; the control operator at the other station is the second party; and anyone else participating in the two-way communications from either station is a third party.

Certain kinds of third-party traffic are strictly prohibited. All amateurs should know that they must not handle the following kinds of third-party traffic: (1) international third-party traffic, except with countries that have assented thereto; (2) third-party traffic involving material compensation, either tangible or intangible, direct or indirect, to a third party, a station licensee, a control operator or any other person; (3) except for an emergency communication, third-party traffic consisting of business communications on behalf of any party. For the purpose of this section, business communication shall mean any transmission or communication, the purpose of which is to facilitate the regular business or commercial affairs of any party. Emergency communication is defined in the amateur rules as "any Amateur Radio communication directly relating to the immediate safety of life of individuals or the immediate protection of property."

The countries that have entered into third-party agreements with the U.S. and Canada are listed in the "DX and DXing" chapter of this book.

U.S. Possessions: Third-party traffic may be handled by U.S. amateur stations with U.S. possessions and territories where amateur stations are also licensed by the FCC. This message handling is under the same conditions as domestic traffic in the States. There must be no pecuniary interest, direct or indirect, to any party.

Military stations: U.S. personnel stationed overseas who are amateurs are no longer licensed by the U.S. Instead, the national governments do the licensing. This means that if there is no third-party agreement between the U.S. and the country involved, third-party traffic is not permitted.

Only amateur stations identified by authorized call signs having a one- or two-letter prefix beginning with A, K, W or N are licensed by the U.S. Government. Third-party traffic may be handled freely with these stations, except for the KA prefix followed by a numeral other than 1 and two letters. These call signs are reserved for U.S. personnel in Japan. Because a current U.S.-Japanese agreement provides that operation by U.S. personnel will be under regulations similar to those existing for Japanese nationals, third-party traffic privileges are denied KA stations in Japan as well as JAs. Traffic with the Federal Republic of Germany, also is normally prohibited, but communications concerning amateur satellites may be handled internationally.

In general, third-party traffic is prohibited with amateur stations in the rest of the world. This is not through reluctance on the part of the U.S. Government, but because of prohibitions on the part of the other governments concerned. The amateur at the other end is commonly forbidden to handle any third-party traffic. U.S. amateurs must abide by this restriction and have no participation in the handling of third-party traffic in such cases.

Active participation by a third party: One of the best ways to get a nonham interested in ham radio is to allow him to speak into a microphone and experience first-hand the excitement of amateur communication. While the U.S. amateur rules allow active participation by third parties, the control operator must make certain that the communications abide by all the restrictions pertaining to third-party traffic. For example, the third party may not talk about business, and he may not communicate with any country that has not signed a third-party agreement with the U.S. Also, the control operator may not leave the controls of the station while the third party is participating in amateur communications. The amateur rules allow third-party participation only if "a control operator is present and continuously monitors and supervises the radio communication to ensure compliance with the rules."

Third-party participation is not the same as designating another licensed amateur to be control operator of a station. A control operator, designated by a station licensee, may operate the station to the extent authorized by his (the control operator's) class of license.

Logging Requirements

Ever since the FCC reduced the amount of information it requires radio amateurs to log, there has been some confusion as to the minimum number of entries necessary for a "legal" log. As a minimum, the following items must be entered in the log of every U.S. Amateur Radio station: (1) the call sign of the station and the signature of the station licensee, or a photocopy of the station license; (2) the locations and dates upon which fixed operation of the station was initiated and terminated; (3) the date and time periods the duty control operator was other than the station licensee, and the signature and primary station call sign of that duty control operator; and (4) a notation of third-party traffic sent or received, including names of all third parties and a brief description of the traffic content.

A station log must be preserved for at least one year following the last date of entry. Copies of the log must be made available to the FCC upon request.

In addition to the minimum logging requirements, a log must contain a notation of the dates and locations upon which portable operation was initiated and terminated. This means that if one is operating his station from a number of locations, he must enter in his log the locations and dates upon which portable operation is initiated and terminated *at each location*. Portable operation is defined in the amateur rules as "radio communication conducted from a specific geographical location other than that shown on the station license."

No special entries have to be made in a station log for mobile operation. This does not mean, however, that one can forget about logging if operating mobile. As a minimum, items 1 through 4, above, must be entered in this log. If you conduct third-party traffic while operating mobile, for example, a notation of the third-party traffic sent or received, including names of all third parties and a brief description of the traffic content, must be entered in the station log. Mobile operation does not "do away with logging," as some people

Guest Operating

Arline, WA1VMC, a Novice, is visiting Jeanie, AB1P, an Extra Class licensee. If Arline would like to operate Jeanie's station, she must use Jeanie's call sign. Jeanie designates Arline to be control operator of AB1P. Since Arline holds a Novice class operator's license, she may not exceed Novice frequency and mode privileges. The fact that she is using AB1P, an Extra Class call sign, *does not* authorize her to exceed the privileges of her own operator license class.

What if the tables in the example above were reversed? Suppose Jeanie, AB1P, the Extra Class licensee, is visiting WA1VMC, the Novice. Does Jeanie have to use WA1VMC? The answer is yes. Arline would have to designate Jeanie control operator of WA1VMC so that she could operate from that station.

Would Jeanie be able to exceed Novice class privileges while operating from WA1VMC? Yes, because Jeanie holds an Extra Class operator's license. But wait a minute! Wouldn't that get Arline into trouble if an FCC monitoring station hears, for example, WA1VMC, in the Extra Class portion of the 75-meter phone band? Yes, it would, but fortunately the FCC has thought about that problem.

Whenever a station is operated within the privileges of the operator's class of license, but they exceed those of the station licensee, station identification shall be made by following the station call sign with the operator's primary station call sign. So if Jeanie is operating WA1VMC on phone, she would say: "This is WA1VMC with AB1P controlling." On cw, this identification would be WA1VMC/AB1P, with the slant bar separating the two call signs.

The second call sign should not be sent if the operator is operating within the limitations of the station license class. If Jeanie were using cw at 3725 kHz, she would send WA1VMC only.

When a ham is operating from a club station, the club station call sign is used. Again, the operator may *never* exceed the privileges of his own operator's license. However, if the control operator holds a license class that exceeds the license class of the station trustee, the operator would identify the station in accordance with the above example.

Table 4
Authorized Effective Radiated Power for Repeater Stations

Antenna height above average terrain	*Maximum effective radiated power for frequency bands above:*			
	52 MHz	*144.5 MHz*	*420 MHz*	*1215 MHz*
Below 50 feet	100 watts	800 watts	1000 watts	1000 watts
50-99 feet	100 watts	400 watts	1000 watts	1000 watts
100-499 feet	50 watts	400 watts	800 watts	1000 watts
500-999 feet	25 watts	200 watts	800 watts	1000 watts
Above 1000 feet	25 watts	100 watts	400 watts	1000 watts

seem to believe; rather, no additional logging entries are required for mobile operation. Mobile operation is defined in the amateur rules as "radio communication conducted while in motion or during halts at unspecified locations."

Logging third-party traffic can be by a means other than written. For example, it is perfectly legal to tape record third-party traffic and keep it as part of your log. If the FCC asks to see the station log,

Table 5
Frequency and Power Restrictions on the 160-Meter Band

Area	*Maximum DC plate input power in watts*							
	1800-1825 kHz	*1825-1850 kHz*	*1850-1875 kHz*	*1875-1900 kHz*	*1900-1925 kHz*	*1925-1950 kHz*	*1950-1975 kHz*	*1975-2000 kHz*
	Day/Night	*Day/Night*	*Day/Night*	*Day/Night*	*Day/Night*	*Day/Night*	*Day/Night*	*Day/Night*
Alabama	500/100	100/25	0	0	0	0	100/25	500/100
Alaska	1000/200	500/100	500/100	100/25	0	0	0	0
Arizona	1000/200	500/100	500/100	0	0	0	0	0
Arkansas	1000/200	500/100	100/25	0	0	100/25	100/25	500/100
California	1000/200	500/100	500/100	100/25	0	0	0	0
Colorado	1000/200	500/100	200/50	0	0	0	0	200/50
Connecticut	500/100	100/25	0	0	0	0	0	0
Delaware	500/100	100/25	0	0	0	0	0	100/25
District of Columbia	500/100	100/25	0	0	0	0	0	100/25
Florida	500/100	100/25	0	0	0	0	100/25	500/100
Georgia	500/100	100/25	0	0	0	0	0	200/50
Hawaii	0	0	0	0	200/50	100/25	100/25	500/100
Idaho	1000/200	500/100	500/100	100/25	100/25	100/25	100/25	500/100
Illinois	1000/200	500/100	100/25	0	0	0	0	200/50
Indiana	1000/200	500/100	100/25	0	0	0	0	200/50
Iowa	1000/200	500/100	200/50	0	0	100/25	100/25	500/100
Kansas	1000/200	500/100	100/25	0	0	100/25	100/25	500/100
Kentucky	1000/200	500/100	100/25	0	0	0	0	200/50
Louisiana	500/100	100/25	0	0	0	0	100/25	500/100
Maine	500/100	100/25	0	0	0	0	0	0
Maryland	500/100	100/25	0	0	0	0	0	100/25
Massachusetts	500/100	100/25	0	0	0	0	0	0
Michigan	1000/200	500/100	100/25	0	0	0	0	100/25
Minnesota	1000/200	500/100	500/100	100/25	100/25	100/25	100/25	500/100
Mississippi	500/100	100/25	0	0	0	0	100/25	500/100
Missouri	1000/200	500/100	100/25	0	0	100/25	100/25	500/100
Montana	1000/200	500/100	500/100	100/25	100/25	100/25	100/25	500/100
Nebraska	1000/200	500/100	200/50	0	0	100/25	100/25	500/100
Nevada	1000/200	500/100	500/100	100/25	0	0	0	0
New Hampshire	500/100	100/25	0	0	0	0	0	0
New Jersey	500/100	100/25	0	0	0	0	0	0
New Mexico	1000/200	500/100	100/25	0	0	100/25	500/100	1000/200
New York	500/100	100/25	0	0	0	0	0	0
North Carolina	500/100	100/25	0	0	0	0	0	100/25
North Dakota	1000/200	500/100	500/100	100/25	100/25	100/25	100/25	500/100
Ohio	1000/200	500/100	100/25	0	0	0	0	100/25
Oklahoma	1000/200	500/100	100/25	0	0	100/25	100/25	500/100
Oregon	1000/200	500/100	500/100	100/25	0	0	0	0
Pennsylvania	500/100	100/25	0	0	0	0	0	0
Rhode Island	500/100	100/25	0	0	0	0	0	0
South Carolina	500/100	100/25	0	0	0	0	0	200/50
South Dakota	1000/200	500/100	500/100	100/25	100/25	100/25	100/25	500/100
Tennessee	1000/200	500/100	100/25	0	0	0	0	200/50
Texas	500/100	100/25	0	0	0	0	0	200/50
Utah	1000/200	500/100	500/100	100/25	100/25	0	0	100/25
Vermont	500/100	100/25	0	0	0	0	0	0
Virginia	500/100	100/25	0	0	0	0	0	100/25
Washington	1000/200	500/100	500/100	100/25	0	0	0	0
West Virginia	1000/200	500/100	100/25	0	0	0	0	100/25
Wisconsin	1000/200	500/100	200/50	0	0	0	0	200/50
Wyoming	1000/200	500/100	500/100	100/25	100/25	0	0	200/50
Puerto Rico	500/100	100/25	0	0	0	0	0	200/50
Virgin Islands	500/100	100/25	0	0	0	0	0	200/50
Swan Island	500/100	100/25	0	0	0	0	100/25	500/100
Serrana Bank	500/100	100/25	0	0	0	0	100/25	500/100
Roncador Key	500/100	100/25	0	0	0	0	100/25	500/100
Navassa Island	500/100	100/25	0	0	0	0	0	200/50
Baker, Canton, Enderbury Howland	100/25	0	0	100/25	100/25	0	0	100/25
Guam, Johnston, Midway, Northern Mariana	0	0	0	0	100/25	0	0	100/25
American Samoa	200/50	0	0	200/50	200/50	0	0	200/50
Wake	100/25	0	0	100/25	0	0	0	0
Palmyra, Jarvis	0	0	0	0	200/50	0	0	200/50

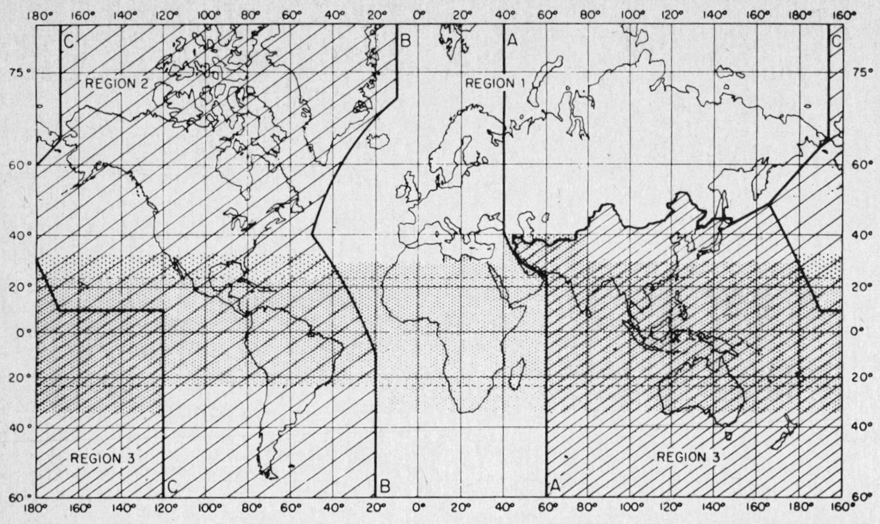

LINE A: Line A extends from the North Pole along meridian 40° East of Greenwich to parallel 40° North; thence by great circle arc to the intersection of meridian 60° East and the Tropic of Cancer; thence along the meridian 60° East to the South Pole.

LINE B: Line B extends from the North Pole along meridian 10° West of Greenwich to its intersection with parallel 72° North; thence parallel 40° North; thence by great circle arc to the intersection of meridian 20° West and parallel 10° South; thence along meridian 20° West to the South Pole.

LINE C: Line C extends from the North Pole by great circle arc to the intersection of parallel 65° 30′ North with the international boundary in Bering Strait; thence by great circle arc to the intersection of meridian 165° East of Greenwich and parallel 50° North; thence by great circle arc to the intersection of meridian 170° West and parallel 10° North; thence along parallel 10° North to its intersection with meridian 120° West; thence along meridian 120° West to the South Pole.

[*Editor's Note:* U.S. amateurs are no longer required to give their ITU region when operating on the high seas, nor do they have to identify their positions with latitude and longitude degrees. However, ITU region location continues to determine the frequencies allowed to U.S. amateurs when outside the territorial jurisdiction of a foreign country and outside ITU Region 2.]

Fig. 2 — Regions as defined in Table of Frequency allocations, Radio Regulations of the ITU. The shaded part represents the Tropical Zone.

however, the licensee is expected to transcribe the tape recordings into written form.

Authorized Power

Generally speaking, an Amateur Radio transmitter may be operated with a power input of up to 1 kilowatt to the plate circuit of the final-amplifier stage of an amplifier oscillator transmitter, or to the plate circuit of an oscillator transmitter. However, there are circumstances under which less power must be used. These are as follows:

(1) Under all circumstances, an amateur is required to use the minimum amount of transmitter power necessary to carry out the desired communications. (2) When operating in the Novice subbands, all amateurs may use no more than 250 watts of power input to the transmitter final-amplifying stage supplying radio-frequency energy to the antenna. (Power used for heating cathodes is excluded from this figure.) The Novice subbands are as follows: 3700-3750 kHz, 7100-7150 kHz (7050-7075 kHz when the station is not within ITU Region 2 — see Fig. 2), 21,100-21,200 kHz and 28,100-28,200 kHz. (3) For stations in repeater operation, the authorized effective radiated power (erp) shall not exceed certain levels, in accordance with Table 4. (4) For stations operating in the 160-meter band, special care must be taken to follow the strict power limitations set forth in Table 5.

In all cases where power exceeding 900 watts is permitted and used, an accurate means for measuring the plate power input to the vacuum tube(s) supplying power to the antenna must be used.

Operating Away from Authorized Fixed Location

An amateur station license authorizes the use, under the control of the licensee, of all transmitting apparatus at the fixed location specified on the station license. It also authorizes portable and mobile operation within the U.S., its territories or possessions (a military recreation station may not be operated portable or mobile).

When portable or mobile operation is conducted outside the territorial limits of the U.S., its territories or possessions, the following rules apply:

1) Operation may not be conducted within the jurisdiction of a foreign government except pursuant to, and in accordance with, express authority granted to the licensee by such foreign governments. When a foreign government permits FCC licensees to operate within its territory, the amateur frequency bands that may be used shall be as prescribed or limited by that government.

2) When outside the jurisdiction of a foreign government, amateur operation may be conducted within ITU Region 2, subject to the limitations of and on those frequency bands listed in §97.61 of the amateur rules.

3) When outside the jurisdiction of a foreign government, amateur operation may be conducted within ITU Regions 1 and 3 on the following frequencies, subject to the limitations and provisions of Section IV of Article 5 of the Radio Regulations of the ITU:

(a)Region 1	Region 3
3.5-3.8 MHz	1.8-2.0 MHz
7.0-7.1 MHz	3.5-3.9 MHz
14.0-14.35 MHz	7.0-7.1 MHz
21.0-21.45 MHz	14.0-14.35 MHz
28.0-29.7 MHz	21.0-21.45 MHz
144-146 MHz	28.0-29.7 MHz
430-440 MHz	50.0-54.0 MHz
1215-1300 MHz	144-148 MHz
2300-2450 MHz	420-450 MHz
	1215-1300 MHz
	2300-2450 MHz

(b) Operation on amateur frequency bands above 2450 MHz may be conducted subject to the limitations and provisions of Section IV of Article 5 of the ITU.

4) Except as otherwise provided, amateur operation conducted outside the jurisdiction of a foreign government shall comply with all requirements of Part 97 of the FCC rules.

Amateur operation in a foreign country: Operation may not be conducted from within the jurisdiction of a foreign government unless that foreign government has granted permission for such operation. Some countries have allowed U.S. amateurs to operate while visiting; however, not all countries permit foreigners this privilege. Every country has the right to determine who may and who may not operate an Amateur Radio station within its jurisdiction. There are severe penalities in some countries for operating transmitting equipment without the proper authority.

Canada-U.S. reciprocal operating: The most liberal agreement between two countries allowing reciprocal operating for their radio amateurs is between Canada and the U.S. No formal application or permit is required for U.S. amateurs to operate amateur equipment in Canada, nor is any formal paperwork required for Canadian amateurs to operate in the United States. U.S. radio amateurs may operate in Canada with the mode and frequency privileges authorized them in the United States, and Canadian radio amateurs may operate in the U.S. with the mode and frequency privileges authorized to them in Canada. In all cases, however, visitors must stay within the band and mode restrictions of the host country.

Other reciprocal operating agreements: Except for Canada, U.S. amateurs must obtain a written permit to be allowed to operate in any foreign country. The administrations of some countries have signed reciprocal operating agreements

Table 6

Countries with which the U.S. shares reciprocal operating agreements:

Argentina	Israel
Austria	Jamaica
Australia	Jordan
Barbados	Kuwait
Belgium	Liberia
Bolivia	Luxembourg
Brazil	Monaco
Canada*	Netherlands
Chile	Netherlands Antilles
Colombia	New Zealand
Costa Rica	Nicaragua
Denmark	Norway
Dominica	Panama
Dominican Republic	Paraguay
Ecuador	Peru
El Salvador	Philippines
Fiji	Portugal
Finland	Sierra Leone
France	Spain
Germany (Federal Rep.)	Suriname
Greece	Sweden
Guatemala	Switzerland
Guyana	Trinidad & Tobago
Honduras	United Kingdom
Iceland	Uruguay
India	Venezuela
Indonesia	
Ireland	

Countries with which Canada shares reciprocal operating agreements:

Austria	Luxembourg
Barbados	Netherlands
Belgium	New Zealand
Bermuda	Nicaragua
Brazil	Norway
Colombia	Panama
Costa Rica	Peru
Denmark	Philippines
Dominica	Poland
Dominican Republic	Portugal
Ecuador	Senegal
Finland	Sweden
France	Switzerland
Germany (Federal Rep.)	U.S.*
Guatemala	Uruguay
Honduras	Venezuela
Iceland	All countries of the British Commonwealth
India	
Indonesia	
Israel	

*Automatic; written permit not required.

Table 7
How did Amateur Radio fare in Geneva at WARC-79?
(Unless otherwise stated, changes should be effective January 1, 1982.)

Bands Below 30 MHz

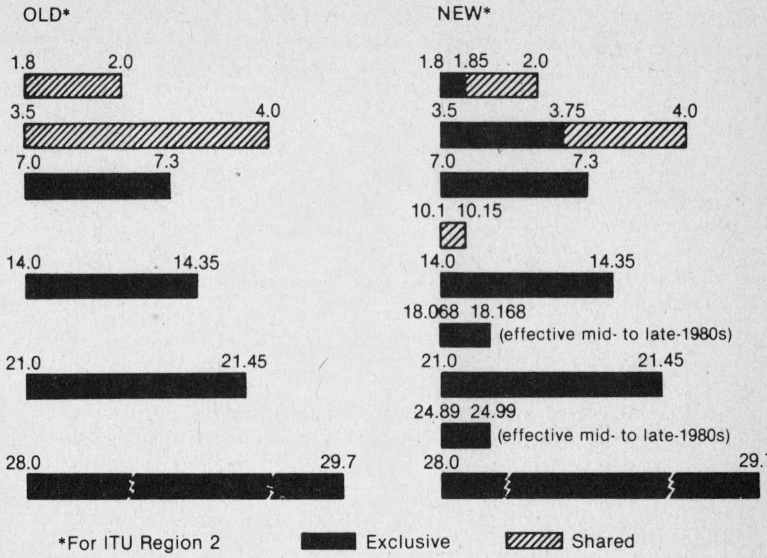

Summary of losses

NONE!

Summary of gains

- Three new bands, spaced between present bands to give us greater ability to adjust to changes in propagation
- Exclusive use of parts of the 1.8- and 3.5-MHz bands, which are now shared
- 250 kHz more spectrum for amateur use
- 500 kHz more exclusive spectrum

Bands Above 30 MHz

Summary of losses**

- 1215-1240 MHz withdrawn to protect new radionavigation-satellite system
- Some additional sharing of certain bands above 220 MHz

**For the USA

Summary of gains**

- New satellite bands at:
  1260-1270 MHz (uplink only)
  2400-2450 MHz
  3400-3410 MHz
  5650-5670 MHz (uplink only)
  5830-5850 MHz (downlink only)
  10.45-10.5 GHz
- New secondary amateur allocation at 902-928 MHz
- New amateur and amateur-satellite bands totalling 4,700 MHz exclusive and 17,040 MHz shared above 40 GHz

with the U.S. Such treaties facilitate the application procedures for getting permission to operate. These agreements, because they are reciprocal, also allow foreign amateurs the opportunity to receive permission to operate while in the U.S. If you want to operate a station in a foreign country, ARRL hq. may be able to help you obtain forms and information. Write to the Membership services Department, International Branch, for details. Many countries require a U.S. licensee to hold a General or higher-class license to qualify for a permit to operate. Table 6 lists the countries that have signed reciprocal operating agreements with the U.S. and Canada.

Radio Frequency Interference (RFI)

The FCC authorizes radio amateurs to use up to 1000 watts input power and many different bands of the radio spectrum. However, these privileges do not come without some effort on the part of the license applicant. A prerequisite for getting on the air is passing an examination testing the applicant's technical knowledge. One area of special importance is knowing about RFI, its causes and its cures. The FCC expects radio amateurs to identify and solve most RFI problems; but RFI is not always a simple matter.

Much has been written about RFI. For example, the ARRL book, *Radio Frequency Interference* examines this subject from both a legal and technical standpoint. Also, the ARRL *Radio Amateur's Handbook* devotes an entire chapter to solving radio frequency interference problems. If you are the recipient of interference complaints, these two books will be of interest to you.

A growing number of amateurs is unjustly accused of causing radio frequency interference. This occurs when an amateur transmitter is operating properly, and the responsibility for the interference falls with the electronic device experiencing the interference. For example, some television sets receive signals they are not supposed to because the manufacturer decided that reducing the per-unit cost was more important than incorporating adequate shielding in the circuit design. In these cases the amateur transmitter is operating with a "clean" signal; it is the television set that must be brought up to today's engineering standards. This usually means that the only effective cure must be performed by adding a filter at the television set. Trying to make your neighbor understand that it is his equipment that is at fault is usually no easy task, however.

Legislation has been introduced into both houses of Congress to require standards for unwanted-signal rejection filtering in electronic home-entertainment devices. Also, some manufacturers have made some progress in dealing with this problem by providing free assistance to owners of their products who experience RFI. However, the onus for causing the interference continues to fall on amateurs.

Citizens Band operators are also increasingly involved in RFI complaints. FCC figures show that only five percent of RFI complaints can be attributed to CB operation. The CB situation is exacerbated because the typical CB operator is usually less-prepared to solve an RFI problem. The FCC figures also show that over 80% of RFI complaints concerning home entertainment devices would never have occurred if the manufacturers had included proper shielding or filtering. (These FCC figures come from *The Extent and Nature of Television Reception Difficulties Associated with CB Radio Transmissions,* Federal Communications Commission, Field Operations Bureau, 1977.)

If You Are Accused of Causing RFI

1) Check your log. Were you operating at that time? (A complete log, although no longer required by the FCC, is very useful in interference situations.)

2) Check with your own nonamateur equipment. If you are not interfering with your own TV set, chances are the problem lies with your neighbor's receiver and not with your transmitter.

3) Solicit the cooperation of your neighbor in testing to determine the exact cause of the interference.

4) Check with your local radio club for a TVI committee or other assistance.

5) Request RFI assistance from the manufacturer of the home-entertainment device.

6) Read the ARRL book, *Radio Frequency Interference* and the appropriate chapter in the *Radio Amateur's Handbook.*

7) Be prompt, courteous and helpful; Amateur Radio's reputation is at stake, as well as your own.

8) For further assistance, write the Technical Information Service, ARRL hq. Include as much detail as you can.

New Bands for the 1980s

In the fall of 1979, the member-nations of the International Telecommunication Union (ITU) held a conference to consider comprehensive changes in the international table of frequency allocations and other important parts of the ITU Radio Regulations. The last such meeting had been held in 1959. The World Administrative Radio Conference (WARC-79) was a major challenge for the Amateur Radio Community because the Conference had the power to make sweeping changes in the privileges enjoyed by radio amateurs.

The world Amateur Radio community was represented by the International Amateur Radio Union (IARU), an organization of Amateur Radio societies from around the world. The IARU's headquarters society is the American Radio Relay League. Over 16 years of preparation brought about a bright picture for Amateur Radio in the years ahead. A complete chronicle of these international efforts can be found in February 1980 *QST* (copies available from ARRL hq.). A summary of how U.S. amateurs fared in the decisions at WARC-79 appears on the opposite page.

How To Apply For an Amateur Radio License

If you are applying for a Novice class license, you must first find a volunteer examiner who is over 18 years of age, holds a U.S. General class or higher amateur license and is not related to you. You must also obtain the FCC's Form 610, which is available from any FCC district office or from ARRL hq. (s.a.s.e., please). If you have difficulty locating a volunteer examiner, drop a note to the Club and Training Department at ARRL hq. for assistance. The volunteer examiner will first give you a five-word-per-minute code test (both sending and receiving). If you pass the volunteer examiner's test, he will fill in the section of your Form 610 certifying that you have successfully completed the code requirement. After you fill out the remaining pertinent sections of the form, the volunteer examiner will mail the application to FCC, Box 1020, Gettysburg, PA 17325. FCC will send the 20-question, multiple-choice test papers to the examiner, who should administer the exam and return the papers to the FCC within 30 days.

If you are applying for a Technician, General, Advanced, or Extra Class license, you must appear before a Commission examiner. If you are physically unable to travel to an FCC examination point because of a protracted disability, however, you may make special arrangements through your FCC district office for taking your exam.

If you live in or near a city with an FCC district office, call or write to find out when amateur exams are given. The FCC also travels to some cities periodically to conduct exams. However, you must file a Form 610 with the district office at least one month in advance of the scheduled examination. Contact the FCC district office nearest you for more information. A list of FCC district offices appears on page 19.

Chapter 4

Traffic Handling

Traffic handling remains a principal Amateur Radio operating activity with a sizable number of amateurs participating in it. But why? Who needs it, modern communications facilities being what they are? You can call Aunt Susie for a song just by picking up the telephone any evening and convey a great deal more information than you can in a 25-word hamgram. So what's the big deal about handling message traffic in standard form? Isn't it antiquated, obsolete?

There is no specific or definite answer to any of these questions. Those amateurs who handle traffic today — and there are perhaps more than one thinks — do so, obviously, because they enjoy it. If they didn't enjoy it, they wouldn't be doing it. The question is, *why* do they enjoy it? What is it that attracts them to an activity that may seem meaningless to the uninitiated? People enjoy doing many different things for many different reasons, and what one enjoys may be anathema to another, or vice versa. Is traffic handling your "bag"? You won't know unless you try it, but do try it before you decide against it. You may find there is more to it than initially meets the eye.

Consider, for example, the public service angle. Basically, traffic handling is doing something for someone else, and the unselfish find pleasure in this. It is a means of utilizing your Amateur Radio operating skill for the benefit of someone else, rather than solely for your own amusement. In this way it also benefits Amateur Radio in that it provides a means of contacting the public on a person-to-person rather than a mass-media basis.

Another facet is the fraternal one. You can't handle traffic by yourself; it takes two to tango and it takes someone on the other end to handle traffic, usually a number of people — hams like yourself who are interested in or at least having a crack at this kind of activity. True, fraternalism exists in all ham activities, but there seems to be a special bond among traffic handlers that does not exist among other operating specialists. Traffic work is 75% skill and 25% equipment, and the skilled have greater respect for one another than pure fraternal groups.

A complete address, including phone number, and clear enunciation or sending, are essential to ensure that a message will get to its destination. (*WA4DSH photo*)

And that leads to the organizational angle. There is probably no group of operating specialists in ham radio so highly organized as the traffic handlers. There are hundreds of organized traffic nets in existence: some 300 of these nets are part of the ARRL National Traffic System, covering the U.S. and Canada like a blanket so that a message originated anywhere will reach its destination systematically in as short a time as possible. More about the details of this later. For the present, suffice it to say that being part of a nationwide system of traffic nets is just another incentive to handle traffic.

Formal traffic in standard form is *record* traffic, which is particularly important in emergencies. It puts something on the record that can be verified, referred to, chronicled, authenticated. Informal (tactical) communications in emergencies have their place, and a very important one, since in the height of an emergency situation it often is not practical to be formal. On the other hand, emergency plans can be and should be set up so that most circumstances can be handled in stride, in a more or less routine fashion. In such cases, formalization of communications is more a help than a hindrance.

These are some of the "whys" in message handling. Now let's get on with the "hows."

Message Form

There are a lot of wrong ways to handle messages, and if you decide to go into this pursuit you will hear many of them. But there is only one right way, according to ARRL-recommended procedures, and this is what we are going to talk about next.

First of all, let's look at the *form* of the message. If all messages are written in the same format and everybody handling message traffic uses it, the procedure will be easier, quicker and more efficient. Elsewhere is shown a typical message in standard ARRL form. Here is an explanation of each part.

1)The *number* identifies the message chronologically. Most traffic operators number these messages consecutively starting with number 1 the first of each year, and this is what we recommend. The number is used to refer to the message, once it is put on the air, should this be necessary.

Handling Instructions

HXA — (Followed by number.) Collect landline delivery authorized by addressee within____ miles. (If no number, authorization is unlimited.)

HXB — (Followed by number.) Cancel message if not delivered within_____ hours of filing time; service originating station.

HXC — Report date and time of delivery (TOD) to originating station.

HXD — Report to originating station the identity of station from which received, plus date, time and method of delivery.

HXE — Delivering station get reply from addressee, originate message back.

HXF — (Followed by number.) Hold delivery until______ (date).

HXG — Delivery by mail or landline toll call not required. If toll or other expense involved, cancel message and service originating station.

This prosign (when used) will be inserted in the message preamble before the station of origin, thus: NR 207 R HXA50 W1AW 12 . . . (etc.). If more than one HX prosign is used, they can be combined if no numbers are to be inserted; otherwise the HX should be repeated, thus: NR 207 R HXAC W1AW . . . (etc.), but: NR 207 R HXA50 HXC W1AW . . . (etc.); On phone, use phonetics for the letter or letters following the HX, to insure accuracy.

2) The *precedence* (accent on first syllable, please) indicates the order of handling, which in turn indicates the importance of the message. To whom? To the originator, but the precedence is determined by the operator of the originating station, not by the person who signs the message. Otherwise, all messages would bear the highest precedence. Furthermore, there are basic rules for determining precedence, as follows:

a) Most messages are *Routine* (R) precedence — in fact, about 99 out of 100 are in this category. You might ask, then why use *any* precedence on routine messages? The reason is that operators should get used to having a precedence on messages so they will be accustomed to it and be alerted in case a message shows up with a different precedence. A *Routine* message is one that has no urgency aspect of any kind, such as a greeting. And that's what most amateur messages are — just greetings.

b) The *Welfare* (W) precedence refers to either an inquiry as to the health and welfare of an individual in the disaster area or an advisory from the disaster area that indicates all is well. Welfare traffic is handled only after all emergency and priority traffic is cleared. The Red Cross equivalent to an incoming Welfare message is DWI (Disaster Welfare Inquiry).

c) The *Priority* (P) precedence is getting into the category of high importance in an emergency situation and is applicable in a number of circumstances: (1) important messages having a specific time limit, (2) official messages not covered in the emergency category, (3) press dispatches and emergency-related traffic not of the *utmost* urgency, (4) notice of death or injury in a disaster area, personal or official.

d) The highest order of precedence is *EMERGENCY* (always spelled out, regardless of mode). This indicates any message having life-and-death urgency to any person or group of persons, which is transmitted by Amateur Radio in the absence of regular commercial facilities. This includes official messages of welfare agencies during emergencies requesting supplies, materials or instructions vital to relief of stricken populace in emergency areas. During normal times, it will be *very rare*.

3) The next part of the message is the *handling instructions* (HX). It is optional with the originating station — but if he includes it, then it must be carried through all subsequent handlings. Ordinarily, it would be omitted, because it is only for the purpose of instructing handling stations to give the message some special handling somewhere along the line or at the delivery point. We recommend that it be used sparingly, certainly not unnecessarily, since it does lengthen the message format somewhat. There are seven HX indicators at present, from A through G. The procedure is simply to add the appropriate letter or letters to the HX to indicate which instruction applies.

4) The *Station of Origin* is also part of the message reference, along with the number. This is the call letters of the station that first puts the message on the air.

5) All messages should have a *Check*. This is the count of the number of words or groups in the text of the message itself. Neither the address nor the signature count, but any salutation or complimentary close should be included in the text and added to the check count. There has been a lot of controversy about how to count words. ARRL's system and recommendations are given in a box elsewhere.

6) The *Place of Origin* of a message is the city or town and state or province in which the message originated. This is usually the place in which the station of origin is located, but quite conceivably it could be a different place. It is best not to use abbreviations for cities, but do use the postal abbreviations for states and commonly used abbreviations for Canadian provinces, such as ON for Ontario, BC for British Columbia, etc. Omit ZIP or postal codes in this part of the message.

7) Next comes another optional part of the message, the time of origin, or *filing time*. Unless this has some significance to the originating person or addressee, it should be omitted. If it is included by the originating station, however, it must be carried through to delivery by all handlers. Since many messages traverse time zones, it is best to use Universal Coordinated Time, the abbreviation for which is UTC and the designation Z. Universal Time always uses the 24-hour system.

Checking the Message

The ARRL check is basically a count of words or groups as *sent* on cw and RTTY, and in accordance with dictionary or specific sender's instructions on phone. Initials, fullstop indicator (X or "X-ray"), hyphenated words, names of cities, etc., all are counted separately if they are sent separately. Here are a few examples:

Rufus T. Firefly — 3 words
San Francisco — 2 words
NYC — 1 word
New York City — 3 words
ARL SIXTY THREE — 3 words
Fifty six — 2 words
6146B — 1 word

It is proper procedure to change the check of a message if the originating station entered the wrong count, but it is improper to change in any way the meaning or content of the message. If you do change the check, do so by retaining the original count and indicating the corrected count by a slant bar on cw or RTTY (/) or the prowords "corrected to" on phone. Do not waste time arguing about how a word or group should be counted. If you are sure you have it correct as sent, "roger" it and get on with the next one.

8) The *date* is the last item on the top line or heading of the message. Amateur procedure is simply to use the month and day (e.g., NOV 1). The year is superfluous (we *hope* the message will be delivered the same year it is originated), so let's omit that.

We have now completed the "preamble," the first of what you might think of as the principal parts of a message. It is composed of informational data and, although very important, it is not the real "meat" of the message itself — whom it is to, what it says, and who sends it. These are the other "principal parts."

9) The *address* comes next. You address a message in much the same way you do a letter, except that punctuation is omitted. The ZIP or Canadian postal code is helpful, but not essential. Also crucially important is the telephone number of the addressee, since most amateur messages are delivered by telephone. In fact, some traffic handlers won't deliver a message unless a phone number is given or listed in the local calling-area directory — and you can't blame them, especially in highly populated metropolitan areas (where directory assistance is no longer free). Messages without telephone numbers addressed to box numbers or rural routes are also be to avoided. Messages without a complete address, even those addressed to other amateurs, should be avoided.

10) The *text* should be limited to 25 words. It is best to avoid salutations (e.g., "Dear John"); they just make the text longer. Use no punctuation in the text. X is used for period or semicolon. The word "Query" denotes a question. Each X (X-ray on phone) counts as a word.

11) The *signature* winds up the message. This is just the name of the

person sending the message, signed just as in a letter. Any complimentary close (e.g., love, yours truly, sincerely, etc.) is a part of the text, not the signature, and let's hope most of them are also avoided. An amateur message is more in the nature of a telegram than a letter; although it is free, it should be composed and sent as though it were a paid service, to improve efficiency.

Sending a Message

Written or formal messages can be sent by any mode used by amateurs, but normally only three are used: code (we usually call it cw), voice (phone) and radioteletype (RTTY). The form of the message does not vary from one mode to the next, but the sending procedure can vary to a degree. So it seems best to divide this section into three so we can consider how a message is sent by each. You might ask, "Which is preferable?" Ah, now you get into an area of considerable controversy, because each mode has its proponents and exponents and each claims it is faster and more efficient than the other. Actually, the speed in any mode, given an equal degree of experience and skill, is about the same. There are many proponents of each mode, who will dispute this, but since messages are recorded or written, the speed of transmission is largely limited, at its fastest, by typing speed coupled with other necessary skills. Let's each do our traffic handling by the mode or modes we like best. But please, let's have some versatility so that we can all communicate with each other in traffic handling and not be divided into separate camps. Much progress has been made along this line in recent years, but more is needed.

As noted above, the procedure of sending by each mode varies, although the basics are the same. Thus, rather than repeat basic information for each mode, we cover it only in discussing cw sending, then cover differences in detail afterward.

Sending the Message by CW

We start with this mode because it is the oldest and was the basis of Amateur Radio traffic handling, starting back in the teens. Of course it has changed considerably along the way, but by its very nature cw requires more abbreviations and procedural signs than other modes. This doesn't make the procedure more complicated, just more detailed. So let's get started.

About the only procedural abbreviation used is NR, which starts every message and stands for "number." It precedes the number of the message and in effect says, "here it comes, start copying." The number itself follows, then the precedence. The precedence is not part of the number and it shouldn't be sent that way. Use normal word spacing. The precedence is normally R for Routine.

THE AMERICAN RADIO RELAY LEAGUE

RADIOGRAM

VIA AMATEUR RADIO

NUMBER	PRECEDENCE	HX	STATION OF ORIGIN	CHECK	PLACE OF ORIGIN	TIME FILED	DATE
280	R		KØDJ	10	COLORADO SPRINGS CO		FEB 5

To

BILL MANN W1KX
13 DIVISION STREET
JAY MAINE 04239

THIS RADIO MESSAGE WAS RECEIVED AT
AMATEUR STATION ________ PHONE ________
OWNER ________
STREET ADDRESS ________
CITY AND STATE ________

TELEPHONE NUMBER 979 7233

WILL ARRIVE PORTLAND AIRPORT MONDAY X SEE YOU THEN 73

BOB POIRIER KØDJ

SENDER'S ADDRESS AND PHONE NUMBER FOR REFERENCE:

	FROM	DATE	TIME		TO	DATE	TIME
REC'D	WA1MUX	2/5	2200Z	SENT			

THIS MESSAGE WAS HANDLED FREE OF CHARGE BY A LICENSED AMATEUR RADIO OPERATOR, WHOSE ADDRESS IS SHOWN IN THE BOX AT RIGHT ABOVE. AS SUCH MESSAGES ARE HANDLED SOLELY FOR THE PLEASURE OF OPERATING, NO COMPENSATION, DIRECT OR INDIRECT, PAID OR PROMISED, CAN BE ACCEPTED BY A STATION OWNER. FOR THE SAME REASON, NEITHER EVENTUAL DELIVERY NOR ACCURACY OF COPY CAN BE GUARANTEED. ANY REPLY MAY BE FILED WITH THE STATION DELIVERING THIS MESSAGE TO YOU. FURTHER INFORMATION ON AMATEUR RADIO MAY BE OBTAINED FROM A.R.R.L. HEADQUARTERS, 225 MAIN STREET, NEWINGTON, CONN. 06111

THE AMERICAN RADIO RELAY LEAGUE, INC., IS THE NATIONAL MEMBERSHIP SOCIETY OF LICENSED RADIO AMATEURS AND THE PUBLISHER OF QST MAGAZINE. ONE OF ITS FUNCTIONS IS PROMOTION OF PUBLIC SERVICE COMMUNICATIONS AMONG AMATEUR OPERATORS. TO THAT END, THE LEAGUE HAS ORGANIZED AN AMATEUR RADIO PUBLIC SERVICE CORPS (ARPSC), CONSISTING OF THE AMATEUR RADIO EMERGENCY SERVICE (ARES) FOR WORK DURING EMERGENCIES, AND THE NATIONAL TRAFFIC SYSTEM (NTS) FOR DAILY NATION-WIDE MESSAGE HANDLING. THE TWO DIVISIONS SUPPLEMENT EACH OTHER IN DAILY OPERATION. MORE INFORMATION IS AVAILABLE FROM A.R.R.L. HEADQUARTERS. LITHO — U.S.A.

Here's an example of a radiogram in proper form. When copying traffic on a typewriter you'll find it helpful to copy 10 words to a line. This will be very helpful in checking the accuracy of the word count. When writing out a message, keep it to five words per line.

Others are W for Welfare, P for Priority and EMERGENCY (never abbreviated).

Filing time comes next. This part of the preamble, like handling instructions, may be absent, since it is optional with the originating station, but if it is there when you get the message from another station, it must remain there. Usually this will be four digits followed by a letter, usually Z indicating Universal Time. Occasionally you will run across an odd-ball filing time, such as 430 PM EST, but experienced traffic handlers use Universal Time, converting it if necessary at the delivery point. Don't change to UTC if the filing time is in some other form. The originator may have had a reason for putting it in another form.

That takes care of the preamble. Now to the address. Incidentally, it is not recommended practice to separate the preamble from the address by a "break" ($\overline{BT}$) sign. The address must follow, so there's really no point in adding this. Along the same line, do not send TO prior to the address, just swing right into it. After the name, insert the procedure sign $\overline{AA}$, to show that the name is completed and you are about to send the next part of the address. Use the prosign $\overline{AA}$ again after the middle part of the address, if any, before the city and state. The city and state/province should include the ZIP or postal code. Then comes another $\overline{AA}$ and the telephone number. Failure to insert the $\overline{AA}$ prosigns between the parts of the address can lead to confusion and possible errors. For example, is 132 PALM AVE EAST MIAMI FL in East Miami, or is EAST part of the street name? After the address is complete and you are ready to start the text, insert the prosign $\overline{BT}$ to indicate this.

The text is really the easiest part of the message to send, because it's in plain language. (Send carefully, giving words and groups proper spacing. Difficult or unusual words can be repeated — after sending it once, use the prosign $\overline{IMI}$ and send it again. If you have a competent operator at the other end and conditions are good, this is usually unnecessary and the receiving operator may be impatient with it.) At the end of the text, another $\overline{BT}$ so indicates, and you are ready for the signature.

Don't use the proword SIG; it isn't necessary and could be confusing. If the signature includes an address or telephone number, use the separator sign $\overline{AA}$ between the parts, just as in the address. After the signature, the prosign $\overline{AR}$ indicates the end of the message. If you have more messages to send to this station, the letter B indicates this; if no more, use the letter N.

So there you have it. You have composed and sent your message. Now, wasn't that fun? If the receiving operator received everything okay, well and good. But he or she may ask for fills, in which case you will want to know what has been asked for. We'll cover that under the heading "Receiving the Message."

Sending the Message by Phone

The principal difference in sending the message on phone is that you don't have to use abbreviations. It is just as easy and a lot more understandable to just *say* the instruction or proword in full. On the other hand, there are many faults in phone sending that are so common that they are not universally recognized as faults at all and some nets consider them standard procedure.

In sending a message on phone, you don't "read" it, you "send" it. In reading, one tries to get expression into his voice. In sending, expression is far less important than articulation. The message should be sent almost in a monotone, without rising and falling inflections but with very careful articulation. The aim, don't forget, is *copy*, not comprehension. Therefore, read one word or group at a time, each one carefully articulated, at a speed slow enough to be written down comfortably. Difficult or unusual words should be spelled out (phonetically if deemed necessary, but plain English words don't usually require this).

Remember, cw abbreviations and Q-signals are not used in phone traffic handling. By the same token, don't name each part of the message as you send it. The receiving operator knows the order of the message parts; naming them at best slows things down, at worst confuses the receiving operator as to what to copy and what not to copy. This is a common fault in phone message sending — even such things as "from," "going to," "the text," and "signature" are unnecessary and should be omitted. If you send each word or group separately, with appropriate pauses to indicate spaces, there will usually be no problem. And above all, *articulate*. Don't slur or omit your consonants, especially fullstops (don't say "Trent'n" when you mean "Tren-ton") and don't drop your voice for unaccented syllables — accent *all* syllables. You aren't talking, you are *sending*, and the other person isn't just listening, he's trying to *copy* your unintelligible mutterings. So open your mouth, pronounce every syllable distinctly. Put yourself in front of the receiving operator's speaker, or between his earphones.

Separate the parts of the message by the proword "break" — that is, after the address and before the signature.

The address is probably the hardest part of the message to send by voice, especially when it comes to initials — and military addresses are just plain murder! There are 26 letters in the English language but only 11 distinctive sounds when pronouncing them, and 14 of them end in an ee sound. F and S are absolutely indistinguishable on phone, as are M and N. B, C, D, G, P, T, V, Y and Z can all sound alike in anything but perfect conditions. So proper and judicious use of phonetics on voice, especially in sending the address, is required if the message is to be received correctly. This in turn requires use of some prowords (procedural words not to be copied) and the standard phonetic alphabet. Usually only initials require phonetics, but often an unusual last name or street name may require them. If you are sending an initial that requires a phonetic (only six English letters are distinctive without a phonetic), say "initial Foxtrot." This tells the receiving operator that you are sending an initial F. If you just say "foxtrot," he may write that down. If you just say F, it's a tossup between F and S, and Old Man Murphy says the guess will be wrong. Use the standard ITU phonetics, not the cute ones of your own devising. There are all kinds of pitfalls in phone sending. Usually, surnames can simply be spelled, unless the spelling is uncommon. Don't, for example, spell Smith, or Miller, or Johnson, and *please* don't say "common spelling." If the spelling *is* uncommon, say "I spell" and do so, using phonetics *if* necessary (phonetics are usually unnecessary on fm). One thing that makes phone sending slower than cw is unnecessary use of phonetics.

Separate the parts of the address by pauses. There is no phone proword equivalent to the cw $\overline{\text{AA}}$.

In sending the text it is sometimes difficult to justify the word count. Make sure the receiving operator knows how the word should be written — either by sending it that way or, in some cases, by actually giving him vocal instructions. Numbers in the text are a prime example. If a number is to be copied as digits, the instruction "figures" will indicate this. If it is to be written out, just spelling it is easiest, quickest and the least confusing.

Saying "break and the signature" is unnecessary and inconsistent with sending the fewest instructional words and prowords possible. Let's skip it — just send the signature, using the same principles as in sending the address.

When the signature is completed, say "end of message" and follow this with "no more" if this is the only or last message you have for this station, or "more" if you have additional.

Sending the Message by RTTY

Proponents of RTTY claim that this mode is ideal for the handling of record traffic, and if properly done this is indeed the case under optimum conditions. Unfortunately, relatively few amateurs equipped for RTTY are interested in handling traffic. As a result, the standards for doing so are not too well established.

The object in sending traffic by RTTY is to give the receiving station a copy of the message that looks exactly like the one you are sending. Because of difference in equipment, this of course is not always possible. The procedure, however, can be much like that used on cw; that is, the same abbreviations and prosigns can be used. We shall concentrate here on the differences.

Normally, the messages are prepared for transmission on paper tape. This minimizes errors and enables use of less air time in transmission. However, it does require time in preparation. There are sophistications that can overcome this, such as using a reperforator in copying, so that the message(s) are punched on tape automatically as received. This tape, provided there are no "gloops" on it caused by receiving conditions or QRM, can be used in sending the message to the next station.

In today's advanced computer technology all manner of things are possible, such as computer printouts of RTTY messages sent at speeds of 100 wpm and faster. Naturally, a readout on a screen is not useful for traffic handling, since the record is temporary. In any case, most amateurs are still using the standard 60-wpm machines that trigger frequency-shift transmitters to operate page printers at the other end. A printed copy of the message is required.

In sending the RTTY message, it is first of all necessary to get tuned up so that the receiving station is making solid copy. A series of RY's (similar to Vs on cw) can accomplish this quickly, and once the transmitting and receiving stations are aligned this procedure need not be repeated. The sending of the message can then begin. It should be preceded by a carriage return, a few line feeds and LTRS to make sure the receiving equipment is properly aligned for reception — and also to allow room for a leader in a reperforated tape, if any. The message starts with NR, just as on cw, and goes completely through the preamble. Then a CR and two LFs put the machine in position for the address. This is sent all on one line, if possible, otherwise continued on the next line with a CR and only one LF. Leave two spaces for each part of the address, or a slant bar. Don't forget that you have to shift back and forth between LTRS and FIGS quite frequently, especially in the address. If you are preparing a tape beforehand, you can correct errors by sending LTRS, which is nonprinting pulse which perforates all five holes. If you are transmitting manually you have to be very careful or your copy will be a mess. Typing on a teletypewriter requires more skill, and more practice, than doing so on a regular typewriter. Use tape if possible. Remember that some machines don't unshift from FIGS back to LTRS when you send a space, so sending LTRS after each numerical group is a required precaution. After the address is completed, a CR and two LFs will provide sufficient separation for the text. The prosign $\overline{\text{BT}}$ is not used on RTTY.

The text should be sent 10 words or groups to the line with extra space between the fifth and sixth word. After 10 on each line, a CR and LF will start a new line. This makes "checking" the message easy. Numbers are sent exactly as originated — using FIGS if they are originated in that form, otherwise spelled out. The signature should be on a separate line following an LF to separate it from the text.

Many RTTY traffic handlers follow this with a "Confirm" line on which all numerals and difficult words are repeated in the order in which they appear in the message. This is all well and good if the confirm line really confirms what was sent above. But if it does not, how is the receiving operator to know whether the error was made in the first or second transmission? It seems better to have the receiving operator *ask* for a specific confirmation or fill if one appears needed; or the sending operator *corrects* the word if he sees that it was sent wrong.

Receiving the Message

Most of us copy messages with a pencil or pen, but those in the higher-echelon nets often use a "mill" (typewriter), because the speed increases and handwriting gets scrawly and difficult to read, even for the writer. So if you type at all, learn to do your code copying by mill. You can do it faster and a great deal more legibly.

Copy the message exactly as sent, very much as in the form shown in the illustration. Sure, ARRL sells message blanks, but you don't have to use them for everyday copying. Use plain sheets of paper. Standard 8-1/2 by 11-inch paper cut in half to make 8-1/2 by 5-1/2-inch sheets are very convenient if you want one message per piece of paper. Amateurs who handle a lot of traffic use scrap paper — that is, paper which has already been used on one side. Some operators use standard sheets, or even legal size (8-1/2 by 14-inch), copying as many messages on the same sheet as its length will allow. Others use 3 × 5 file cards. We have even known hotshot traffic operators who mount a roll of Teletype paper on their mills and copy consecutively, tearing a message off now and then. At the other extreme, some whose traffic-handling activities are not extensive just copy the messages on the blank side of their log sheets. Any way you do it is okay, provided you comply with the regulations which will be discussed later.

It is best to copy the preamble all in one line, with normal separation between the parts. When you come to the address you can copy each part on a separate line, as though you are addressing an envelope. This may seem wasteful of paper space, however, so another method is to copy the address consecutively, putting some kind of separation mark between the parts. Probably the slant bar (/) is easiest and quickest. Other options are the equals sign (=) or just a space.

The text is copied separately from the rest of the message, of course, and the signature is added underneath it. Copy the text five words to a line if you are writing longhand, 10 words to a line if typing. With a little practice, as your copying proficiency increases it is easy to do this. When typing 10 to a line, copy five words, leave two spaces, then copy five more. When the message is finished you can tell at a glance how many words it contains. If this is the number reflected in the check, you're in. If not, you have a problem; either the check count is wrong or you missed something or copied wrong. You should agree on the check count before you "roger" the message, to make sure that nothing is amiss.

Maybe you think you copied certain words correctly but are not sure. It will be necessary to make sure they are correct. If you missed part of the message, you will need to get "fills." The procedure varies with mode and will be discussed separately.

Receiving Messages by CW

A very useful device in cw is called "break-in." This means that the receiving operator can interrupt the sending operator while the message is being transmitted. Thus, if a word is missed because of a crash of static or QRM, fading or other phenomena, transmission can be interrupted and the missed word repeated on the spot. "Break-in" is not keying through a VOX relay, unless such action is instantaneous. To be effective, the transmitting operator must be able to hear signals while he is transmitting — not just between words or pauses, but between characters and parts of characters. When he hears the receiving station "break" (BK), he stops sending instantly. The receiving operator then sends the last word received. The sending operator repeats that word and continues on, or sends $\overline{\text{AA}}$ (all after) and continues with the next word.

To indicate that he is equipped for break-in, the sending operator sends QSK (I can work break-in) before he starts transmitting his first message. Some operators indicate they have no break-in by sending NO QSK. If the sending operator does not indicate either way, try breaking if you miss something — just once. If it doesn't work, forget it and get fills at the end of the message.

The way to get fills (missed parts) in a message is simply to send $\overline{\text{IMI}}$ BN (last word received before the missed portion) $\overline{\text{IMI}}$ (next word received). This translates to, "What is between *blank* and *blank*?" The sending operator should then send the missing portion. If you know that only one word or group was missed you simply ask for WA (word after) or WB (word before), whichever is shorter.

Sometimes you think you copied a word correctly but want to make sure. (Be careful not to guess wrong!) You can ask for a fill on that word just as though you missed it entirely, but the average operator will ask for a "confirm" (CFM). *Example*: you received the word JEWELL but are not sure if the other operator meant to send it with two L's, so you say CFM JEWELL (the way you received it). If he says CFM or C (correct), you're in business. If he says N (negative), he will then repeat the word correctly. You can also use this procedure if you know you received it correctly but suspect he may have sent it wrong. It is not necesary to use a question mark in getting a confirmation.

If your word count does not agree with the message check, you have a problem that must be resolved before the message can be "rogered" (confirmed as received correctly). For example, say you count only nine words but the preamble check was 10. You ask the sending operator to CFM CK 9. If he disagrees, you send QTB (I disagree with your word count) and send the first letter or digit of each word or group in the text. He may break you in the middle of this procedure and fill in a missing word, or wait until you have completed the QTB procedure, or tell you a certain word (week end is a good example) should count as two, or he may agree with your count. In any case, the result will be assurance that the message content is correct and complete. That's what the "check" is for. Once you are sure you have it correct, don't waste any more time arguing how a word or group should be counted; just get on with the business at hand. If the check must be corrected, you indicate this by a slant bar and the correct check, like 10/9. The check that the message started with always remains in the preamble, even if it was wrong. Resolving a disagreement on word count need take minimal time in handling a message, and is worthwhile from an accuracy standpoint. If it takes more than a few seconds, you are wasting time.

Receiving Messages on Phone

There is no break-in on phone, except via the quick-VOX or fast push-to-talk method, and even this is not true break-in because the receiving operator cannot interrupt while the sender is talking. However, it is practical if done properly — sending one group or letter at a time, pausing very briefly between each one. The effect is somewhat like a telephone conversation, but unless adequate frequency separation is used it is not identical. In most cases, the sending operator pauses after the preamble and the address to allow the receiving operator to break if he missed something.

Otherwise, the receiving procedure is very similar to that used on cw, without the abbreviations. Although phone operation allows the use of many more words with little loss in time, it is still not good procedure to use unnecessary words. Amenities may show that your mother brought you up right, but they have no place in traffic handling; they are just so much extraneous procedure. Ask for "all before," "all after," "word before," "word after," "between blank and blank," "signature," and so forth. On the other hand, one of the advantages of phone operation is that it permits you to explain your dilemma, if you have one, without using up a lot of time in doing so.

Resolving a disagreement on word count is also easier on phone, if perhaps a bit more frequent. If you count 15 words and the check was 16, just say "I count 15." Usually, the sending operator can quickly cover possible error spots. If not, the receiving operator quickly reads (he doesn't "send" it in this case) the text as he received it — not the first letter of each word or group — and the sending operator quickly fills in the missing part. If the check was wrong to begin with, change it to "16 corrected to 15" and write it 16/15 — but the sending operator should have done this before he sent the message.

Don't say "I QSL your message number so-and-so." Just "Roger" is sufficient. Remember, "Roger" means "received okay and in full." It is *not* an affirmative and does *not* indicate concurrence, agreement or compliance. *So* many operators use this proword incorrectly. "That's a Roger" does not mean "that's correct" or "I agree." It means only "I have received your transmission (or message) correctly and in full" — and just "Roger" is enough. "Roger" is used in message handling to acknowledge receipt of a message. Q-signals should not be used on phone.

Receiving Messages by RTTY

The procedure for receiving on RTTY is pretty much the same as on cw. RTTY operators proficient on cw may even prefer to get their "fills" by this mode. Ask for WA, WB, $\overline{AA}$, AB, ? BN blank ? blank, etc., just as on cw. The procedure may be a little slower on RTTY than on cw because break-in operation is not usually available, but with practice it can become quite efficient. Confirming and correcting check count uses the same procedure as on cw — CFM, QTB, etc. For more information on RTTY, see the "Visual Communications" chapter.

Originating Messages

In Amateur Radio traffic parlance, the word "origination" does not refer to just any message that is sent for the first time ("originated") from your station. It refers *only* to those messages that originate with a member of the general public other than yourself, and are subsequently sent from your station. "Origination," in this sense, is therefore entirely an off-the-air function, entirely separate from sending the message.

We hope this distinction is not confusing. We make it because it is important that we keep ourselves in the public eye, and to do this we offer an extra traffic-handling "count" for obtaining a message from a nonamateur to be transmitted to its destination by Amateur Radio. The operator gets a count for "originating" the message, then another count when he puts the same message on the air. More about counting messages later. The thing to remember at this point is that an "Origination" involves contact with the public.

How to obtain such messages? Up to a point, it's easy. Members of your family qualify as originators, and so do your relatives, neighbors and close friends. Your free service (for noncommercial purposes) can be offered to hotels, civic groups, at public gathering places such as fairs, expositions, conventions, shopping malls, etc. There is no end of possibilities. Occasional originations via your station can keep Amateur Radio face-to-face with the general public at the same time it helps keep those interested in traffic handling occupied and happy. Too much is bad. Too little is also bad. A moderate amount of traffic flowing through our nets and traffic system will keep us all in practice to handle really important traffic when or if the occasion requires, meanwhile performing a public service during normal times. These are the objectives of Amateur Radio traffic handling.

Delivering Messages

Once the message is received at a point from which it can be delivered, the actual delivery process begins. This is another off-the-air function quite separate from the act of receiving a message and likewise receives extra credit in message counting. You cannot deliver a message to yourself, but you can deliver one to a member of your family or any other person, including other amateurs, as long as it is off the air. Again, the principle is contact with the public.

Most deliveries are made by telephone. Consequently, the demeanor of the delivering person is most important to the image of Amateur Radio. You are calling, in most cases, a total stranger. His sole impression of Amateur Radio may lie in your methods in conveying the message to him. You can say, "Hey, I gotta message for ya" and let Amateur Radio take the consequences for this kind of an approach, or you can introduce yourself politely and explain the circumstances in a manner which will make the recipient feel that amateurs are ladies and gentlemen and that a worthwhile service is being performed. If the manner of the person at the other end is brusque or even offensive, all the more reason for you to be conciliatory and polite. Usually, the reception is one of delighted surprise and gratitude. If it is not, shrug it off as part of the game. Things don't always go as we want them.

Delivery by mail is a controversial topic and as in most such topics there is logic and lack of it on both sides. You receive a message without a telephone number and none can be obtained. What do you do with it? First of all, one thing you do *not* do with it is throw it in the wastebasket. Under any circumstances, this is a strict no-no in Amateur Radio. The handling of the message must be completed, one way or another. Some amateurs refuse to accept messages for delivery if they do not contain a local telephone number. This is not a reasonable attitude. An attempt should be made to deliver any message destined to your delivery area, telephone number or not. Even in this day, not everyone has a telephone. In amateur message handling, not all deliveries are easy.

If all attempts at telephone delivery fail, then the message should be mailed on a postcard or in a letter. ARRL has delivery cards, and radiogram blanks available for this purpose at a nominal cost. These are designed to present Amateur Radio in the most favorable light.

An amateur in a heavily populated metropolitan area for which many messages are destined has a legitimate gripe about handling traffic without telephone numbers. The mailing of such traffic can run into money — and besides, the argument goes, the originator in most cases could have mailed a postcard or letter in the first place, often to arrive before the message would have. These are valid arguments, and the only counter-argument is that while we cannot and have no intent to compete with the mail or commercial communications companies, in most cases, amateur messages are faster than the U.S. and Canadian mail. For those which are not, it is nevertheless incumbent on us to carry through to completion and delivery every possible message. The mailed message should contain some note that no other type of delivery was possible so that the recipient will not get the impression that delivery by mail is standard procedure.

The expense of delivering mail messages can be considerable if many deliveries must be made in this manner. Amateurs not wishing to deliver messages in this manner should *always* service the originating station so he will know his message was not delivered. As a matter of perspective, however, stations never cancel messages; the most they do is file them undelivered. Same thing? Perhaps, in effect, but they are entirely different in perspective. Only the originating operator can cancel a message. We recommmend that messages undeliverable by telephone be mailed. All of us spend a little money doing what we like to do most. Under normal circumstances, the cost of mailing the occasional undeliverable message will be negligible. Some conscientious message handlers even offer to mail copies of messages delivered by telephone, as a gesture of good will. Some recipients may like to have such copies as mementos. This is just another way in which you can introduce ham radio service to the public.

Book Messages

Occasionally, especially in originating messages at public places, you will run into a situation in which messages have

Book Messages

Assume you have four messages that are all identical except for the number and address. These can be transmitted as a "book" as follows (cw form is used):

BOOK OF 4 R K1XA 19 NEWINGTON CT OCT 1 $\overline{BT}$

HOPE YOU WILL FIND AN OPPORTUNITY TO HANDLE SOME TRAFFIC DURING THE SIMULATED EMERGENCY TEST IN OCTOBER X 73 $\overline{BT}$
HALPRIN K1XA $\overline{BT}$

NR 1 BILL KELLER W2RQ $\overline{AA}$ 27 ALBRIGHT CIR $\overline{AA}$
MADISON NJ $\overline{AA}$ 888 1436 $\overline{BT}$
NR 2 NICK ELIAS N3AIU $\overline{AA}$ 1454 MAIN ST $\overline{AA}$
BETHLEHEM PA $\overline{AA}$ 656 3322 $\overline{BT}$
NR 3 WOODY HUDDLESTON K4SCL $\overline{AA}$ 219 DRIFTWOOD
LN $\overline{AA}$ LARGO FL $\overline{AA}$ 791 6524 $\overline{BT}$
NR 4 ART SMITH W6INI $\overline{AA}$ 4515 MELISA WAY $\overline{AA}$
SAN DIEGO CA $\overline{AA}$ 555 2731 $\overline{AR}$ END BOOK $\overline{AR}$ N

On phone, of course, the cw abbreviations and prosigns are not used. One says "break" for $\overline{BT}$, pauses in speech for $\overline{AA}$ and says "end of message" for $\overline{AR}$.

Transmitted in this form, K1XA would get credit for two *Sent*, and the station receiving would get credit for two *Received*. If the receiving station relays two of the messages in book form but sends the other two single to different stations, he would get three *Sent* points. This counting system discourages the practice of originating messages in book form to get easy points at the same time it makes the form available for quicker handling efficiency.

identical parts. For example, let's say you originate messages for a national club organization sending inter-chapter greetings on anniversaries. Each such message has the same text and the same signature, and much the same preamble. Only the address is different. This is the classical example of a "book" message, for which a format has been provided so that such messages can be handled with greater efficiency and dispatch than by sending them each individually. It is basically very simple. All the parts common to each message are sent first. After this is completed, then the parts different in each message are sent. We'll discuss details in a moment, but first some important perspective on the handling of book messages.

The "booking" of messages is strictly an option with the originating, handling and delivering operators. Every operator has the right to combine messages into books or to "decombine" them if he feels this is indicated — as indeed it often will be when the routing sends parts of the book to different places. Every message originated and handled by Amateur Radio is important enough to receive separate and complete handling. The booking procedure is merely a convenience to step up the procedure. Just because you receive a message in book form does not mean it must be sent that way; or just because you have for origination many messages with the same text or other common parts does not mean you must book them. Please remember, however, that the booking of messages does not demean their importance. The book form has one object only: speed and convenience in handling. If this objective is not met, then it is preferable to handle each message completely separately. Do *not* use the word "same" for any part of a message.

Most messages that are booked have the following parts in common: precedence, station of origin, check, place of origin, date, text and signature. The number and address are usually different. The sending procedure, therefore, is to send first an indication that a book of messages follows and the number of messages the book contains — on cw or RTTY, BOOK OF 4; on phone, "book of four." You then skip the number, send the rest of the preamble in standard order followed by a break ($\overline{BT}$). The address is omitted and the text follows, after which follows another break sign and the signature. At the end of the signature you do not send the "end of message" sign ($\overline{AR}$), but another break sign and then go into the separate numbers and addresses. Each number and address, as it is sent, is followed by a break sign until the last one, which is followed by "end of book" ($\overline{AR}$ END BOOK) and then "end of message" ($\overline{AR}$). This is followed by the customary indication of "more" (B) or "no more" (N).

Some of the confusion arising from the sending of book messages has been caused by variations from the above example, such as a situation in which the preamble, text and signature are not the common parts and the others are not the different parts. The basic principle of sending the common parts first (whatever they are) and the different parts last (whatever *they* are) still applies; any confusion will lie in the details of sending the message, not in the basic principles. You can work out the details for yourself, but bear in mind that in such cases it may be deemed more prudent to send each message entirely separately. The sending operator must exercise his judgment in such a case and make a decision on how the message or book is to be handled.

Servicing the Message

Every message should contain servicing data, and should be kept on file somewhere in your station for not less than a year. The servicing data on an *Originated* message should indicate the name and address of the sender. A *Sent* message should show the call of the station to which sent, along with the date and time. A *Received* message should show the call of the station from which received, along with the date and time. A *Delivered* message should show the method of delivery, along with the date and time. These data are essential in case any question arises later regarding the message.

Handing of Service Messages

A Service Message is a message regarding the handling of another message. It is amateur-to-amateur only and usually relates to delivery problems. But it can also relate to other discrepancies in the message itself or in its handling. The form and procedure are exactly like that of any other message, except that some amateurs like to precede it with the proword "service" (SVC) before the number (*not* in place of the check, which should be entered as usual). On cw or RTTY it is permissible to use common abbreviations in the text. Otherwise, it is a standard message with all the trimmings as above. Note that the use of "service" (SVC) and abbreviations is permitted, but not required.

There is no excuse for a service message with a long text. One hears a lot of them. Stick to the essential basic facts. Some cw abbreviations that can be used are GBA (give better address), RE UR (reference your), ADX (address), ADEE (addressee). The updated ARRL Numbered Radiogram list now includes a service-message text — ARL SIXTY SEVEN.

Keeping It Legal

The following is quoted directly from the U.S. Amateur Regulations:

97.114 Third Party Traffic

The transmission or delivery of the following amateur radiocommunication is prohibited:

a) International third-party traffic except with countries which have assented thereto;

b) Third-party traffic involving material compensation, either tangible or intangible, direct or indirect, to a third party, a station licensee, a control operator or any other person;

c) Except for an emergency communication as defined in this Part, third-party traffic consisting of business communications on behalf of any party. For the purpose of this section, business communication shall mean any transmission or communication the purpose of which is to facilitate the regular business of commercial affairs of any party.

It's rather self-explanatory. Every amateur should be familiar with 97.114. Also, while third-party traffic is permitted in the U.S. and Canada, this is not so for many other nations. A special treaty is required to make such traffic permissible. The current list of third-party agreements appears in the DX section of this book. Check this list before agreeing to handle any kind of international traffic.

Such agreements specify that only unimportant, personal, nonbusiness communications be handled — things that ordinarily would not utilize commercial facilities. (In an emergency situation, we handle traffic first, then face the possible consequences later. It is not unusual for a special limited third-party agreement to be instituted during an overseas disaster.) If we value our privileges, we must take care not to abuse any regulations, whether it be on the national or international level. Furthermore, FCC regulations require that a notation of all third-party traffic be kept in the station log, including the names and addresses of the third parties involved, for a period of one year. In the case of formal traffic handling, it is quite simple to keep a file of each actual message handled.

Responsibility

Amateurs who originate messages for transmission or who receive messages for relay or delivery are accepting the responsibility for clearing such messages from their stations on their way to their destination or delivery in the shortest possible time. No amateur is required to accept or handle a message. Once you have accepted one, however, it is your moral responsibility to see that it is relayed onward or delivered. It is unethical in amateur practice to "dead end" any formal message. Any amateur who does so is playing fast and loose with a reputation for integrity and responsibility that other amateurs have worked for decades to establish.

Message Count

In order to add a little of the spice of competition to message handling, a count is kept each month of the number of messages handled by each station. This count is reported to your ARRL Section Communications Manager (see page 8 of any *QST*) near the first of each month, and he includes it in his monthly report to ARRL headquarters. These reports are published in *QST* under the heading of "Section Activities." Stations that accumulate a large number of traffic points during the month are listed in an Honor Roll called the Brass Pounder's League (BPL). More on this elsewhere.

You may or may not be interested in trying for BPL or in having your monthly traffic total listed by the SCM. Most traffic handlers are. In any case, here are brief definitions of each type of message handling.

Originated: One point for each message obtained from a third party for sending via your station. This "extra" credit is given for an off-the-air handling because of the value of contact with the general public.

Sent: Every message sent over the air from your station to another amateur receives a point in this category. Thus, a message that is eligible for an *originated* point as above receives another point when it is sent on the air. Likewise, a message that is received on the air gets a *Sent* point when it is transmitted to another station. A message that you initiate yourself, while it gets no *Originated* point, gets a *Sent* point when cleared. All *Sent* points require *on-the-air* sending.

Received: A message received over the air gets a *Received* point, whether received for relaying (sending) or for delivery to the addressee. Any message received which is not eligible for a *Delivery* point (i.e., one addressed to yourself) is nevertheless eligible for a *Received* point. This is also strictly an on-the-air handling.

Delivered: The act of delivery of a message to a third party receives a point in this category, in addition to a *Received* point. This is strictly an *off-the-air* function and must be coupled with receipt of the message at your station. Thus you cannot get a *Delivered* point unless you first get a *Received* point.

Brass Pounders League

To qualify for the Brass Pounders League listing in *QST*, a station must have a total traffic count of 500 or more or a total of Originations-plus-Deliveries of 100 or more, for a calendar month. Traffic totals should be reported to the SCM, who will issue a card certificate to each amateur who makes BPL. Any station who makes BPL for the third time will receive from headquarters a medallion with the call letters engraved on the back thereof.

The BPL is an old and dear ARRL tradition. Who pounds brass anymore? Hardly anyone, literally speaking, except maybe on Straight Key Night. The BPL is open to all traffic-handling amateurs, regardless of mode used. The name hangs on, like the *Relay* in ARRL, a symbol of a long and glorious history.

Counting Book Messages

Counting of book messages has also been a matter of controversy among traffic handlers through the years. The present method is to count one handling point for each three messages in the book, plus another point for anything over an exact multiple of three. For example, a book of four messages would count two *Received* points if received in that form and an additional count of two points when *Sent* the same way. A book of six would also count two, but a book of seven, eight or nine would count three, etc. This count applies only to sending or receiving. Messages cannot be originated or delivered in books, so each message originated from a nonamateur gets one *Originated* point whether it is booked for sending or not. Similarly, each message delivered to a nonamateur third party gets a *Delivered* point whether received in book form or not.

Messages received as a book and split up for sending or delivering count exactly as above. The received book is counted according to the number of messages in the book. If it is then split up for sending or delivery, it counts according to the form in which the parts were handled. The box contains a typical example.

Remember that booking of messages is a device used solely for the purpose of speeding the handling of messages with common parts. It is not intended to enhance any individual operator's traffic count; in fact, it will usually have the opposite effect. Whether to book or not book messages should not be based on count, but on whether it will have a desirable effect on speed and efficiency of handling.

ARL Text

An aid to originating traffic is the list of standard numbered radiogram texts. This makes it possible for a third person (nonamateur) to select a suitable text from a list provided in ARRL Form CD-3. If one of these texts is used, the prosign ARL is used prior to the check of the message in the preamble. This prosign then also precedes the number of the text used for the text of the message. This number is always *completely spelled out*. This minimizes errors. More than one such text can be used as a message text, if desired by the originating person, and additional words can also be added, keeping in mind the desirability to limit texts to 25 words or less. The check of such a message is not the total number of words when the complete text is substituted for the ARL number, but the actual number of words or groups in the message as sent from one station to another. In making delivery, however, the delivering operator makes no mention of the ARL number (which probably would mean nothing to the recipient) but delivers the text that the ARL number stands for. Using ARL texts, consequently, it is possible to formulate quite a lengthy message which would still fall within the limit of 25 words or groups.

Before sending a message with an ARL text, make sure the other operator is familiar with such texts and has a list of them. If you are operating in an ARRL net, this usually can be taken for granted. Otherwise, before sending the message, inquire ARL? If the answer is negative or the other operator doesn't know what you mean, send the whole text and forget about ARL. The ARRL numbered text list appears in the reference section of this book.

Traffic-Handling Appointments

ARRL's Field Organization was originally based on traffic handling, so even in its modern state it has quite a bit of traffic handling flavor however, contrary to the belief of many, the Field Organization is not based solely, or even mostly, on traffic handling. It is headed by the Section Communications Manager, who is elected by the ARRL full members in his section for a term of two years. The SCM is not necessarily a traffic handler. The SCM is, nevertheless, the administrative head of all ARRL-sponsored operating activities insofar as the particular section is concerned. A number of leadership appointees assist the SCM in the various operating fields.

The major SCM appointee in the traffic-handling field is the Section Traffic Manager. This official is in charge of all traffic-handling activities within the section. He reports to and is responsible to the SCM. There are two other traffic-handling appointments in the Field Organization — Net Manager and Official Traffic Station. The Net Manager (NM) is the official in charge of a section traffic net sponsored by ARRL and recognized by the SCM as an ARRL net. The Official Traffic Stations (OTS) constitute the best traffic-handling stations in the section and are appointed only after proving their merit as traffic handlers.

So get the "chain of command" — in quotes, of course, because all is voluntary. The SCM is the elected "boss" of all ARRL operating activities in the section and makes *all* appointments, although he usually relies heavily on the leadership people for advice in making many of them. There are no appointee-appointed appointees. The STM is the "assistant SCM in charge of traffic-handling activities," and has jurisdiction over all ARRL traffic activities of a sectionwide nature, under the SCM. The Net Manager is appointed by the SCM, usually at the recommendation of the STM, to manage a particular ARRL section or local ne . While there is only one STM per section, there may be as many NMs as there are nets, both at section and local level, as long as they are traffic nets and recognized by the SCM as part of the ARRL Field Organization.

Then there are the OTS appointees, also appointed by the SCM, usually at the recommendation of the NM or STM. These are the section's crack traffic handlers, the amateurs who set the example of efficient traffic handling for the rest of the amateur fraternity. You must be a full ARRL member to hold any SCM appointment, but being a member any participating in an ARRL net do not automatically qualify you for OTS appointment. For this you must prove yourself as a traffic handler. You can be a member of an ARRL net without being a League member or an OTS, but you cannot be an OTS without being both a League member and a regular ARRL net participant.

Traffic Nets

At one time all traffic was handled on a station-to-station basis. This was the basis on which traffic handling formed the origins of the American Radio Relay League. Details on how messages were handled during the first 30 years or so of Amateur Radio make fascinating reading. Today, however, traffic handling is accomplished mostly on spot-frequency nets.

All nets don't handle traffic, or course. Some of them are simply in preparation for emergencies, with routine traffic handling as a side issue. Others are held for social or fraternal purposes and even avoid handling third-party traffic. We shall be speaking here solely of nets organized for the purpose of handling third-party traffic as a public service. In passing, however, let this policy be noted: Every traffic net should be prepared to handle traffic in an emergency just as efficiently, so should be familiar with any special procedures required for so doing. Conversely, every emergency net should be familiar with standard message form and how to handle record messages. In other words, every traffic net is to some extent an emergency net, and every emergency net is to some extent a traffic net. During an emergency situation, the two types of nets should be able to work together in harmony and with maximum efficiency.

But back to traffic nets. There are many different types of traffic nets, some of them part of an organized traffic system, some of them independent of any central organization. A good comparison is the independent telephone companies and the gigantic Bell System. The ARRL National Traffic System is the Amateur Radio traffic-handling equivalent of the Bell System and its many local companies. As in our nationwide telephone system (by far the best in the world), the local independent nets rely mostly on NTS to handle their long-haul traffic. And, like the same system, NTS often relies on independent local nets to be the source of supply and delivery of much of its long-haul traffic.

A traffic net is a group of amateur operators gathered on a spot frequency for the purpose of exchanging third-party traffic on an organized basis. For each net session there is a net control station, who is a sort of master of ceremonies, and one or more representatives designated to take messages destined for points out of the net's normal coverage area, or to bring traffic in from outside the net's area destined for points the net normally covers. Other stations in the net are either at random or representing a particular city or town or other designated part of the net's coverage.

The net control (NCS) has the function of directing the flow of message traffic among members of the net. Usually, this is his sole function and usually it is more than enough to keep him busy, although in some nets short of qualified operators or with little traffic to handle the net control may handle some of the traffic himself. During the time the net is in directed session, the station designated net control for that session is the boss. His directions are to be complied with immediately and without question if they are understood.

Most traffic nets meet daily at a fixed time on a designated frequency. They start with a "call-up" by the NCS, using a pre-designated abbreviation on cw or RTTY or the name of the net on phone. The net control is usually not the same station every session, allowing more operators to accumulate experience in net controlling. Net management policy is to pass the NCS duty among regular personnel, as soon as any operator has accumulated enough net savvy and know-how to qualify for this duty.

This brings us to the net manager. In ARRL nets, the net manager is appointed by the SCM on recommendation of the STM, as already noted. In the case of High-level NTS nets, the appointment is made by the office of the ARRL communications manager. In the case of nets not part of the ARRL system, the net manager may be self-appointed or elected by the membership of the net. In any case, the manager is the operator who designates net control stations for the various sessions of the net and also designates "liaison" stations to conduct regular liaison with other nets to insure the flow of traffic to and from places beyond the net's coverage area.

We have mentioned coverage area several times without explaining just what it means, although the term is mostly self-explanatory. Every net has a finite area or designated points in which its stations are located or can cover by local telephone. As long as all traffic is originated within that area or among those points and destined to the same, no liaison with other nets is required; but there are few if any nets of this nature. Thus, every net must conduct liaison with other nets having different coverage areas or points, and this requires the availability of stations who can take traffic from one net to another. Sometimes a net control will merely ask for a volunteer to handle a message outside the net's area, or to liaison with another net. The well-organized net, however, has "thru" stations already designated for each net session (a different station or stations for each session, if available), and designation of such station is a function of the net manager. The net manager has the function of keeping net records and rosters and, in the case of ARRL nets, of reporting monthly to the STM.

Reporting Into Your Section Net

There are so many different kinds of nets with so many different procedures that the best way to handle this procedure is to take a typical example and stipulate that most reporting procedures will approximate it. The example we take is an ARRL National Traffic System section net, of which there are hundreds in operation on a daily basis. Some sections have

ARRL QN Signals For CW Net Use

*QNA**	Answer in prearranged order.
*QNB**	Act as relay *B*etween____ and____.
QNC	All net stations *C*opy. I have a message for all net stations.
QND*	Net is *D*irected (controlled by net control station.)
*QNE**	*E*ntire net stand by.
QNF	Net is *F*ree (not controlled.)
QNG	Take over as net control station.
QNH	Your net frequency is *H*igh.
QNI	Net stations report *I*n. I am reporting into the net. (Follow with a list of traffic or QRU.)
QNJ	Can you copy me? Can you copy____?
*QNK**	Transmit messages for____ to____.
QNL	Your net frequency is *L*ow.
*QNM**	You are QRMing the net. Stand by.
QNN	*N*et control station is____. What station has net control?
QNO	Station is leaving the net.
QNP	Unable to copy you. Unable to copy____.
*QNQ**	Move frequency to____ and wait for ____ to finish handling traffic. Then send him traffic for____.
*QNR**	Answer____ and *R*eceive traffic.
QNS	Following *S*tations are in the net.* (Follow with list.) Request list of stations in the net.
QNT	I request permission to leave the net for____ minutes.
*QNU**	The net has traffic for *you*. Stand by.
*QNV**	Establish contact with____ on this frequency. If successful, move to and send him traffic for____.
QNW	How do I route messages for____?
QNX	You are excused from the net.* Request to be excused from the net.
*QNY**	Shift to another frequency (or to ____ kHz) to clear traffic with____.
QNZ	*Z*ero beat your signal with mine.

*For use only by the Net Control Station.

Notes on Use of QN Signals

The QN signals listed above are special ARRL signals for use in amateur cw nets *only*. They are not for use in casual amateur conversation. Other meanings that may be used in other services do not apply. Do not use QN signals on phone nets. *Say it with words.* QN signals need not be followed by a question mark, even though the meaning may be interrogatory.

both phone and cw nets and many of them conduct more than one session daily in conformity with NTS principles.

So okay, you are the average amateur who has had a license long enough to know what amateur operating is all about, and you decide to take the plunge into traffic handling by reporting into your section net. Some sections have special nets for beginners; if you are lucky enough to live in such a section you may want to start with the Section Novice or Training Net. In any case, the question is, how do you go about it?

Let's assume first that you are reporting into a cw net. To do this, you must have a list of special QN signals, because these are widely used, as are a number of special cw net abbreviations. Armed with this material, listen to a few net sessions to get the hang of it. Even among NTS section nets the procedure may vary considerably, but the average would go something like this:

NCS: PSN PSN PSN DE K9NCS K9NCS K9NCS QNN QNI K

(Translation: This is the Podunk Section Net being called to order by Net Control [QNN] K9NCS. Stations will report into the net [QNI]).

There is also a recommended variation in having stations answer the call-up. Either they can answer one at a time and be acknowledged in turn by NCS, or they answer consecutively, being careful not to interfere with each other. This can be accomplished on cw by sending DE and then listening to see if anyone else is trying to QNI, or on phone by saying "this is" and listening. If no one else is transmitting, then you can go ahead and QNI. After it has appeared that there are no more QNI, NCS acknowledges them all in a group (QNS). In some nets, the reporting-in is done in alphabetical order of the "regulars," or in alphabet groups, to minimize the possibility of reporting stations "doubling" on each other. In the case of the Podunk Net, we assume reporting-in is at random, and so the first station to report is W9NET. He sends N (first digit of his call suffix) and NCS answers with the same letter. W9NET then knows that NCS is alerted to his QNI.

W9NET: K9NCS DE W9NET QNI QRU $\overline{\text{AR}}$

(Translation: W9NET reports in with no traffic.)

K9NCS: W9NET DE K9NCS R $\overline{\text{AS}}$

(Translation: NCS acknowledges the QNI of W9NET, directs him to stand by.)

WB9ABC: A

K9NCS: A

WB9ABC: K9NCS DE WB9ABC QTC AVON 1 $\overline{\text{AR}}$

(Translation: WB9ABC signals the first letter of his call suffix, gets a return signal, says he has one message for Avon.)

K9NCS: WB9ABC W9NET UP 5 AVON K

(Translation: Net control directs WB9ABC and W9NET to move up 5 kHz and clear the message to Avon. Acknowledgement of WB9ABC's QNI is, of course, understood. Using side frequencies is the usual method, especially on cw nets where crowding is less a problem.)

Traffic is cleared on the net frequency only when there are no further QNI to attend to and only when further dispatching is not required. Note also that as soon as two stations with traffic to be handled are QNI, they are dispatched to clear the traffic immediately, without waiting for further QNI. Stations reporting in always indicate the destination of their traffic, if any, by sending the *destination first* followed by the number of messages for that destination.

When two stations are dispatched to a side frequency to clear traffic, the station to receive the traffic calls first. The number of kHz up or down is only approximate; the receiving station does not call on top of an ongoing QSO, even a casual ragchew. He calls as close to the designated frequency as possible without causing QRM. The station designated to transmit should look around until he finds the receiving station, then zero on him, whether this is exactly on the designated frequency or not. It is logical for the receiving station to pick the frequency on which he will have to copy the message.

The QN Signals

The "Special QN Signals for Net Use" originated in the late 1940s in the Michigan QMN Net, and were first known to Headquarters through the then head traffic honcho W8FX. Ev Battey, W1UE, then assistant communications manager, thought enough of them to print them in *QST* and later to make them standard for ARRL nets, with a few modifications. The original list was designed to make them easy to remember by association. For example, QNA meant "Answer in *A*lphabetical order," QNB meant "Act as relay *B*etween ...," QNC meant "All *N*et Copy," QND meant "*N*et is *D*irected," etc. Subsequent modifications have tended away from this very principle, however, in order that some of the less-used signals could be changed to another, more needed use.

Since the QN signals started being used by us amateurs, International QN signals having entirely different meanings have been adopted. Concerned that this might make our use of QN signals with our own meanings at best obsolete, at worst illegal, ARRL informally queried FCC's legal branch. The opinion then was that no difficulty was foreseen as long as we continued to use them only in amateur nets. Occasionally a purist will insist that our use of QN signals for net purposes is illegal. Should anyone feel so strongly about it as to make a legal test, we might find ourselves deprived of their use. Until or unless we reach such an eventuality, however, let's continue to use them, and use them right. After all, we were using them first.

When the message has been cleared, both stations (after a very short pause to make sure NCS has not sent another station up to intercept one of them) return to net frequency and, at the first break in procedure, check back in by sending just the suffix of their calls. NCS will acknowledge by repeating the suffix or just sending R. Stations dispatched to a side frequency do not chew the rag with each other while NCS is perhaps waiting for them to come back.

W9ILN: I

K9NCS: I

W9ILN: K9NCS DE W9ILN TO 9RN QRU $\overline{\text{AR}}$

K9NCS: W9ILN DE K9NCS R $\overline{\text{AS}}$

(Translation: W9ILN indicates he is reporting in as the station designated by the net manager to take any "THRU" traffic, destined to points outside the net's direct coverage area (to the ninth region net), and that he has no traffic. NCS directs him to stand by. As other stations report in with "THRU" traffic, they will be dispatched to a side frequency with W9ILN to send him such traffic.)

Let's assume that while WB9ABC is up

Rules for Net Discipline

A good net is a disciplined net, and discipline is something that must be enforced by the net control station and observed by all net stations. To the extent it is not, to this extent does the efficiency of the net suffer. Here are some hints to help you help the NCS observe proper circuit discipline.

1) *Zero beat your signal with that of the NCS.* The net frequency is where the NCS is. Your station should be equipped so that you can zero beat NCS without putting a signal on the air. Don't make it necessary for him or her to tell you that you are too low (QNL) or too high (QNH). Make sure you are right on before you QNI.

2) *Report in on time.* Be ready to QNI at the appointed time. A net cannot operate smoothly if net stations report in three, five or 10 minutes late.

3) *List your traffic properly,* if you have any. The proper destination first, number of messages second. A prerequisite to this rule is to be familiar with the check-in procedure of the net.

4) *Never transmit without the NCS's permission,* in a directed net, except in an emergency or when returning from a side frequency assignment. In such cases, make your transmissions ultra-brief.

5) *Call the NCS only if you must.* This is in line with the above, but occasionally you have a situation that requires you to notify the NCS. Avoid doing this frivolously. If you must be excused, disregard the reason; just ask to be excused. Never call the NCS to advise him he is doing something wrong.

6) On the net frequency, *transmit* only to the NCS unless he instructs you otherwise. Unauthorized side comments with or to another station are strictly out of order.

7) *Answer promptly when the NCS calls you.* Stay alert. Don't fall asleep and be caught napping. In a busy net, the NCS may call you only once and if you don't respond promptly he may assume you have left the net and cross you off the list.

8) *Don't leave the net until you are excused by the NCS,* either specifically or by a general QNF ("the net is free"). If you *must* leave the net temporarily, be sure to get the net control's permission first, then report back in as soon as you can.

9) *Answer the NCS's questions or inquiries briefly.* Avoid long explanations; just answer the question. If he asks if you will volunteer for a liaison job, just say yes (C) or no (N). If the latter, don't explain the reason.

10) *Know your procedure.* Nothing is more frustrating for an NCS than to have to explain procedure to someone who should be familiar with it. The more thoroughly familiar you are with the procedure the more valuable you are as a net station.

five sending his Avon message to W9NET, WA9XYZ reports in with another message to Avon. NCS might wait until W9NET returns to the net frequency, then dispatch him off frequency with WA9XYZ. Or, to speed things up, he might direct him as follows: WA9XYZ QNQ UP 5 AVON W9NET AFT WB9ABC, meaning "move up 5 kHz and send your Avon message to W9NET after he finishes with WB9ABC." Sometimes NCS uses only call suffixes instead of sending the entire call each time.

NCS instructions should always be acknowledged. The simplest way of doing this is just to depress your key. Some net stations say R (instructions received) or G (going to another frequency), but simply a dit or a dah is sufficient.

During the course of the net, after it appears that all stations have reported in, NCS may start excusing or releasing (QNX) those for whom there is no traffic and who have no traffic to send. The usual procedure is to call each such station and say QNX, at which the excused station will identify and leave the net. After all traffic has been cleared and the net has no further business, NCS will close the net. He may simply call the net and say QNF (net is free, no longer directed) or he may QNX each station individually. In any case, after all stations are excused and/or the net is free, NCS will say PSN DE K9NCS QNX QNF $\overline{SK}$, meaning "All stations are excused, the net is free, end of communication." Now stations may chat among themselves, if they wish. Up to that point, the only communication among stations is at the direction of the NCS.

Take a look at the list of QN signals. The use of some of them is self-explanatory. In most cw nets, some are never used because they are understood, but in beginner nets or nets in which there are many beginners, you may find much use of such signals as QNK, QNR, QND, QNO and QNY. In "savvy" nets, such as those at region level and above, they are seldom used. However, let us touch briefly on use of some of the less-obvious QN signals.

QNB: Often two net stations cannot copy each other (QNP), so NCS sends a third station off frequency with them to act as relay. The instruction might go like this: WD9DEF QNB W9NET ES WA9ABC UP 5, meaning "WD9DEF go up 5 kHz with W9NET and WA9ABC and act as relay for clearing their traffic." QNB is sometimes used by a net station to indicate he needs help, such as QNP WA9ABC NEED QNB, meaning "I cannot copy WA9ABC, I need a relay." This is a slight misuse of QNB, but it usually gets the job done.

QNC: When a net station has a message he wants to send to the entire net, such as a bulletin of general information, he may report it in his QNI as QNC 1. NCS may or may not inquire into the nature of the bulletin, depending on who the station is. In any case, when he tells that station to QNK QNC (send your message to all net stations), the station will send QNC several times, then send the general message in standard ARRL form. Usually he will end the transmission with NO QSL, meaning "No acknowledgement of receipt is required." If this is the case, the sending station gets one *Sent* point, but receiving stations do not get *Received* points. If the message ends with QSL (acknowledgement required), NCS will then call the roll and each net station will acknowledge receipt of the message, getting fills if necessary. In this case, all receiving stations get one *Received* point.

QNG: This signal has infrequent use, but it is important. Perhaps NCS does not show up at the net beginning time. It is customary to wait three minutes, then the alternate NCS may take over or, if there is none, the net manager may break in and direct someone to take over (e.g., WA9ANC QNG). During the course of net operation the NCS may be called away by an emergency, or he may have transmitter trouble, in which case he could ask a net station to take over. These things do not often happen, but the signal is there for use if they do. Once an alternate NCS takes charge, he keeps it; he does not give it back to the regular NCS should that individual show up late. Changing NCS in the middle of a net is a time-wasting and confusion-causing procedure.

QNQ and QNV are signals frequently used. We have already illustrated use of the former. QNV is used when NCS has some reason to think that two stations he wishes to pair to exchange traffic cannot copy each other. They establish contact on net frequency and, if copy is okay, move to the designated side frequency to exchange their traffic. NCS will direct the instruction (QNV) to the station who is to do the transmitting. If the station to receive cannot copy, he says QNP; otherwise, he says okay and off they go.

QNW may be used either by net control or a net station. NCS may use it any time during the course of the net, but the net station should use it only when reporting in, so he knows how to designate his traffic.

There are a great many variations in cw net procedure from the above, but this gives you an idea of the basic procedure used. After you have QNI'd a few times, you will begin to get the hang of it and most likely you will find it an exhilarating example of teamwork.

A section net on voice would be conducted in much the same fashion as above, but of course without the QN signals and other abbreviations. K9NCS would come on net frequency at the appointed time with a call-up something like this:

"Calling the Podunk Section Net, this is K9NCS, K9 November Charlie Sierra, net control, calling the Podunk Section Net." (This can be repeated several times so that all net stations can zero on the frequency.)

Some nets include a long, drawn out

explanation of the net's purpose, how often it meets, its operating procedure, etc., but this is unnecessary, a time-waster and not likely to impress anyone. After a short call-up as above, NCS will announce that stations will now report in. Usually he calls for stations with traffic first, or liaison stations. As each station with traffic reports in, he inquires if any station able to handle such traffic is on frequency, if not already reported in.

Dispatching of traffic off frequency is not so easy on phone, because stations occupy wider channels, there are more stations, and therefore there is less leeway for doing this. Add to this the fact that on 75 meters, on which most phone traffic nets meet, only the top 100 kHz of the band is available for net operation, since the remainder of the band is not open to General class licensees. This portion is therefore greatly overcrowded to begin with, so QRM becomes a decided problem on 75. Consequently, much traffic on 75-meter nets may be handled on net frequency. This makes the need for efficient operation even greater. But side frequencies should be used whenever possible.

Phone traffic nets are usually more relaxed and informal than their cw equivalents because of the very nature of the mode; but this informality can lead to laxity if not kept within limits. On RTTY, cw procedure can be used for the most part, with adaptations as necessary to accommodate circumstances peculiar to RTTY.

Pointers for the Net Control

It is commonly accepted that no net is any better than its control station and little better than the worst of its participating stations. To be a good net control, one must first have the right temperament and the proper perspective. One should be calm, self-assured, highly tolerant and patient, but still able to make proper decisions quickly. It's not an easy combination to come by, and by no means all net controls will possess all of them. The perspective must take into account that controlling a net is not a time or place to show off, to exert one's authority, to "throw your weight around." The NCS is doing a job, and it's a hard job. If it is done well, it can be a source of much pride, but it requires skill and experience and the utmost cooperation of all participants. Here are some factors an NCS should observe to "run a tight net."

1) *The net control is entirely responsible for the performance of his or her net.* He is appointed by the net manager because he is felt to have the necessary qualifications, and his performance will be compared with that of other net controls and with that of other nets. The net's performance will perforce suffer if participants are untrained or undisciplined, but the training and the maintenance of discipline are part of the NCS's job.

2) *NCS is the absolute boss* while the net is in directed session. This authority pertains only to a participant's performance in the net and does not extend to the maintenance of the station itself or the person of its operator. Within its scope, however, the NCS's authority is absolute, his instructions to be carried out promptly and without question if they are understood.

3) *NCS must call the net promptly at the appointed time.* Traffic reported in should be dispatched just as soon as two stations with traffic for each other are in the net. NCS must consider liaison schedules, perhaps calling for liaison stations to report in first (QNA). He must observe precedences and see that participants also do so. He should know his net, its members and their coverages, and be familiar with coverages of other nets with which liaison is conducted.

4) *NCS should have a commanding signal and superior operating ability.* A weak signal cannot adequately control a net, and a net control with a poor fist or faltering, stammering speech is usually an inadequate NCS.

5) *The NCS establishes the exact net frequency.* The published frequency is only for the purpose of helping participants find the net. NCS "fires up" on a clear or relatively clear adjacent frequency if the published net frequency is occupied at net time. Attempts to clear the frequency or keep it cleared are usually ill-advised. QRM is a way of life in Amateur Radio and we must learn to put up with it and work through it. All stations should be required to zero beat the NCS frequency.

6) *NCS should keep an accurate and complete log,* far beyond FCC requirements, showing the identity of each reporting station, traffic handled by each, from whom and to whom, and the beginning and ending time of the net. These data are usually reported to the net manager.

7) *NCS gets traffic moving as quickly as possible.* If stations can be dispatched to side frequencies to handle traffic, it can start flowing as soon as two stations report in with traffic for each other. Should traffic with EMERGENCY precedence be reported in (this will seldom happen in normal operation), all net business stops until that traffic is cleared. Priority traffic is the first order of business. Welfare and Routine traffic are then handled, with consideration being given to clearing those stations with the fewest messages first, or those who must liaison with other nets.

8) *Handling of "informal" traffic does not take place while the net is in directed session,* and is not considered traffic at all. It is the practice in many nets to permit such informal exchanges. An NCS who does not discourage this, if not outright prohibit it, is running the risk of having his net deteriorate into a bunch of individual ragchewers.

9) *NCS excuses stations promptly* as soon as their business is cleared. No station likes to sit idly and listen until the net is over, and requiring anybody to do so can greatly discourage attendance. The excused (QNX) station may continue to listen, but he is also free to leave.

10) *NCS closes the net promptly after all net business is cleared.* He should not engage in ragchewing with net members just because traffic is light. This can take place after the net is clear of all traffic and NCS has declared QNF.

Repeater Nets

Repeaters deserve special treatment, as they do indeed have a vitally important application to traffic- and emergency-net operation for local coverage purposes. In some cases, they can even effect section-wide coverage.

Since few traffic-net organizations own repeaters, a principal problem is to make contact with a repeater group and get exclusive use of the repeater during the period in which the net will operate. Most repeater clubs are willing to do this if the public service angle is emphasized properly. It is sometimes difficult to set the time during normal local and/or section net meeting times, however, since such times, occurring in early evening, are considered "prime."

The Repeater Net control station can be that of any operator with the qualifications who is "full quieting" into the repeater; thus the restriction to one with a commanding signal is not so compelling. Operation can then be conducted just about the same as any phone net. Any two stations that can hear each other direct can be dispatched to a simplex frequency. Others may be dispatched to another repeater if one or more are available. This puts an extra burden of knowledge on the control station and also requires more prearrangements. The nice thing about a repeater net is that while some stations operate only certain modes and certain bands, an increasing number of stations are equipped for 2-meter fm, since equipment cost is comparatively low and operation so simple. Furthermore, the day is fast approaching when repeaters in space or on the moon, and linked repeaters, will make wide-coverage nets practical on vhf and uhf. When this occurs, and it will, high-frequency nets may become history, like spark and the Trunk Lines.

Higher-Level Independent Nets

There are quite a number of traffic nets that operate at levels "higher" than section. This does not necessarily mean that their operators are more skilled, but only that the coverage is wider. You will find some of these nets on 80 cw and 75 phone, but most of them operate on 40- or 20-meter phone and cover vast areas; many are international in scope. Because of their special circumstances, most of these nets have their own special operating procedure. If you intend to participate in such a net, our best advice is to listen long and carefully before taking the plunge. Much of the procedure will be standard, as above, but some of it may be peculiar to the particular net and should be learned before reporting in.

The ARRL National Traffic System

For the first 30 years of Amateur Radio, the League's sponsored traffic organization consisted of Trunk Lines. At its peak, 14 of them criss-crossed the U.S. and Canada, not to mention a number of independent lines organized and operated by individuals not affiliated with ARRL. Trunk Lines were the last word in traffic routing in the twenties and thirties, with fixed-frequency operation using crystals (supplied by ARRL) and individual schedules between stations in the same

NATIONAL TRAFFIC SYSTEM AREA AND REGION MAP

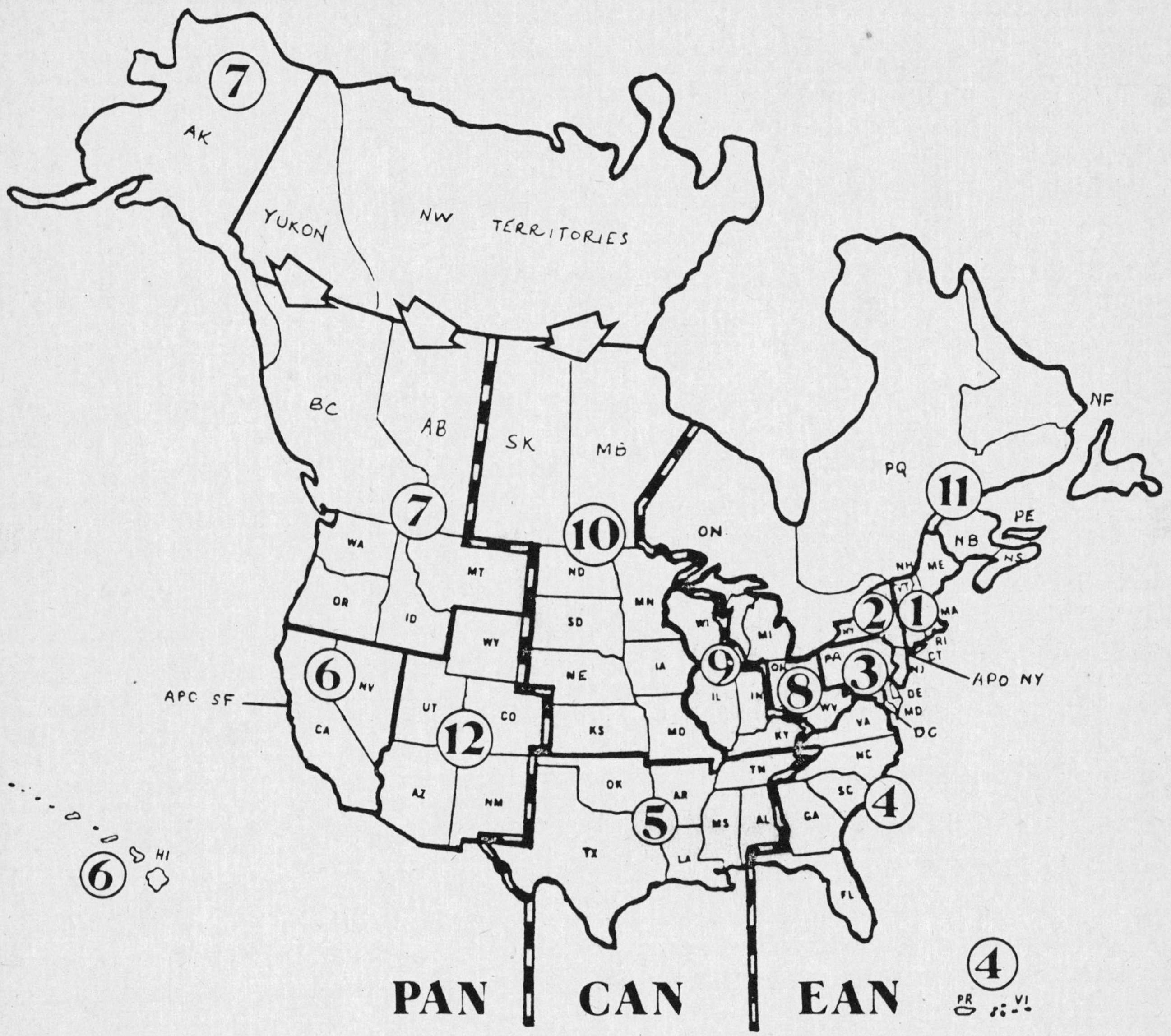

line. The system was reinstated after World War II, but the postwar breed of traffic operators discovered that things had changed. The Trunk Lines started operating like regional nets, and the concept of traffic flow from station to station along a prescribed route was gradually destroyed as stations from far outside the line's route started to take part. Still, the tradition hung on. As the system became a hodge-podge of overlapping regional nets, interline disputes arose concerning coverage and interconnections.

So in 1949, without totally abandoning the Trunk Line concept and all who espoused it, ARRL organized and started operating the National Traffic System, consisting of nets at three levels interconnected by designated liaison stations. The concept faltered, as trunk liners were slow to or refused to adopt it, but it finally caught on and now, after 30 years, is still the official traffic system of ARRL.

The NTS of today bears little resemblance in detail of operation to that which started on October 1, 1949, but the basis is still the same. The traffic stills flows from section to region to area net, whence it is transferred to its destination area net and comes back down through region and section nets to delivery. Now consisting of over 300 organized traffic nets, NTS is by far the largest, if not the only, amateur traffic net organization in existence and is known as the tightest-operating organization in all of Amateur Radio. The fine details of its operation are contained in the ARRL *Public Service Communiations Manual,* so will not be covered here, but some of the broader principles and procedures merit coverage here.

General Principles of Operation

From its very beginning, NTS abandoned the tradition of the "iron man" traffic handler who manned a Trunk Line post night after night (aside from weekends and summers). The new system was set up to operate every day throughout the year, including weekends and holidays, but it spread the work among as many qualified amateurs as were available. NTS nets exist at four levels — local, section, region and area. Local nets cover a comparatively small area, such as a city, county or designated "zone" within a section. They usually operate on vhf and often serve as both traffic and emergency nets. Section nets operate at the ARRL section level, each one covering a section or, in some cases where interest or availability of traffic is insufficient to support a section net, two or more sections. The next level up is the region net. The basis for the coverage of region nets is a call area, but this has been complicated by time differences and other considerations within some regions. The

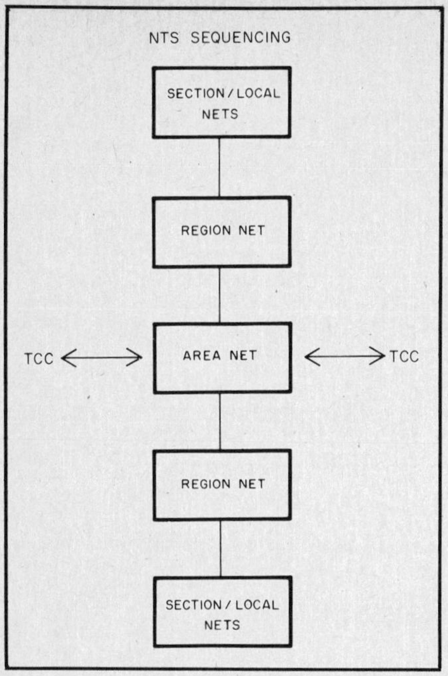

National Traffic System Routing Guide

State/Province	*Abbrev.*	*Region*	*Area*
Alaska	AK	7	PAN
Alabama	AL	5	CAN
Alberta	AB	7	PAN
Arizona	AZ	12	PAN
Arkansas	AR	5	CAN
British Columbia	BC	7	PAN
California	CA	6	PAN
Colorado	CO	12	PAN
Connecticut	CT	1	EAN
Delaware	DE	3	EAN
District of Columbia	DC	3	EAN
Florida	FL	4	EAN
Georgia	GA	4	EAN
Guam	GU	6	PAN
Hawaii	HI	6	PAN
Idaho	ID	7	PAN
Illinois	IL	9	CAN
Indiana	IN	9	CAN
Iowa	IA	10	CAN
Kansas	KS	10	CAN
Kentucky	KY	9	CAN
Labrador		11	EAN
Louisiana	LA	5	CAN
Maine	ME	1	EAN
Manitoba	MB	10	CAN
Maryland	MD	3	EAN
Massachusetts	MA	1	EAN
Michigan	MI	8	EAN
Minnesota	MN	10	CAN
Mississippi	MS	5	CAN
Missouri	MO	10	CAN
Montana	MT	7	PAN
Nebraska	NE	10	CAN
Nevada	NV	6	PAN
New Brunswick	NB	11	EAN
New Hampshire	NH	1	EAN
New Jersey	NJ	2	EAN
New Mexico	NM	12	PAN
New York	NY	2	EAN
Newfoundland	NF	11	EAN
North Carolina	NC	4	EAN
North Dakota	ND	10	CAN
Nova Scotia	NS	11	EAN
Ohio	OH	8	EAN
Oklahoma	OK	5	CAN
Ontario	ON	11	EAN
Oregon	OR	7	PAN
Pennsylvania	PA	3	EAN
Prince Edward Island	PEI	11	EAN
Puerto Rico	PR	4	EAN
Quebec	PQ	11	EAN
Rhode Island	RI	1	EAN
Saskatchewan	SK	10	CAN
South Carolina	SC	4	EAN
South Dakota	SD	10	CAN
Tennessee	TN	5	CAN
Texas	TX	5	CAN
Utah	UT	12	PAN
Vermont	VT	1	EAN
Virginia	VA	4	EAN
Washington	WA	7	PAN
West Virginia	WV	8	EAN
Wisconsin	WI	9	CAN
Wyoming	WY	12	PAN
Virgin Islands	VI	4	EAN
APO New York	APO NY	2	EAN
APO San Francisco	APO SF	6	PAN

only regions that now observe call area boundaries are the first, second, third and eighth. The map shows the NTS regions as they now exist.

The highest level of NTS net is the area net. There are three areas at present. The original basis for area nets was the four time zones in the contiguous U.S., but sparse amateur populations in the Mountain Area made it impossible to operate an area net there, and this became part of the Pacific Area. Even so, the time zone boundaries are not followed faithfully in designating NTS areas, since it is more practical to do this than to divide states or sections through which the time zone boundary passes.

All these NTS nets are bound together for traffic interchange by liaison stations designated by the various net managers. The identity of these liaison stations, varies from day to day insofar as the availability of qualified personnel permits. There are still a few "iron men" and "iron women" left, but they are a vanishing breed. The average NTS liaison or net control now does his stuff once a week, on a designated day. At other times, he reports into the section and/or local net to assist as possible at that level, or directs his energies toward other Amateur Radio pursuits. Versatility is a great thing. More amateurs should practice it.

But what makes NTS a system rather than a loose group of nets operating at various levels is the time sequence in which these nets operate. This principle also made initial organization so difficult that the venture almost collapsed before it took hold. For here it became necessary to find the stations that could or would fit into the system's operating timetable rather than arrange the system to suit the convenience of a majority of traffic-interested amateurs. The timing of net operations is crucial to NTS's success. Without nets meeting in proper sequence, traffic flowing through the system will be delayed until it was "slower than walking."

The timetable calls for local nets to meet first, followed by section nets. At section level, liaison stations are designated to go up to region level, with traffic destined for points outside the section. Region nets meet after section nets to clear intersectional traffic, and here also stations are designated to go to the area net with traffic designated for points outside the region. At the area level, region net representatives exchange traffic, and then, with each region representative clutching the traffic for his own region, the region nets meet again to distribute the traffic among section net representatives. Finally, the section nets meet the second time to further direct the traffic to stations from which it can be delivered. This completes the NTS cycle in a single area.

But what of the large volume of traffic destined from one area to another? This is where the NTS Transcontinental Corps (TCC) comes into the picture. The TCC is not a net, but a corps of elite operators specifically recruited for the purpose of transferring traffic from one area to another so that it can receive subsequent handling in its area of destination. This is done by means of liaison stations in each area net who are there for the purpose of clearing traffic for a specific other area and seeing that the traffic gets to that area in the shortest possible time. The various TCC functions are designated by letters. For example, the station having the function of reporting into the Eastern Area Net (evening session) for the purpose of taking traffic for the Central Area is designated A, the station to take traffic for the Pacific Area is B, etc., with different functions all the way up through V. If the specific identity of such stations changes every day of the week, this means that 151 stations would be serving in TCC capacities alone. So NTS has plenty of room for qualified skilled operators.

But how do these stations get this traffic to its destination area? The obvious way is by reporting it into the destination area net. However, this method has one big drawback: The distances to be covered are often too great on the net's frequency. The TCC functionary is thus often unable to hear his contact well enough to copy, his contact cannot hear him; or both. To solve this dilemma, the traffic is sent to another station is the destination area in an out-of-net schedule, and *that* station (who also receives a TCC letter-designation) reports it into his own area net. What's the difference? The difference is that an out-of-net schedule has some leeway in time and maximum leeway in selection of frequency band. The two stations can arrange to meet on a band on which the path is optimum at that time, and often can even select an optimum time, whereas direct QNI requires the reporting station to report into the area net at the time and on the frequency on which that net meets. This frequency and time are not optional. They must conform to system standards.

NTS Cycles

We mentioned above that the

NTS and The Phase III Amateur Satellite

Phase III has important ramifications for the future of the Amateur Service; unlike the present amateur satellites, Phase III will offer an average of eight hours of reliable propagation per orbit (there will be approximately two orbits per day). Also unique is the concept of specialized service channels at the outer edges of the passband. There will be a 150-kHz passband, with special cw channels at the low end, 124 kHz of mixed-mode general amateur operation in the middle, and special ssb channels at the high end. In an emergency, all routine communications will cease; specific emergency notification procedures will be formulated and disseminated at a later date (an emergency-calling frequency may also be shared with another service on one of the ssb channels). Phase III will utilize a Mode B transponder; that is, you transmit on 70 centimeters and receive on 2 meters.

One of the cw channels is to be allocated as the National Traffic System channel, for the daily handling of authorized formal traffic. It should be an exciting challenge.

It is envisioned that the primary use of the NTS segment will be for TCC (transcontinental) schedules. Traffic *nets*, as such, won't be pushed because the satellite will not be in the same place at the same time each day. Nor will TCC-type functions be all the action, because the traffic handling potential or utility is unlimited. If capabilities and conditions permit, net controls can direct stations on NTS area, region and section nets to QSY to the spacecraft, instead of down five, to handle their traffic. Most importantly, traffic "hotlines," for dispatching emergency/priority traffic during crises, should be relatively easy to set up. A 24-hour NTS schedule is very much in the hearts and minds of the people in the traffic community also; Phase III may cause or contribute to the ultimate expansion.

On the subject of expansion: Remember the war protesters who chanted, "The whole world is watching"? On this new satellite, the whole world will be *listening*. Amateurs from many different countries will get exposure to daily, efficient NTS operations (a network they may have heard or read about, but never actually *heard*) so we may see the day when the National Traffic System becomes the *Inter*National Traffic System.

The satellite will give us a more reliable means of contact with South and Central America, where communications disruptions aren't uncommon and third-party agreements are common. Though we don't have third-party traffic treaties with European nations, it's no secret that amateurs there have a great interest in formal traffic handling. Let's hope that Phase III helps secure those third-party agreements. A worldwide traffic system beckons.

As a comparison with another interest area, the traditional problem with RTTY traffic handling is that RTTY freaks don't care about traffic and traffickers don't care to acquire RTTY gear, even though it's the most efficient way to handle messages, especially in high-volume situations. This is a similar case. At present, there are few NTSers involved in satellite work and vice versa. But for the NTS channel to be a success, we need coordination, cooperation and *participation*. For traffic handlers, forthcoming issues of *QST* will detail the equipment needed to access the spacecraft (briefly, 1000 watts erp on 435 MHz — 10 watts into a 10-dB amplifier into a 10-dB antenna with circular polarization). And we'd certainly appreciate the old hands at the satellite game trying to get involved in organized public service activities. The NTS channel can't suffer from neglect; otherwise it will cease to be the NTS channel.

It will require much planning and work to make this particular channel what it should be and we'd like to hear from those interested in taking part in the effort. The intent is not to get too heavy, though, as AMSATer WA2LQQ points out: "The prospects for accomplishment are truly enthralling. And the promise of productive public service, scientific advancement, technical innovation, international cooperation and just plain fun in 'hamming' makes these exciting times indeed for Amateur Radio." Traffic devotees in particular should keep in mind that North America is no longer the only game in town. For more on Phase III, refer to the satellite section of this book.

completion of the net sequence from local to area and back to section completes one cycle of NTS operation in a single area. It then follows that the completion of this sequence in all three areas constitutes a full cycle of NTS operation throughout the system. However, there are *two* cycles of NTS operating daily, one in the evening and one during the day. The evening cycle has been standard since 1949, while the daytime cycle was implemented in 1973.

The basic evening schedule has section nets meeting between 1630 local and 1900 local, followed by the region net at 1945, the area net at 2030, the second session of the region net at 2130 and section nets at 2200. Each of the three areas (eastern, central and pacific) adhere to this basic schedule. Similarly, in the daytime, section nets meet at 1200 local, regions at 1330, area nets at 1430, regions at 1530; the late daytime section nets can be separate nets or can merge into the first evening section nets in the 1630-1900 local time period. The areas, of course, are tied together by TCC schedules.

The starting and ending times of nets within the evening portion of NTS are fairly tightly constrained (say ± 15 minutes) to their present values by the limited hours available. The daytime cycle does not need to be so restricted, however. One of the advantages to participants in the daytime cycle is greater flexibility in the times that section and region nets meet, as well as breathing space between net sessions. There is adequate provision adjacent to all section nets for a layer of local nets. Since the daytime and evening cycles of the system are part of one unified NTS, linkage of the two cycles is a primary function of TCC.

In emergencies or overload situations, extra cycles of NTS can easily be held at three-hour intervals. Therefore, in addition to the normal area net sessions at 2:30 P.M. and 8:30 P.M., new area net sessions at 11:30 and 5:30 P.M. can be held, along with their associated region and section/local nets. The system is symetrical, regular and repeatable. All this sounds very complicated but it is actually quite simple. In NTS, things are done systematically; we don't just "play it by ear," so the methods are standardized. Full details of NTS operation can be found in the *Public Service Communications Manual*, available from ARRL headquarters. Specific information on net frequencies can be found in *The Net Directory*, also available from Hq.

NTS detractors are fond of saying that the many relays necessitated by following the system only invite message garbling. This is true to some extent, and it is also true that many messages going through the system are garbled, but these are human errors, not system defects — efforts which only training and experience can minimize, never eliminate, in NTS or any other traffic system. It is also true that for this or that single message or group of messages it may indeed be faster to transmit them directly to their destination locality — provided a station willing, capable and responsible can be found to receive them. NTS is geared to systematic handling of message traffic, all traffic handled in accordance with its precedence. It relies not on any single station or net, nor does it require extended daily effort on the part of anyone. This is the measure of its success throughout the years and as long as amateurs are permitted to offer this public service.

The Upper Echelon of NTS

Local and section NTS nets are under the direction of the SCM and his leadership appointees, STMs and NMs. But

Who Owns the Frequency?

Traffic nets often have a problem when it comes time for the call-up and a ragchew is taking place on the published net frequency. What to do? Well, you could break in on the ragchew and ask politely if the participants would mind relinquishing the frequency. This usually works, but what if it doesn't? The net has no more right to the frequency than the stations occupying it at net time, and the ragchew stations would be perfectly within their rights to refuse to relinquish it.

The best thing to do in such a case is to call the net near, but not directly on, the stations — far enough (hopefully) to avoid causing QRM, but not so far that net stations can't find the net. Usually, the ragchewers will hear the net and move a bit further away — or even if they don't, the net can usually live with the situation until the ragchew is over.

It is possible to conceive of a situation, especially on 75-meter phone, in which the net frequency is occupied and the entire segment from 3900 to 4000 kHz is wall-to-wall stations. In such a case, the net will just have to do the best it can and take its chances on QRMing and being QRM'd. In any case, it is *not* progressive to argue about who has the most right to the frequency. Common courtesy says that the first occupants have, but there are many extenuating circumstances. Try to avoid such controversies, especially on the air. They make us look bad.

The Net Control Sheet

Net controlling is no easy task, requiring much talent on the part of the operator. A useful "prop" is a set system for keeping track of net operation so that you don't get balled up and start to lose control. This can best be effected by a sheet of paper on which you record who has what traffic, who covers what, and who is on what side frequency. Trying to keep this information in your head is a losing battle unless you have a remarkable memory.

There are many methods for doing this, depending to a great extent on what net is being controlled and the exact procedure used. In general, however, the best method is to list the calls of stations reporting in vertically down the page, followed by their coverage, if known. The coverage may be unnecessary if NCS knows his stations, but it is a good idea to leave space for it in case an unfamiliar station reports in and you have to ask his location.

Next, list horizontally across the page the traffic reported by each station coming into the net, using destination (abbreviated) followed by the number of messages. From this you can see at a glance when traffic flow can start. In most nets it is best to start it right away, not wait until all stations have reported in. As traffic is cleared, it can be crossed off. Whenever you get a station who has no traffic and for whom there is none, that station can be excused (QNX) and crossed off the list. As stations clear traffic and there is none for them, they also can be excused.

If side frequency (QNY) procedure is used, net controlling is a bit more complicated, but the use of such frequencies vastly speeds up the process. In this case you will need to keep track of who is on which side frequency clearing which traffic, and you will be kept busy dispatching them, sending station up or down to meet stations already on side frequencies, checking stations back into the net as they return, etc. Both your fingers and your key or mic button will be kept going, and it can be a nightmare if not handled properly.

Probably the best method is to keep the side frequencies on a separate sheet, each side frequency utilizing a separate column, labeled up 5, up 10, down 5, down 10, or whatever spacing intervals you find practical. As two stations are dispatched to a side frequency, enter the suffixes of their calls in the appropriate column, at the same time crossing out the traffic they are sent there to handle. When they return, cross them out of the side-frequency column; this side frequency can then accommodate two other stations. Of course, if you dispatch a third station to the side frequency to wait to clear traffic to one of the stations already there, enter him in the column also, and then only one of the two originally dispatched will return, so just cross off that one. When all your listed traffic is crossed off, the net is ready to secure (QNX QNF).

There are a number of refinements to this method, but the above is basic and a good way to start. Experience will soon indicate better ways to do it. For instance, K2KIR's novel approach follows.

Net Controlling EAN

At the NTS Area and Region levels, the complexity of the Net Control task often suggests the desirability of using a matrix form of log sheet. Many of the EAN Net Control Stations use the form shown in Fig. 1. A similar form for region nets can be made by replacing the RN column headings with the Section Net designations and the CAN/PAN columns with single column titled THRU (or EAN, CAN or PAN). Although this form can be used with no other accessories than a pencil, the use of moveable objects such as 6-32 hex nuts, buttons, push pins, etc., materially aids the visualization of which stations are on the net frequency and hence which stations are available for pairing off and passing traffic.

As stations are sent off frequency, their hex nuts are moved from the NET column to the appropriate side frequency and a single diagonal line is drawn through the traffic being cleared. If, for any reason, the traffic is *not* cleared, a circle is drawn around the traffic total as a reminder to the NCS that it still must be cleared (in other words, the circle overrides the slash). Assuming the traffic is cleared on the second attempt, an opposite diagonal is drawn, thus totally crossing out the traffic quantity.

For the example shown, the following notes should explain how the various features of this NCS sheet are utilized:

Halfway through the Eastern Area Net session shown in Fig. 1:

W1EFW and VE3GOL are both clear and have been QNXed at 0055 and 0057, respectively.

W8PMJ has cleared his CAN and PAN traffic, as well as 8RN traffic from W1TN. He has yet to clear his 3RN, plus receive 8RN from K3NGN. He is presently DOWN TEN with W2CS.

W2CS was previously sent off to clear his 4RN with KB4N, but they were unable to complete the pairing. NCS will try again later, perhaps by using a relay station.

W4SQQ has a net report for K2KIR, to be sent direct to him if he QNIs.

WB2TOM and K1XA are UP TEN, clearing 2RN traffic.

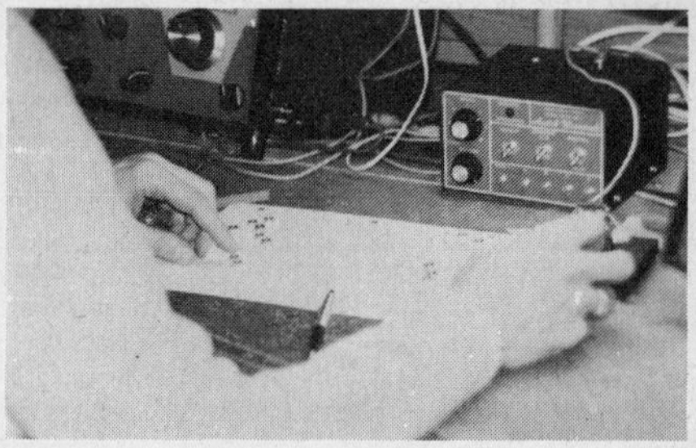

WA3PXA, KB4N and K4SCL are standing by on the net frequency, waiting for new assignments.

There are many techniques for determining in what order to clear the traffic listed, but a couple of fairly common algorithms are worth mentioning.

1) Assign highest priorities to stations having the largest totals. Thus, in Fig. 1, W1NJM (PAN RX), W8PMJ (8RN TX and RX) and K3NGN (3RN TX) should be kept waiting as little as possible. If conditions are good, QNQing other stations to these three will speed things up immensely.

2) Tackle the smaller individual destination totals first. This allows early excusing of stations having small traffic totals to clear, and avoids a last-minute panic near the end of the net.

3) Clear short-haul pairings first. In the winter near a sunspot minimum, short-haul communications are most likely to be successful early in the net session before the skip has gone out. At all other times, especially in the summer with nets on Daylight Savings Time, it delays long-haul pairings until band conditions have improved for those paths.

Obviously, there are times when these (or any other algorithms) will conflict with each other. In the final analysis, a good NCS bases his decisions on the specifics of a given net session: the band conditions, the operators, the traffic distribution and the time remaining.

— *Bud Hippisley, K2KIR*

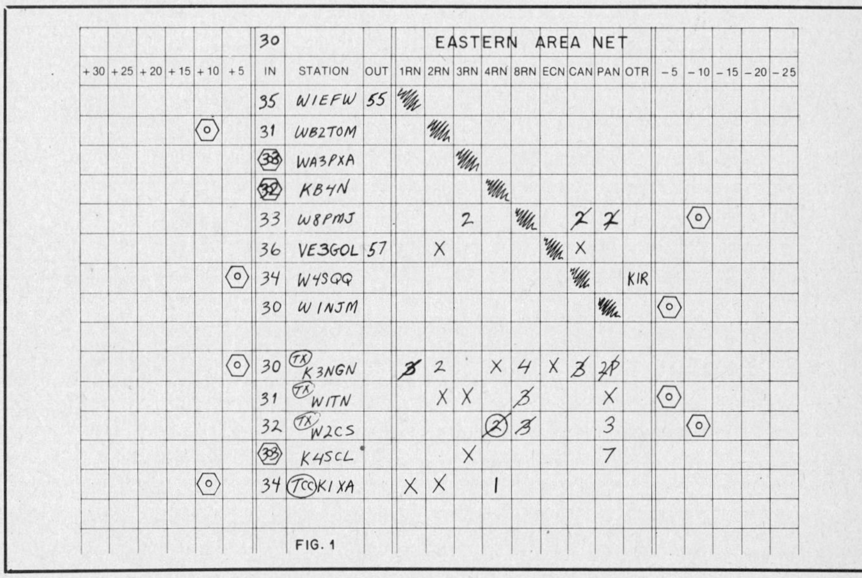

FIG. 1

since region and area nets and the TCC far transcend section boundaries, the appointments at this level are handled directly from ARRL headquarters, under the supervision of the communications manager. There are 36 such appointments — 24 region net managers, 6 area net managers and 6 TCC directors. An assistant communications manager at headquarters is delerated the function of handling these appointments.

At such a distance from many of the appointees, headquarters' personnel are not always in a good position to pick the most qualified person, and so a system of advisory committees has been set up called the NTS Area Staffs. There is one Area Staff in each NTS area. Each staff con-

MARS

It is not generally known by modern-day traffic handlers that our standard message preamble is largely fashioned after the form used in the Army Amateur Radio System, which had its heyday in the thirties. The ARRL standard preamble prior to that time was quite different, but the AARS form was adopted because it had advantages and was so widely used. AARS nets were numerous in the thirties, using WL calls on two frequencies (3497.5 and 6990 kc.) outside the amateur bands and many frequencies, using amateur calls and amateur participants, inside the bands.

MARS, the post-World War II successor to AARS, encompasses all three armed services. Although it operates numerous nets, all of which are outside the amateur bands on military frequencies, thousands of amateurs participate in MARS. This service performs some traffic coverage that we amateurs are not permitted to perform. MARS, which stands for "Military Affiliate Radio System" is conducted in three different organizations under the direction of the Army, Navy and Air Force. Therefore, MARS is not in the strictest sense Amateur Radio.

Nevertheless, many amateur messages find their way into MARS circuits, and MARS messages find their way into amateur nets. In fact, NTS has a semi-formal liaison with MARS on the West Coast to handle the many messages from families in the states to their sons and daughters serving overseas. A similar liaison on the East Coast also exists.

Traffic for some points overseas at which U.S. military personnel are stationed can be handled via MARS, provided a complete military address is given, even though some of these points cannot be covered by Amateur Radio or NTS. The traffic is originated in standard ARRL form and refiled into MARS form (now quite a bit different from ours) when it is introduced into a MARS circuit for transmission overseas. In this manner, traffic may be exchanged with military personnel in Germany, Japan and a few other countries which do not otherwise permit the handling of international third-party traffic.

Traffic coming from MARS circuits into amateur nets for delivery by Amateur Radio are converted from MARS to amateur form and handled as any other amateur message. The *exception* occurs when traffic originates overseas. In this case, the name of the country in which it originates, followed by "via MARS," should appear as the place of origin, so it does not appear that such messages were handled illegally by Amateur Radio. There are places in which U.S. military personnel are stationed with which even MARS cannot handle traffic, presumably because of objections by the host country. This information is in the hands of MARS "gateway" stations and changes from time to time.

The amount of MARS traffic appearing on amateur nets is not great, since MARS has a pretty good system of handling it on MARS frequencies, but it is important that we maintain liaison as closely as possible since all civilian MARS members are U.S.-licensed amateurs.

Information concerning MARS may be obtained directly from the individual branches at these addresses:

Air Force MARS:
Chief, USAF MARS
HQ AFCS/DOYR
Scott AFB, IL 62225

Army MARS:
Commander
U.S. Army Communications Command
Attn: CC OPS OM
Fort Huachuca, AZ 85613

Navy-Marine Corps MARS:
Chief, Navy-Marine Corps MARS
Bldg. 13, NAVCOMMU Washington
Washington, DC 20390

sists of the region net managers, the area net managers, the TCC directors and three "members-at-large" (MAL) selected by the rest of the staff. Thus, all 45 of the NTS higher-echelon appointees serve in an advisory capacity to the communications manager.

The area staffs meet in person at infrequent intervals, whenever it is possible to get them together, and discuss problems being experienced in their respective areas in particular and in the entire NTS in general. Headquarters relies heavily on area staffs to recommend replacements for managership vacancies at region, area and TCC level. Recommendations are drawn up and submitted to the communications manager. The area staffs, like other ARRL advisory committees, can only recommend; they have no directive powers. The communications manager, in considering an area staff recommendation, must decide whether any proposal is feasible as applied to the entire system, not just the area from which it originated. Naturally, the problems of one area are not always the same problems as those of another area, but system procedures must be standard throughout for best system effectiveness.

Alternate NTS Routings

Things don't always go according to plan. If a liaison station doesn't show up, or if an outlet to a certain destination is unavailable, NCS may have to do some fancy improvising. He can always ask QNW (How do I route traffic for . . .?) and hope to get a volunteer, or he can route the traffic through a station known to be able to take it to another net.

In the average NTS net, the normal NTS routing is sacred only when it is available. If not available (temporarily, of course), NCS may route the traffic any way that *is* available, even if it is irregular.

Here's an example. The manager of a region net finds that he gets more than enough liaisons from and to Section A, but Section B often doesn't show. Nothing wrong with assigning a Section A station to handle Section B traffic (and take it to the Section B net) — temporarily, until a Section B station becomes available.

NTS QNI Policy

National Traffic System nets at local and section level are open to all amateurs in the net's coverage area. The object is coverage, and the more stations QNI the better. At region and area levels, participation is normally restricted to representatives of sections and designated liaison stations. However, stations from outside the coverage area of the net concerned, or other not-regularly-designated stations who report in *with traffic* will be cleared, provided they can maintain the pace of the net as to procedure, speed and general net "savvy." Such stations reporting in *without* traffic will immediately be excused by the NCS unless they can supply outlets not at that time available through normal NTS channels. Visitors to NTS nets should bear in mind that they operate on a time schedule and that no offense is intended by observance of this QNI policy.

Handling Traffic in Emergencies

Contrary to popular belief, traffic-handling procedures and rules do not go out the window in an emergency. There is still need for formal traffic and operators who know how to handle it. In the early hours of a communications emergency there is likely to be some EMERGENCY and Priority traffic to handle, and once commercial communication is reestablished to take care of such traffic there is sure to be much Welfare traffic. On such occasions, nets will operate extra sessions or continuously, in accordance with the apparent needs and in coordination with ARES personnel in charge of the amateur phase of operation.

In all amateur traffic operating, EMERGENCY precedence messages are handled *first*; they can bypass any normal NTS channels if this will get them delivered more quickly. Remember, an EMERGENCY precedence message signifies a life-or-death situation affecting the welfare of groups of people or the general public. As soon as an EMERGENCY precedence message gets out of the disaster zone to a place where normal telephone service is available, it should be sped to delivery by that means. Emergency operations are treated in detail in the next chapter.

Chapter 5

Emergency Communications

The laws and regulations supporting the existence of the Amateur Radio Service justify Amateur Radio as a noncommercial service whose value (in part) is to provide emergency communications. Another aspect of the justification is to provide a pool of skilled communicators available to the country in time of need.

The ARRL recognizes the need for emergency preparedness and emergency communications training through the Amateur Radio Emergency Service. ARES will be discussed in detail since it is the organization *directly* sponsored by ARRL. Other organizations and their relation to ARES will be discussed later in this chapter.

Communications Emergencies

There is a tendency in modern society to lose perspective concerning the base it is organized on. For example, we do not recognize our dependence on telephones and commercial electrical power until something causes disruption of these services. Most local governmental jurisdictions have provided police, sheriffs, fire departments, etc., with radio communications and thus feel relatively independent of the telephone and power services but have not really considered the impact of the loss of those services on the radio links they have. Virtually all state governments have emergency communications plans and these are tied together at the federal level by the Federal Emergency Management Administration. Some of these systems are quite sophisticated and extensive. However, they are *all* susceptible to overload and malfunction. It is economically impossible to provide total backup for the telephone service. When the general population cannot do its own communicating, the requirement exists for a service that can do it for them. Very few government officials at any level are trained as communicators, so even when the backup systems are put into effect there is a shortage of adequately trained personnel.

When a situation occurs that requires handling a large amount of information in a short period of time, we have a communications emergency. This is when ARES can provide valuable support to the community. It should be noted that the telephone lines do not have to be disrupted for such an emergency to occur. All that is required is that the amount of information desired to be transmitted exceeds the capacity of the existing telephone and radio services. This happens frequently in any situation that affects a large number of the population in a given area. Further, there are still many areas where the number of telephone lines barely support the existing daily loads.

Tactical Communciations

In an emergency situation that affects only a small number of people in a small area for a short period time, the need for communications will most likely be of the

The Pasadena, California, Radio Club and the Pasadena chapter of the Red Cross teamed up for operation during ARRL's annual Simulated Emergency Test. From left W6HAH, W6ZH, Joshua Potter of Red Cross, and N6UA provide a communications link from the chapter house. *(Walt Mancini, Pasadena* Star-News *photo)*

WB5YJO was one of more than a hundred ARES members involved in the Wichita Falls tornado emergency. *(K5HGL photo)*

type which we will label "tactical." The requirement will be to get the information in and out of the area in the time frame of seconds or minutes at most. The information will usually be in the form of instructions or inquiries. Situations covered by this could include automobile accidents, lost-persons searches, fires, train or airplane wrecks, tornadoes and local flooding. These communications are normally best carried out in network fashion utilizing frequencies where the propagation characteristics are such as to confine the signals to the local area. Vhf, uhf or 10-meter networks are typical, although 10 and 6 meters can be a problem if the bands are open to other areas.

Because of the growth of the use of 2-meter fm and repeaters, this band has become a natural for such communications. The quality of fm signals allows for more-precise verbal communication. The only drawback is in another characteristic of fm: The station with the strongest signal will "capture" the frequency. This can lead to communciations difficulties when operators desire to help and are not properly trained in emergency procedures. There is a tendency to feel that such communications "come natural" to the amateur and that we will be able to handle it without much preparation. Experience shows otherwise and such training is one of the major functions of a local ARES organization.

Tactical communications put a great burden on the amateur since there is no written text involved. Care must be taken to keep the information accurate and simple. A close relationship between the amateurs and local authorities is necessary to prevent misunderstandings and the dissemination of misleading information. A rumor or false report spread through an amateur network can undo in a few moments what has been accomplished by months of planning and cooperation.

Formal Message Communications

When the emergency situation affects a large number of people over an extended area and will last for a long time, there will be need for more-permanent type of communications. This need is most-readily filled by use of the standard ARRL message form in networks set up for the purpose. Formal traffic also has the advantage that there is less likelihood of misunderstandings since the civil authorities "sign off" on the outgoing message and receive a copy of the delivered message in most cases. Further, the messages are capable of being sent throughout the U.S. and Canada and most parts of the world where such messages are allowed and they can be routed via the National Traffic System. A great number of public agencies and the general public are already familiar with such messages, which also facilitates their origination and delivery. The ARRL also has established good relationships with services such as the American Red Cross and the Salvation Army to expedite the handling of such information when these agencies become involved in an emergency situation.

Emergencies that would give rise to the requirement for formal message communications could include hurricanes, snowstorms, major flooding and earthquakes. Many of the disasters in this country are related to adverse weather conditions and one of the uses for amateur communications is the collection of weather information in situations that could lead to a disaster. In several areas of the country, collection of the weather information in conjunction with the National Weather Service has proven extremely beneficial. In many cases the information is conveyed by formal messages. Another advantage of this system is that the amateurs are already in place and operating so that they can easily shift to handling emergency and priority traffic should the situation get worse.

Emergency Networks

Whether the need for communications is tactical or by formal message, it becomes apparent that the basis of amateur assistance will be a group of amateurs communicating using a series of mutually agreed-upon frequencies. Once we have the amateurs at the necessary locations and the equipment operating, the next problem is to maintain an orderly flow of information between the necessary points. Without some kind of formal procedures that are mutually followed, the result would be chaos instead of the intended public service.

Emergency nets have a special problem in that they have to handle both tactical and formal messages under conditions where there will be greatly increased participation resulting from everyone's desire to help. This last factor has caused great problems in many emergency situations. The answer appears to be to have designated net control stations who are trained in "crowd control" and to have a network plan that is known to many in an area and is frequently exercised. One of the main duties of the ARES emergency coordinators, district emergency coordinators and section emergency coordinators is to develop such communications plans and arrange for groups of ARES members to test out such plans periodically. The net control stations with experience can sometimes be recruited from operators who are regulars in traffic nets. The procedures can also be outlined as modifications of traffic handling procedure. These procedures are spelled out in the ARRL *Public Service Communications Manual.* While it seems simple to participate in such a net, the necessary procedures cover a wide range of problems and practice is probably the best teacher.

2-Meter Fm Nets

The 2-meter fm net is currently the basis for local emergency communications because of the widespread use and coverage of repeaters. Repeaters are especially helpful because of the use of mobile and portable stations in the typical local emergency. When the repeaters have emergency-power capabilities they become the backbone of the local system. Repeaters without emergency power are still useful but should not be relied upon in any realistic emergency-communications plan.

The issue of appropriate procedures is extremely important on 2-meter fm. Knowing when to talk and when to listen is probably the single biggest problem. The rule of thumb is to listen unless specifically called. Recall that a strong station checking in simply to find out what is going on will block a weaker station who may be at the emergency site. The Net Control Station must attempt to keep the frequency as clear as possible for incoming calls. The participating stations can assist by following directions as carefully as possible. In some localities fortunate enough to have two or more repeaters or with good simplex coverage, routine formal traffic or non-emergency informal communications can be sent off frequency. In a true emergency situation, nothing other than strictly emergency traffic may be allowed on the frequency. The issue of the priority of the information is a critical one and amateurs must carefully consider the importance of their information compared with other information on the circuit. Where there is close liaison between the amateur group and local civil authorities, this decision can be a bit easier. As communicators, amateurs are not usually in a position to determine the relative importance of two pieces of information. When this decision can be made by the officials who have that responsibility, the amateur role is much clearer and the emergency nets usually function more smoothly.

Another important function of the local net is to provide liaison to hf nets with wider coverage for the purpose of getting messages into and out of the local area. This liaison function is an important one and needs to be practiced and also made part of the plans. The liaison will usually involve getting formal messages in high volume to a section or higher-level net that will route such messages to their destination. The National Traffic System can play a major role here, as the routing already exists and is practiced daily. This means that the liaison stations should have experience in operating in NTS nets.

Hf Nets

In major emergencies it is likely that formal traffic, especially health-and-welfare messages, will become a significant factor in the amateur workload. These messages will usually find their way into the nets of section-level (or more) coverage operating predominantly on the 80/75- and 40-meter bands. Most of the NTS traffic nets operate on these frequencies. These nets may also end up with tactical communications requirements in some cases. While many of the problems are the same, the hf nets increase the "crowd control" burden on NCS because of their increased coverage. Further, in an emergency extending over a period that exceeds the usual time the net occupies a frequency (assuming it is a net which meets often), problems with frequency coordination can occur. Other nets may use the same or adjacent frequencies

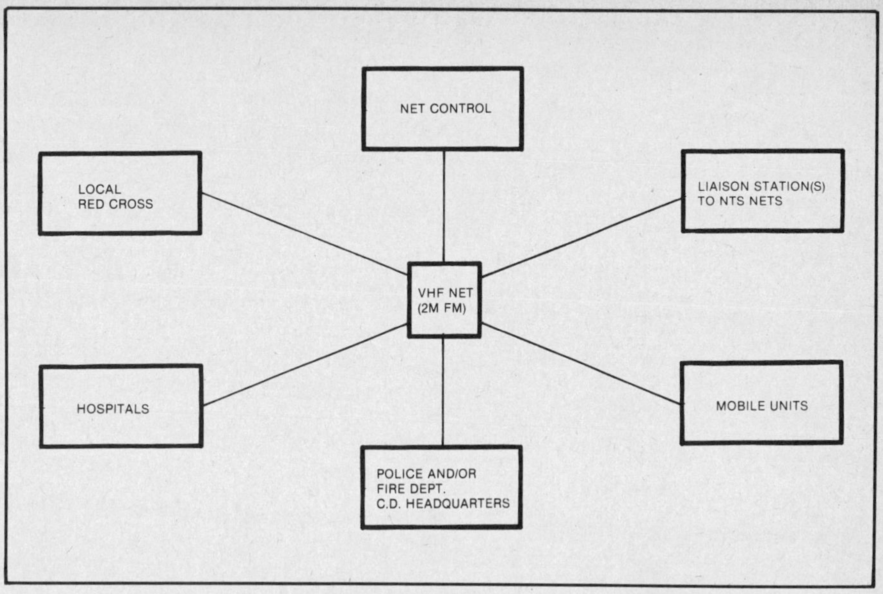

Fig. 1 — Typical station deployment for local ARES net coverage in an emergency.

during scheduled times with no interference between operations that during an emergency cannot operate without some change in frequencies, times or methods.

NTS training will usually be of great value during these types of operations since the traffic loads can become quite high. Nets that are used to 20 to 30 messages in a normal session may be surprised to find hundreds of messages listed. This should not be surprising if one considers that even in a small community a significant number of residents may wish to contact friends and relatives concerning the situation. The hf traffic nets should be prepared to cope with high volumes of traffic for extended periods.

General Emergency Net Structure

While many of the operating procedures for an emergency net can be taken directly from traffic-net procedures, some additional items should be considered. Of greatest importance is the distribution of the net members and arrangement of the participating stations. The two situations most amateurs are likely to become involved in concern the use of emergency nets at the local and ARRL section (or state) level.

Fig. 1 shows a network arrangement for a local emergency situation such as a flood, hurricane or tornado. Although the stations filling the various functions will rotate depending on the length of the emergency situation, the functions shown are typically those that will need to be covered. Stations should be placed where the lack of communications is most likely to disrupt vital community services. The figure shows hospitals, the local Red Cross chapter and the police or fire departments as likely locations. Obviously, operation from such locations has to be arranged in advance of an actual emergency to avoid confusion and interference with the operation of these agencies. This point will be covered in more detail later.

Assuming these arrangements have been made, there will usually also be a need for mobile units to check out various areas of the community and provide relief communications as the situation develops. An experienced traffic handler, or better yet a group of traffic handlers that can "spell" each other, should be used for liaison to Section or other nets to get formal traffic in and out of the locality. Generally, there will be a need for two such operators in the local net and two in the Section or State net at the same time. One of the operators in the local net is picking up outgoing traffic to take to the higher level net and the other has come in from the higher-level net with traffic for the locality. This would occur after the emergency had been in progress for several hours, assuming the situation was one that would generate health-and-welfare traffic. If the Section or State Net were running at the same time, there would be the next shift of these two operators in that net. One of them would be picking up the traffic to return to the local net and the other would be relaying the traffic from the local net. Obviously, when traffic is not great the same station could perform both functions. This latter situation is the one that occurs most often and is the way most local nets that meet regularly practice for liaison with the Section Nets.

In times of emergency the Section or State Nets can also take on a slightly different composition. Fig. 2 shows a typical structure for an hf net during an emergency. The liaison stations to the local nets have been covered above. The local independent outlets for traffic are simply the normal check-ins that such nets would

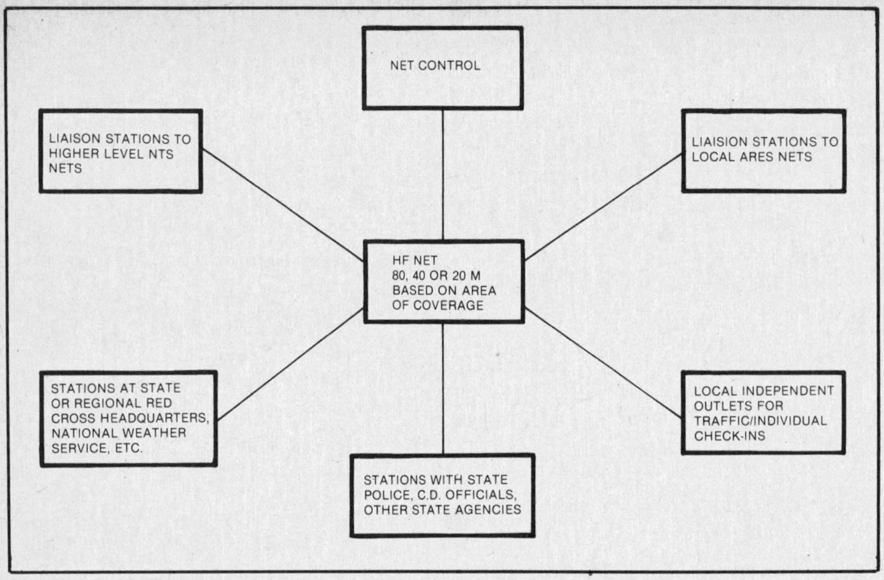

Fig. 2 — Typical structure of hf network for emergency communications.

usually have, i.e., stations that are the usual net members. Most Section Nets maintain liaison to the NTS Region Nets so this part of structure is nothing new. The station locations that represent a departure from the routine would be those located at strategic places likely to have a communications need resulting in formal traffic. Such places could include the state police, the state emergency preparedness center (sometimes also doubling as the civil defense center) and other state agencies. Another set of such locations could be the state headquarters of agencies that normally function in emergencies including the Red Cross, Salvation Army and National Weather Service.

The exact disposition of the stations in either case will depend on the situation in your location and the plans that have been made prior to the need for such communications.

Operating in an Emergency

Operating during an emergency requires some thought beforehand if the activity of your station is to be used to full advantage. The requirements are different if you are operating from the emergency area than if you are outside the area and simply assisting area stations. Since the latter case will usually involve assisting with the routing of formal traffic, it does not require special emphasis beyond your skill as a traffic handler. Traffic handling is covered in the previous chapter. If you are able to assist with "informal" traffic that might arise and are outside the emergency area, remember the basic law of emergency operation, *"listen and follow the instructions of the Net Control Stations."*

One can never really tell when one will be *in* the disaster area so the guidelines to be mentioned here are probably valid for all stations that expect, at some time in their amateur activities, to become involved in emergency communications.

Since most emergencies of any scope will involve loss of power and possible loss of telephone communications, it stands to reason that the amateur equipped for emergency communication will have emergency-power capability. While gasoline generators are highly desirable, the advent of solid state equipment makes operation from 12-volt batteries both feasible and economically attractive. A battery that can be charged by your automobile while you are using the automobile for normal driving is a practical solution to the problem of emergency power. Most heavy-duty automotive batteries will sustain the operation of a 200-watt transceiver for several hours with little difficulty. While some emphasis has been placed on the possibilities of using low-power (QRP) operation during an emergency, it should be noted that this is a last resort. One should not *plan* to use a low-powered station because time generally means lives and property and it will take you longer to get information through with low power. The low-power station would be a good backup to the main station equipment.

Many emergency situations are accompanied by the type of weather that is likely to take antennas down. An amateur well equipped for emergencies will have a spare antenna, usually for 80 and 40 meters, ready to be strung between any two convenient supports. Restricted space (electrically shortened) antennas, either dipoles or verticals, fit this requirement nicely. The vertical antenna is probably a little easier for one person to erect in a hurry but the dipole will give more consistent results regardless of the electrical ground in the vicinity of the antenna.

Since 2-meter fm is so widely used today and has the advantage of easy portability, a 2-meter rig is essential for an amateur who is serious about emergency communications. It should be noted, however, that 2 meters will not necessarily be of much use in getting information in and out of the disaster area unless a sizable fraction of the operators is also equipped with both hf and 2-meter equipment. With respect to getting the information in and out of the disaster area, it should be recognized that cw offers several advantages over other modes. Not only does it deliver the "most signal" for a given power in adverse band conditions, but it also provides some security from the general public. This last point is important when handling messages for civil officials that could cause panic or lead to more problems, if prematurely disclosed.

In further planning emergency-station capabilities, it should also be recognized that record-keeping is going to be an important part of the emergency activity. Not only is there the FCC requirement for documenting third-party traffic (both formal and informal) but in an emergency, the times at which messages are received and sent, where they were sent, from which station and current status, become very important. There is no simple system that appeals to everyone, but some thought ahead of time will at least cover some of the possibilities. One of the benefits in participating in routine tests and drills is to better define your own operating system for such cases. A generous supply of paper, sharpened pencils (or pens) and the possible addition of a tape recorder to the station operating, will assist the bookkeeping chores. With the paperwork correctly done, the "trail" of messages going through your station will be easily followed. To complete your participation in an emergency situation or drill, send a report to your ARES emergency coordinator or section emergency coordinator after the event. It will be appreciated, since this type of documentation is important for the continued justification of our Amateur Radio operating privileges.

Distress Procedures

Occasionally we are called upon to be the key station in providing emergency communication service. Initiating such communications can be a troublesome as well as rewarding experience. In situations like automobile collisions or other local events, the basic idea is to pass as much information as possible very quickly. Unfortunately, public services like the Coast Guard and fire departments are sometimes plagued by false reports. It is advisable for you to begin your report by identifying yourself completely, with name, call, location and home location if you are operating mobile or portable. This will not only provide the authorities with a rapid means of validating your report, but will increase their appreciation

Distress Calling

An amateur who finds himself in a situation where *immediate* emergency assistance is required (at sea, in a remote location, etc.) would call "MAYDAY" on whatever frequency seems to offer the best chance of getting a useful answer. "MAYDAY" is from the French *m'aidez* (help me). On cw, use SOS to call for help. QRRR has been previously discarded by ARRL. The distress call should be repeated over and over again for as long as possible until answered. The amateur involved should be prepared to supply the following information to the stations who respond to an SOS or "MAYDAY":

The location of the emergency, with enough detail to permit rescuers to locate it without difficulty.

The nature of the distress.

The type of assistance required (medical aid, evacuation, food, clothing, etc.).

Any other information that might be helpful in locating the emergency area or in sending assistance.

of the seriousness of the situation. This prologue should be followed by the pertinent facts at your disposal with as little added commentary as possible. It is best to state the facts as succinctly as possible and then stand by to answer questions. This is generally more efficient than trying to anticipate their requirements.

In situations that have a broader scope and which will continue for a while, the initial reports are best done as formal messages (if being relayed through other amateurs) or read from a text (if sent direct). In these situations, getting a civil authority involved as soon as possible should be a prime objective. In either of these cases we should remember that the FCC rules are still in full force during emergency operations including proper identification, modes, band privileges, and so on.

ARES Section Structure

As with most ARRL activities, the Amateur Radio Emergency Service is based on the section organization. To see how you might be able to fit into the structure of ARES, it would be helpful to understand how it is set up. The section communications manager appoints a section emergency coordinator (SEC) to take care of all matters pertaining to emergency communications. Although some SCMs will carry out the work of the SEC themselves, most feel that it takes the full-time attention and energy of an individual to handle this important function. There is only one SEC appointed in each section. His or her duties are:

1) To encourage all groups of community amateurs to establish a local emergency organization.

2) To recommend to the SCM action on all section emergency policy and planning, including the development of a section emergency-communications plan.

3) To cooperate and coordinate with the section traffic manger so that emergency nets and traffic nets in the section present a united public-service front.

4) To recommend to the SCM candidates for emergency coordinator and district emergency coordinator appointments (and cancellations), and determination of areas of jurisdiction of each amateur so appointed. Note that the SEC does not make (or cancel) appointments; he or she simply recommends them to the SCM.

5) To promote ARES membership drives, meetings, activities, tests, procedures, etc., at the section level.

6) To collect and consolidate emergency coordinator (or district emergency coordinator) monthly reports and submit monthly progress summaries to ARRL hq.

7) To maintain contact with other communications services and liaison at the section level with all agencies served in the public interest, particularly in connection with state and local government, civil defense, Red Cross, Salvation Army and the National Weather Service.

Fig. 3 shows the general organization structure for ARES as a part of the Public Service Corps within the Section. The concept of dividing a section into districts is relatively new to ARES, although it has been used by many sections informally over the years. In many sections, either the size of the areas to be covered or the numbers of amateurs involved (or both) are so large that a single individual (the SEC) cannot effectively cover it. This is especially true in times of emergency when the SEC's presence may be required at some central headquarters location to deal with state officials or other agencies, and a senior ARES official may also be needed in the disaster area to coordinate local ARES groups. In sections of only a few counties, this level in the organization is not required and thus is not used. To understand the full difference in the organization at different levels it is well to understand what these ARES appointments are all about.

Let us pass over the district emergency coordinator for a moment and consider the local emergency-communications group. This is the group that most amateurs will come in contact with and it is lead by the emergency coordinator. The job of the EC can be contrasted with that given for the SEC to see how the nature of the job changes as one gets to the local level. The emergency coordinator shall:

1) Coordinate the training, organization and emergency participation of interested amateurs in the area of jurisdiction.

2) Establish an emergency-communications plan for the area that will effectively utilize the group to cover the needs for both tactical and formal message traffic requirements.

3) Establish a viable working relationship with governmental and other agencies who might need the service of ARES

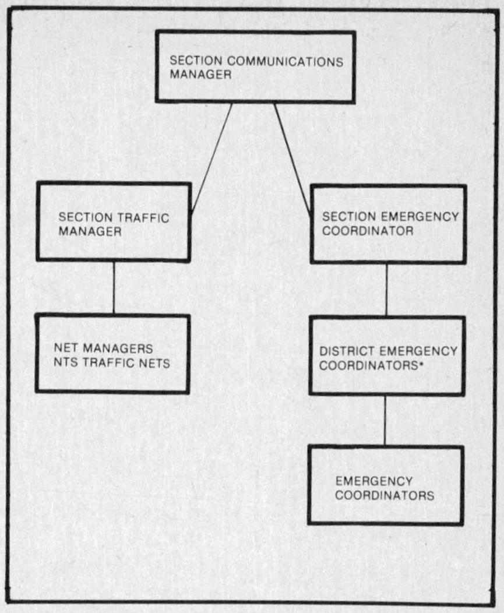

Fig. 3 — Section structure for ARES.

(e.g., Red Cross, Salvation Army, Rescue Squads, Weather Service, Hospitals, Police Department and Fire Department).

4) Establish local communications networks run on a regular basis and periodically test these networks by drills.

5) Establish an emergency traffic plan utilizing NTS to the extent possible and coordinate liaison with NTS nets.

6) In times of disaster, evaluate the communications needs of the jurisdiction and respond quickly to those needs.

7) Do all that is possible to further the image of Amateur Radio by dedication to purpose, reliability and a thorough understanding of the mission of Amateur Radio.

If the EC were actually expected to carry out all of these items by him or herself, it would greatly restrict ARES leadership. However, it should be realized that the EC has the responsibility to see that they are done, but not necessarily through doing it. A special appointment, the assistant emergency coordinator (AEC), was created to help carry out the functions that form the responsibility of the EC. Unlike other appointments, the AEC is not a SCM appointment; it can be made by the EC. The idea is to find amateurs in the local group who have the skills to help carry out the job. These assistants then take care of their area. A typical ARES structure for a group of 30 to 50 amateurs is shown in Fig. 4. This one uses four assistants to cover four functions that are part of the EC's responsibilities. Note that depending on the size, interests and qualifications of the group, the nature and number of assistants can change. For example, where 6-meter and 10-meter nets were in use there could well be assistants to coordinate the activity on these bands, also. There could also be an assistant to help with planning and reports, and possibly one to interface with the local clubs if there were more than one club in the area. In any case the idea is to

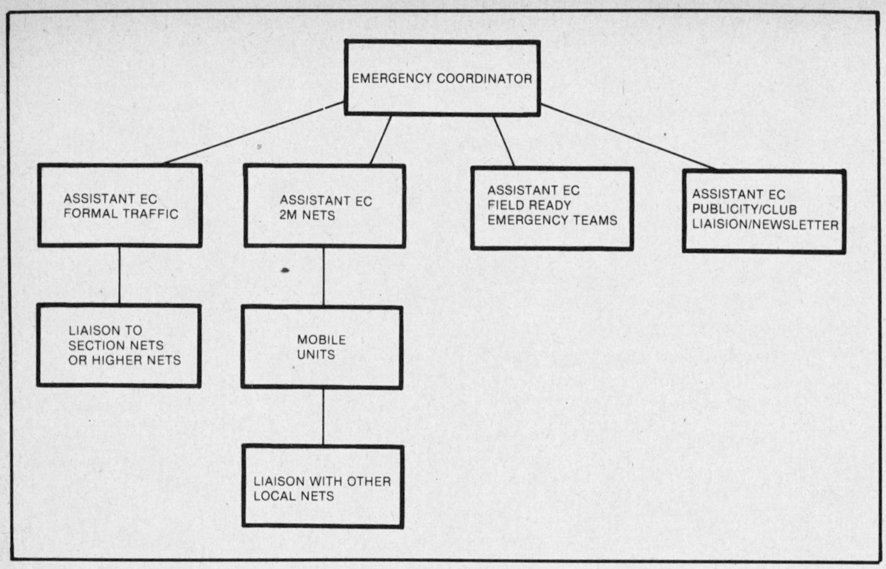

Fig. 4 — Local ARES structure for a county, city or other area of coverage.

spread the workload to get the job done without creating a tremendous burden on any one individual.

It should be clear from this that there is a place in the ARES organization for anyone with an Amateur Radio license and an interest in public service communications. Your present skills will be put to work, while you will also be exposed to the opportunity to develop new ones. Your area of interest could become an area of help to your local ARES group.

Now that we have looked at the top of the section organization and the local level where the actual communications take place, it would be well to recognize that in large sections the local groups could proliferate to the point where simply keeping track of them would be more than a full-time chore not to mention the idea of trying to coordinate them during an actual emergency. To this end, many sections started using the concept of grouping their EC jurisdictions into logical units or "districts" and appointing an EC to coordinate the others in that district. In some cases the districts conformed to the boundaries of governmental planning or emergency-operations districts, while in others they were simply based on repeater coverage or geographical boundaries.

In recognition of the advantage of these types of logical groupings in the ARES organization, ARRL has recently officially added the district emergency coordinator (DEC) appointment to the structure. The DEC is generally an amateur experienced in emergency communications who can assist the SEC by taking charge in the area of jurisdiction, especially during an emergency. The DEC shall:

1) Coordinate the training, organization and emergency participation of emergency coordinators in the area of jurisdiction.

2) Make local decisions in the absence of the SEC or through coordination with the SEC concerning the allotment of available amateurs and equipment during an emergency.

3) Coordinate the interrelationship between local emergency plans and between communications networks within the area of jurisdiction.

4) Act as backup for local areas without an emergency coordinator and assist in maintaining contact with governmental and other agencies in the area of jurisdiction.

5) Provide direction in the routing and handling of emergency communications of either a formal or tactical nature.

6) Recommend EC appointments to the SEC and SCM and advise on OES appointments.

7) Coordinate the reporting and documentation of ARES activities in the area of jurisdiction.

8) Act as a model emergency communicator as evidenced by dedication to purpose, reliability and understanding of emergency communications.

As can be seen from the description, the DEC position requires more experience, more work and a better understanding of the larger-scale communications needs of the section. Ideally, the DEC would be fully conversant in NTS routing and procedures as well as have a thorough understanding of the locale and role of all vital government and volunteer agencies that could be involved in an emergency.

The ARES structure is based on the realization that we must build on the strengths of the available amateurs as well as encourage members to continue to build their skills in areas vital to emergency communications. In addition to the appointments outlined above, there is also the official emergency station (OES) appointment to acknowledge the efforts and skills of those stations vital to the maintenance of a successful emergency-communications service. Qualifications for this appointment include:

1) Possession of full ARRL membership and a Technician class license or higher.

2) Regular participation in the local ARES, if any, including all drills and tests, emergency nets and, of course, real emergency situations.

3) Ability to operate independently of commercial mains including mobile capability on at least one band.

4) Being fully acquainted with standard ARRL message form and being capable of using it in any third-party messages.

5) Reporting monthly to the SCM on form CD-210 or equivalent.

The OES appointment is the next level in the structure of ARES. Coupled with general membership in ARES, it provides the basis for the public-service function that justifies the privileges granted with the amateur license.

The backbone of the ARES organization is the general member. ARES members normally participate to the level that their personal circumstances allow.

Organizing the Group

The above discussion gives you an idea of how the ARES structure can work. Unfortunately, in many local areas of the country, it is difficult to easily become part of an effective ARES program because of a lack of knowledge of who the "right" amateurs are to contact. To obtain information on the existing organizations in your area, you can contact your SEC, SCM or ARRL hq. in that recommended order. The names and addresses of the SECs are usually given in *QST* articles that announce the yearly Simulated Emergency Test. The names and addresses of the SCMs are given on page 8 of each *QST*.

If it turns out that there is no active group in your area, you may wish to find some interested amateurs to form a group. Even if you are alone in the area, you are still needed and can participate in the section activities. If an interested group of amateurs can be found, it is a simple matter to arrange a meeting, sort out various capabilities and abilities, and then make your group known to your SEC or SCM. The diagram in Fig. 4 gives an idea of the functions to be covered. Your SEC/SCM can help by providing guidance in the selection of the EC for the group. Frequently, the group can recommend their most qualified member to the SEC/SCM for their consideration as EC.

The important point is not to let a perceived lack of activity keep you from participating or being involved in beginning an organization. The local 2-meter repeater or simplex frequency is an ideal place to meet amateurs who may share your interest.

Emergency Plans

In any emergency situation, the ability

of amateurs to respond effectively will only be as good as their preparation and training. This is not to say that amateurs who are relatively unprepared have not made tremendous contributions to disasters requiring energy communications. There is ample evidence of contributions from amateurs who just happened to be on the spot with the results saving life or property. The point is that when the amateurs are prepared and trained, they can accomplish much more in the same time and it is *time* that is at a premium in an emergency.

The basis for better preparation can usually be found in formal plans that are made, practiced and distributed well in advance of their need in an emergency. These plans do not have to be elaborate but they should cover the pertinent points.

The Section Plan

Fig. 2 gives some idea of how a typical hf network could be organized to cover a section. A section plan would be valuable if it simply documented what frequencies were to be used and which stations or groups were going to perform the function indicated. For example, the agencies that are involved in handling official emergency preparedness are likely to be located within the jurisdiction of a particular ARES group. The plan could spell out whether that group will cover those agencies or whether a separate group will handle that function. The distribution of communications on ssb, cw, fm and RTTY could also be covered with appropriate frequencies and liaisons spelled out. The plan could also cover the alerting procedure (see below) to be used and provide some definition of the functions of SCM, SEC, STM, etc., during an emergency.

The Local Plan

In a similar manner, the details of how the local group is going to respond should be placed in a plan well before it is needed. This plan should also include (probably as an appendix that can be easily updated) a full list of members, including phone numbers, equipment and operating capabilities. The local plan is the most detailed availability and action guide in ARES.

While it is obvious that the local plan has to tie in closely with the larger-scale section plan, it is not so obvious that the section plan depends on the existence of local plans. The details of which operator(s) will show up to cover initial points is contained in local plans. The SCM, SEC, DEC or even EC cannot be at every critical location. It is imperative that operators understand their assignments.

Developing a plan of this nature seems to be among the least-popular aspect of ARES. When a situation arises requiring such detailed organization, local groups often do not use the plan to its full potential. The local plan is similar to military planning in many respects. This may be the reason that some of the best ARES groups have been led by, or have had their plans drawn up by, retired military personnel.

While all amateurs should be rightfully proud of what we have accomplished in past emergencies, it is only through planning and practice that we will be able to improve that record in the future.

Alerting Procedures

Much has been written in recent years about the need for various ways of alerting ARES membership in times of emergency. There have been suggestions that special receivers be built along the lines of "pocket pagers." Simpler solutions have been suggested along the lines of using repeater monitor radios or using local radio broadcast stations. Another suggestion is the use of National Oceanographic and Atmospheric Administration (NOAA) weather radio stations, since so many emergencies are weather-related.

In the last analysis the simplest (and possibly most effective) means of alerting is the alerting "tree." For a given local group, this can usually be done such that no one has to make a long-distance phone call. Of course, to rely exclusively on the telephones is dangerous indeed, since they may become overloaded or go out of service before the ARES group has been alerted. This is precisely the place where 2-meter fm communications (on the local level) can save the day. Everyone in the group should know exactly what frequency to monitor as soon as they became aware that the telephones were out of service. This is precisely the kind of information that should be covered in the local emergency plan. Repeaters are fine for this purpose but should be backed up by a simplex frequency should the repeater be out of service because of the emergency situation.

Fig. 5 simply depicts the classical "tree" concept. Note that the EC may not be the first person to obtain true information concerning the need for the alert. In this case, the person should attempt to contact the EC *but also* continue the alert down the tree (and horizontally if time permits). The idea is to get as much of the organization into action as possible. Where a local plan exists, the tree is set up so that each "branch" contains amateurs covering all needed skills and areas. Thus, one should not put all the amateurs with one skill or capability (e.g. traffic handlers, those with generators, etc.) in one branch of the tree!

A similar concept can be used at the section level, as shown in Fig. 6. Note here that it is assumed that the emergency begins in a local ARES area or District and that an EC, DEC or assigned liaison station performs the beginning alerting function. If it is the SEC, SCM or STM that first obtains the alerting message (as is likely where these ARRL officials head calling lists), the situation is even simpler since the SEC should have his or her own "tree" for bringing up the districts or local areas. Note that the concept of district emergency coordinator fits nicely into this "chain of command" structure.

Assistant emergency coordinator for district nine in Kentucky, Fred Jones, WA4SWF, is QRV for hf emergency power operation.

Depending on your location the telephone may not be of much use as a sectionwide basis. One must carefully analyze the size of the area to be covered to decide on appropriate frequencies to be used. In some of the smaller, highly populated sections, 2 meters may suffice for section communications. In the larger sections, however, it is virtually certain that hf communications will be needed. The frequencies to be used will be based on distances and propagation conditions. In some cases it may be necessary to have a daytime and nighttime frequency. Remember that one goes to these frequencies as soon as there is any inkling of a potential emergency situation.

The choice of frequencies is usually made easier because over the years, traffic nets covering ARRL sections have migrated to frequencies that have been proven to give reliable communications over that section. These frequencies would seem to be an ideal spot for the alerting function and, in fact, have served exactly that purpose in the vast majority of cases. These same frequencies are then brought "into action" to carry the traffic load into and out of the disaster area.

Relation to NTS

It has been stated often that the National Traffic System (NTS) is available to

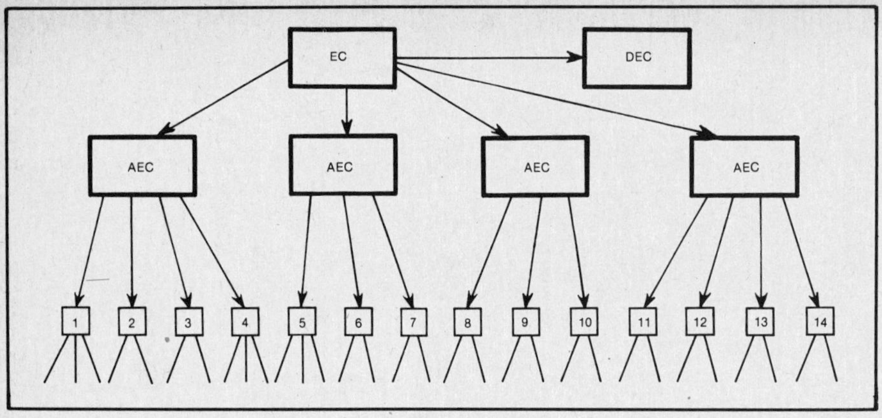

Fig. 5 — Alerting procedure using telephone/2M tree.

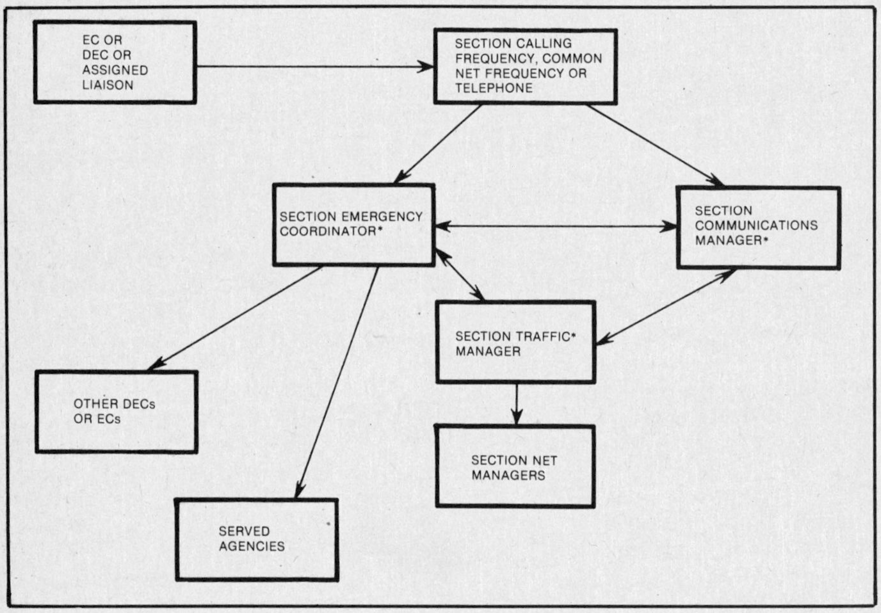

Fig. 6 — Alerting procedure (section wide) using radio or telephone.

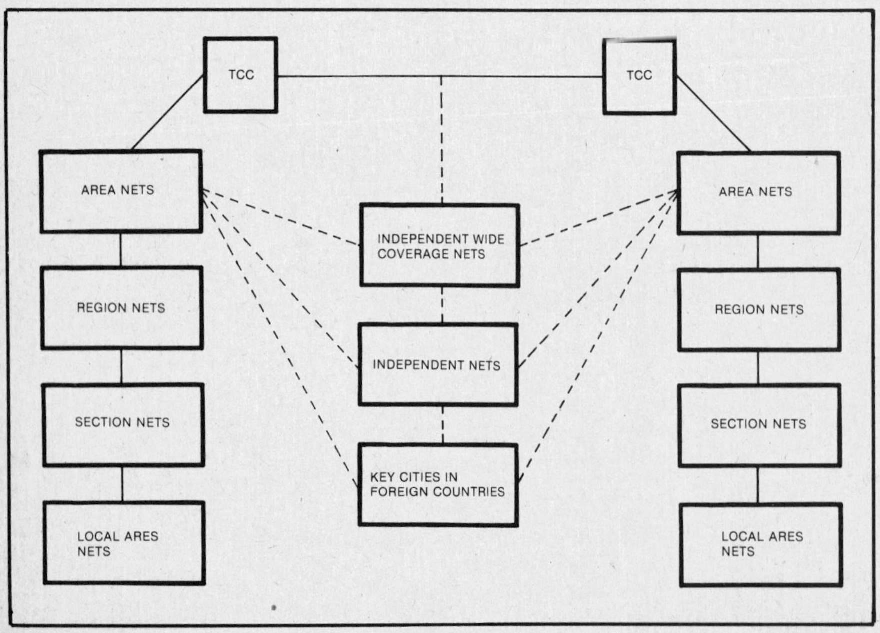

Fig. 7 — Emergency communications through linked nets.

serve ARES in times of emergency. One point of constant confusion is that the goals of emergency and traffic nets are somewhat different. As noted earlier, emergency nets exist to handle both tactical and formal traffic, while traffic nets exist to handle formal written traffic. This is particularly true of NTS nets where it is strictly traffic first.

Since the operation of NTS is already covered in another chapter, it will not be repeated here. Changes have recently been established in NTS scheduling and procedures to greatly facilitate moving traffic during emergencies, increasing the number of "cycles" as needed up to continuous operation.

The information given in Fig. 7 depicts one "cycle" if one starts at one corner (local ARES net) and moves up the network schedule to section net, region net, area net, Transcontinental Corps (TCC) representative to TCC representative and back "down" the cycle in another part of the country. Once the traffic is placed in the section net, it begins its journey and the local ARES group has fulfilled its obligation (see Fig. 1 and 2 again for review).

In addition to the NTS routing system, wide-coverage independent nets, independent regions nets (covering more than one section) and direct connections to key cities in foreign countries are also available. These key cities, usually accessed through the independent nets, have been especially valuable in assisting emergency communications in Central and South America. The independent nets take on a wide variety of types and forms with many of the most active heard daily on 40 and 20 meters. The 20-meter nets are especially important in covering the areas of the Caribbean and South America.

Fig. 7 shows the traffic reaching these independent nets through the higher levels of NTS, i.e. from liaison out of either the area nets or TCC. The reasoning for this is simply that stations who operate in the area nets and in TCC have the most "savvy" and experience. They know which nets meet when and the best modes of routing traffic into or out of certain areas. Further, in times of emergency, the net managers at this level maintain contact with ARRL hq. and are thus aware of any special arrangements that have been made regarding traffic routing.

One last unanswered question relates to the gap between the section and higher-level nets in times of emergency. How does the section get the help of the NTS? It has been agreed by NTS officials meeting at ARRL hq. that the persons in the section most familiar with the NTS (STM, SCM, SEC) should request such aid from their region net manager(s) who will then alert the other sections and the area net as needed. As the higher-level net managers become involved they will develop the needed extra schedules and contacts.

In some cases it will be most expedient and efficient to set up "hot-line" communications directly with the disaster area. This goes along with the principles of NTS and anything that has been developed concerning emergency communications. When the traffic load and/or situation demands such a response, it is simply "woven into the fabric" of the communications available. This is simply treated as another "key city" in the scheme of Fig. 7, only in this case the key city does not have to be in a foreign country or even a city at all! It is simply a location within the disaster area capable of receiving and sending large volumes of traffic quickly. This is now quite often accomplished using RTTY equipment brought to the site for the purpose.

All things considered, the resources available for emergency communications by amateurs are quite vast. With the appropriate availability of people, the combined networks currently available should be able to handle virtually any type of emergency.

After the Emergency

So far we have focused on preparing for and operating in an emergency. It should be realized that much of the work still remains to be done *after* the emergency is over. This is especially true if one has handled any significant volume of third-party traffic. Not only must this record be kept for a year as per FCC regulations, but there could be future inquiries from state and local governmental agencies. One should make sure the disposition of each piece is clearly marked and that it is sorted in some sensible manner in case it is needed for reference.

Additionally, one should consider the possibility of publicizing the effort. As noted at the beginning of this chapter, Amateur Radio is a service and the privilege of holding a license is contingent upon fulfilling this service. It is important that as many people as possible realize that the service has been fulfilled when, in fact, this is the case. Your local newspapers and radio/TV stations may find your story of interest and, of course, ARRL is interested in having a record of the effort. The information is usually forwarded to ARRL through your EC but anyone is invited to submit such information "for the record."

In some cases it may be advisable to make notes of your involvement for your own information and to help others in the future. It is through such documentation that we learn how to improve our efforts the next time an emergency requires our communications expertise.

RACES

Since there has been some misunderstanding concerning the role of the Radio Amateur Civil Emergency Service (RACES) and its relation to ARES, this chapter would not be complete without bringing the RACES story up to date.

After World War II, it was recognized that there was a need for an organized effort to use Amateur Radio to support the Civil Defense (cd) program. Based on this, the FCC authorized RACES in 1951. While it was originally based on potential use for wartime, it has evolved over the years, as has the meaning of civil defense, to encompass all types of emergencies. This has occurred as the civil defense role has been combined with general emergency-preparedness operations in most states and localities.

The expansion of RACES into the general arena of emergency communications began to conflict somewhat with the role played by AREC (now called ARES). This conflict led to a restructuring of RACES which culminated in 1976 with an FCC order on the subject. This order had the following effects:

1) It "deregulated" RACES by streamlining its administration.

2) It discontinued the requirement that RACES communications plans be filed with FCC.

3) It authorized a civil defense entity to hold its own RACES Amateur Radio station license.

4) Except in those cases when the President invokes his War Emergency Power, it provided for shared use of amateur frequencies between amateur and RACES stations.

5) It prohibited all, other than licensed amateurs, from operating in RACES and made operating privileges identical to those held in the Amateur Radio Service.

6) It restricted amateurs operating in RACES to communicating only with other licensed amateurs in RACES.

7) It limited *routine* RACES usage of amateur frequencies for such matters as tests and drills to not more than one hour per week.

8) It abolished all tactical call signs and required all RACES operators to identify by regular assigned calls.

9) It reaffirmed and recognized that RACES is in fact and name an Amateur Radio Service.

All this made it possible and simple for an ARES whose members are all enrolled in and certified by RACES to operate in an emergency with great flexibility. Using the same operators and the same frequencies, an ARES group enrolled also as RACES can "switch hats" from ARES to RACES and RACES to ARES to meet the requirements of the situation as it develops. For example, during a "nondeclared" emergency ARES can operate as ARES, but when an emergency or disaster is officially declared by a state or federal authority, the operation can become RACES with no change in personnel or frequencies.

This situation is still not well understood and accepted throughout the country; both ARES and RACES organizations still exist, separately, in many areas. You will have to determine the situation in your area since the differences are too great to catalog here. Where there is currently no RACES, it would be a simple matter for an ARES group to enroll in that capacity. In cases where both exist, it is possible to join both or to be involved in either. Hopefully, as time progresses, they will be merged into one strong organization with coordination between the ARES and RACES officials using the same groups of amateurs. In several sections in the U.S. today, the ARES structure has also been accepted as the RACES structure.

Those wishing further details should consult the FCC Rules and Regulations. Part 97 (Amateur Radio Service), available in the *Radio Amateur's License Manual* or from the Government Printing Office, Washington, DC 20402. See also a discussion in May 1976 *QST*, page 52.

Chapter 6

DX and DXing

Bet you can't work just one DX station. If not addictive, DXing is at least a satisfying and self-perpetuating aspect of that diverse hobby we know as Amateur Radio. DXing can be the embodiment of all the finer points of hamdom. We have the friendly rivalry and competition to be "the first to work that new DXCC Country." We have individuals (DXers) from different countries and cultures meeting each other on a one-to-one basis with a common bond: The dentist from the United States meets the student from Japan, the Canadian athlete gets to know a little about the Caribbean hotel manager, a Polynesian homemaker chats with a Jordanian monarch. The possible combinations are endless.

Can you envision yourself in far-off exotic lands? See gently swaying palm trees? Long to hear the rhythm of foreign tongues? It's a pretty sure bet that you'll succumb to the lure of DX. A vicarious trip to foreign lands is just as close as your rig and a DX QSO. DXing can be every bit as pleasant as you care to make it, run up your DXCC totals or just drift off to faraway places via the ether. Your Amateur Radio license is your ticket to the exciting world of DXing.

Who is the DXer?

The DXer is the newly licensed Amateur for whom the first DX QSO is only a "CQ" away. The DXer is also the oldtimer who is on the DXCC Honor Roll and is about to jump into the pile-up to work the brand new country that has just shown up on the bands.

We don't expect to be able to tell you *everything* about DX and DXing in this one chapter. But with references to additional related material, we hope to provide a solid starting place for those interested in taking up DXing, and pass along a few tips to the more-established DXer on how to climb a little closer to that DXCC Honor Roll.

DXing can be compared to riding a bicycle. Once you've learned how you'll never forget, but to learn how, you've got to jump on and give it a try. Reading about DXing, like reading how to ride a bike, will give you a sense of the fundamentals, but only jumping into those DX pile-ups and giving it your best shot can make you an accomplished DXer.

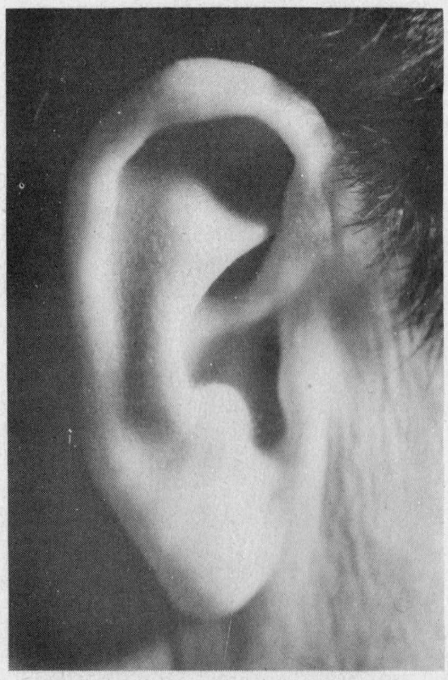

This is about the most important tool that the DXer has at his disposal. Of course the other half of this matched set and the "gray matter" that lives between them are pretty important also.

What is DX?

DX is one of those ambiguous entities that is correctly defined in different ways by different people. DX can be that 150-mile contact into the next state on 2 meters via tropospheric ducting. DX can also be a transcontinental QSO on 20 meters using your homebrew QRP peanut-whistle. In the beginning, to most amateurs, DX is the lure of seeing how far you can get. DX can be a personal achievement, bettering some previous "best distance worked," involving a set of self-imposed parameters. DX can also be competing on a larger scale, trying to break the pile-ups on a rare DX station and aiming for one of the DX-oriented awards. The type of DXing dealt with here is working foreign Amateur Radio stations.

Where to Find DX

Like gold and happiness, DX is where you find it. Granted, the law of averages dictates that the person who CQs long or often enough in segments of the bands DX stations are known to frequent, is likely at some time or another to snag one or more of the rarer DX stations. But the DXer who knows where to find (and how to apply) the readily available information on propagation and the operating habits of certain DX stations has eliminated some of the uncertainties in finding and working the DX stations. At the same time, this operator has considerably increased his chances of successfully adding to his total number of countries worked.

Let's agree, then, that knowing when and where a given DX station is likely to be operating are valuable pieces of information to the DXer. When and where are parameters that depend on two factors:

1) Radio-wave propagation, which depends on cyclic variations of sunspot activity and the time of year and determines which parts of the frequency spectrum will support radio communication between any two given points on earth, and 2) The operating habits/preferences of the individual DX station.

Radio-wave propagation can be a rather interesting and involved study in itself. Luckily, the ARRL *Radio Amateur's Handbook* contains an excellent chapter devoted entirely to a discussion of propagation.

The monthly propagation charts, provided in each issue of *QST* (the monthly journal of the ARRL), make fairly accurate predictions, based on the latest scientifically obtained data, on which bands are most likely to support radio communications between two given areas on earth. If you have a receiver that will cover 2.5, 5, 10, 15, 20, or 25 MHz, you will be able to listen to WWV and WWVH. The regularly scheduled transmissions from these stations have more uses than just checking the time or aligning the receiver. At 14 minutes past the hour, the WWV/WWVH broadcasts contain information (updated several times each day) that indicates solar activity and the condition of the earth's magnetic field, two factors that affect propagation.

A detailed discussion of both the *QST* charts and the WWV/WWVH broadcasts is contained in "The Art and Science of DXing," reproduced in condensed form in this chapter.

Stew, W1BB, calls 160 meters "home." He holds 160 DXCC no. 1 and is living proof that there *is* life below 80 meters.

On the Bands

You can choose the mode on which you intend to chase DX: ssb, cw or mixed. There will be stations that only show up on the bands on one mode (ssb is the more-popular mode because of the faster QSO rates possible). The serious DXer will be proficient in both modes to maximize his chances to "work 'em all." And that brings us to the problem of who can go where to work those choice pieces of DX.

On cw, those privileged 25 kHz at the bottom of most amateur bands are a sanctuary for the DX station whose call sign can cause those really huge pile-ups. This is not to say that you can't work DX above that bottom 25 kHz, because more than just a few Novice class operators have a DXCC Certificate (DXCC, administered by ARRL, is the premier DX operating award in the world) hanging on the wall to show for their DXing efforts. What it boils down to is that having an Extra Class (in the U.S.) license is a definite bonus, but not a necessity, when hunting the rare ones on cw.

It is almost a necessity for the phone (ssb) DXer to have at least an Advanced class amateur license. Without fail, the most desirable DX stations show up on phone in the 20-meter band, transmit below the U.S. portion of 20 and listen in the Advanced portion of the U.S. phone band.

Moving up the ranks in the licensing structure should have at least as high a priority to the serious DXer as upgrading the station. The higher-class license is every bit as much a tool to the DXer as the station hardware.

160 Meters (1.8-2.0 MHz)

Top band: 160 has a character all its own. Since it is presently shared with LORAN — long range navigation radio system — power levels differ in various areas of the U.S. and Canada depending on time of day. A chart showing power levels and frequencies for 160 meters appears in chapter 3.

Like the other low bands (80 and 40 meters), 160 is considered a nighttime band. DX signals peak approximately one hour before sunset (local time) and one hour after sunrise. Optimum conditions for DXing occur when your sunset coincides with the DX station's sunrise, or the other way around. Most of your DX on 160 will be worked during the hours of darkness. The countries that have amateur 160 allocations do not necessarily have mode segments (cw or ssb) that correspond with the majority of worldwide allocations. Thus, you can expect to work some split frequency (listening on one portion of the band while transmitting in another portion) and should have a rig capable of split operation.

Working DX on 160 "ain't all that easy." Not all that many 160-meter WAC (Worked All Continents) certificates have been issued. In fact, until recently, 160 was a very sparsely used band. Kept alive by stalwarts such as Stew Perry, W1BB (who, by the way, holds 160 DXCC Certificate number 1), Top Band has seen a revival in popularity, mainly because most rigs built commercially in the last five years have 160-meter capabilities.

W1BB publishes a 160-meter newsletter that is a definite bonus to the 160-meter DXer. A self-addressed stamped envelope to Stew's *Callbook* address will get you a copy.

An interesting and informative book on 160-meter operation and equipment, *The Challenge of 160!*, by Si Dunn, K5JRN, published in 1977 by 73 Inc., Peterborough, NH, is available through the publisher or Ham Radio Bookstore, Greenville, NH 03048. Send an s.a.s.e. for price and order form. The book is complete and takes the reader from a "I've never even heard of 160 meters" state through the construction of transmitters for Top Band.

80/75 Meters (3.5-4.0 MHz)

It's called 80 meters in the cw portion of the band and 75 meters in the phone part of the band. We'll refer to it simply as "80."

This band is not commonly recognized as a DX band. This is underscored in times of increasing sunspot numbers, especially as we approach the peak of solar cycle 21, when the high bands (20, 15 and 10 meters) are open to somewhere around the clock.

Here, as on 160 meters, 80 is a nighttime band with best DX times being an hour before sunset and an hour after sunrise. Those hours between sunset and sunrise are not a vast wasteland, either. More-than-adequate possibilities to work some medium- and long-haul DX exist

The Art and Science of DXing

The big questions are: Which is the right band, and when is it open?

Whether you are a gung-ho DXer or a casual weekend ragchewer, the answer can be found in the propagation charts that appear monthly in *QST*'s "How's DX?" column. These provide a monthly update of the latest data from the Department of Commerce Institute for Telecommunication Sciences in Boulder, CO. They are easy to use and can help you fill in the who, what, when and where of successful long-haul operating, especially when used with the latest bulletins from W1AW and WWV.

How to Use the Charts

No propagation prediction can ever be perfect, so these charts show the *probability* of ionospheric support between two points at different frequencies and times. A total of 30 paths have been selected as being typical of those of interest to the DX-minded amateur. Table 1 lists the exact location the generalized areas are based on. Naturally, a path will be reciprocal. Assume, for example, you are in southern California and want to work the Caribbean. The Puerto Rico-to-West Coast chart would be the one for you, just as if one had been separately prepared for West Coast to Puerto Rico.

Three factors strongly influence propagation: the season of the year, solar activity level, and station location. The season affects the sunrise and sunset times at both ends of the circuit and, therefore, the amount of daylight and darkness along the signal path. In turn this influences the daily ionization levels, especially at higher latitudes.

But predicting the solar activity level for each individual day is much more difficult. This becomes especially true for *QST* because of a publishing deadline of about two months ahead of when you, the reader, see it. Should you be interested in predicting precisely what the bands are going to be like for an approaching weekend, you can obtain a much more accurate prediction through W1AW and WWV bulletins than *QST* could ever hope to publish. On the other hand, it is possible to make a reasonably accurate prediction of the *median* value of solar activity for the month as a whole. These charts reflect those values.

Three Curves

Appearing on each chart are three curves: optimum traffic frequency (fot), maximum usable frequency (muf), and highest possible frequency (hpf). A computer program predicts each on the basis of seasonal and long- and short-term trends, as well as propagation by the F2 layer, E layer and sporadic E.

The lower curve shows the fot, which is an estimate of the frequency having ionospheric support on 90 percent or more of the days of the month.

In the middle is the muf. Conditions will allow useful radio communication at this frequency for at least 50 percent of the days and will be exceeded half the time.

At the top is the hpf. For about 10 percent of the time, three days of the month, communications can be supported at the given frequency.

Table 1

In the monthly feature, the locations of the points on which the curves are based are generalized so as to emphasize that they are useful for more than just one city or state. Here is the list of actual points which are used in the computer run. They were selected as being representative of the amateur population in a geographical area. Some corrections can be made on the basis of latitude; the general rule is, the higher the latitude, the lower the muf. If the muf curve is marginal for the band in question for Washington, stations in Miami have a better chance and stations in Montreal a poorer chance for a band opening.

Generalized Location	*Actual Location*
East Coast	Washington, DC
Midwest	Kansas City, MO
West Coast	San Francisco, CA
West Europe	London, England
Eastern Europe	Kiev, Ukraine
Japan	Tokyo
Australia	Melbourne
South Pacific	Pago Pago, Am. Samoa
South America	Asuncion, Paraguay
Central Asia	New Delhi, India
Southern Africa	Lusaka, Zambia

You may notice that all three lines merge into one at certain times of the day. When a line becomes near-vertical, it means that local sunrise or sunset is at one end of the path or another, causing rapidly changing ionization levels. The other single-line condition occurs as a horizontal trace at the top of a given chart. The Boulder computer does not provide graphs above 30 MHz, and today's conditions often exceed that!

What Bands Are Open?

A general rule of thumb for selecting a band is to first find the muf, then take the closest band below that. Signals near the muf have the advantages of (1) being reflected at a shallower angle than at lower frequencies, giving them a longer skip distance; and (2) requiring less effective radiated power (erp) for a given received level, thanks to minimal absorption. The muf may rise even higher than the median value, thereby providing yet more enhancement of signals, but the probability drops off to 10 percent at the hpf point.

Although the fot presents the highest reliability, it may drop to such a low frequency that *absorption* becomes an important consideration. The further below the muf you operate, the weaker signals will be, since radio waves relinquish energy while traveling through the ionosphere by setting ionized particles in motion. This absorption is greater at lower frequencies, especially 1.8, 3.5 and 7 MHz. It can sometimes be overcome with the brute force approach of an amplifier and gain antenna. Commercial stations often do this, as do some amateurs. Nevertheless, the various bands can be useful to most amateurs seeking DX.

For instance, say you can be on the air only from 1300 to 1400 UTC. From Fig. 1, your best band would be 20 meters. If you are a Novice, you might have a chance on 40 meters, through the muf would climb to 15 meters within an hour.

If, on the other hand, you can be on the air at almost any time, pick the point where the muf curve approaches or crosses a band you can use, then read at the bottom of the chart the time you should throw the switch and start transmitting. According to Fig. 1, the best time to be on 40 for that path would be from 0800 to 1200 UTC; you could work the circuit on 15 from 1400 to 0100.

WWV Information

At 18 minutes after every hour, National Bureau of Standards station WWV, Fort Collins, CO, transmits bulletins on solar activity and the condition of the earth's magnetic field. This information is essentially current and therefore invaluable to anyone concerned with radio propagation. WWV is on the air continuously, using constant power levels and omnidirectional antennas. It should be noted that WWVH, Kekaha, HI, does not carry the bulletin service, but its signals nevertheless provide propagation data. Other time and frequency staions, some of which are listed in Table 2, can be used similarly.

In order of their presentation, the WWV bulletins give the solar flux and A-index for the previous Universal Coordinated Time (UTC) day; the K-index (changed every 6 hours); the state of solar activity and the condition of the earth's magnetic field for the previous 24 hours, and the same factors predicted for the coming 24 hours. Information on all WWV services is available from the National Bureau of Standards, Boulder, CO 80302.

Solar Flux

The *solar flux* is a measure of the sun's radiation at 2800 MHz, taken daily at 1700 UTC in Ottawa. Similar measurements are made on many different frequencies at other observatories. The 2800-MHz flux is given because this value correlates well with the ionization density of the ionospheric F region, with the Zurich sunspot number and the muf.

A steady rise or fall in solar flux will show clearly in radio propagation effects, and also in the observer's view of sunspot activity. Trends either way are important. They often run for several days, during which the associated changes in muf, and in the duration and geographical distribution of openings on frequencies above about 15 MHz, are easily observed. Flux readings of 80 or higher will make the 21- and 28-MHz bands come alive, even near the bottom of the solar cycle.

In the intermediate years of the cycle, as in 1977 to 1979, flux values tend to range between about 100 and 200, the latter being high enough to make even the 50-MHz band worth watching for worldwide communication on the more favored paths. When values above 200 come, the 21- and 28-MHz bands are open almost around the clock during the cooler months of the year. Sustained periods above 250, particularly in spring and fall, will bring widespread 50-MHz openings.

A- and K-Indices

The A-index is a 24-hour figure for the activity of the geomagnetic field. It is measured on a scale of 0 to 400 or so, though values of 100 or more are rare. A quiet field (A-indices of 10 or lower) is characteristic of the best propagation conditions. Absorption of wave energy is low at such times, so signals are strong on long paths. This effect is most noticeable on circuits crossing the higher latitudes, where very low K values must prevail or propagation will be very poor or nonexistent. The effect of geomagnetic activity is very slight on low-latitude paths or any circuit crossing the equator. In fact, it is observed that north-south or transequatorial propagation may *improve* briefly at the onset of a high A- (and K-) indices, especially on the highest frequency that is usable at all.

The K-index is similar to the A-index, except that it is as near to a current figure as can be given in bulletin form. It is also given on a different mathematical scale in order to make short-term changes more apparent. The information given by WWV is for Fredericksburg, VA. It is likely to be higher for Boston, and lower for New Orleans, as the data are directly related to geomagnetic latitude.

The K-index is given for six-hour periods beginning at 0000, 0600, 1200 and 1800 UTC. It represents the conditions during the last three hours just before the bulletin's time of issue. Thus, it is close to a *now* statement of a factor of vital importance to any user of the hf radio spectrum. It may interest vhf operators as well when the values go above 3. The *trend* is important — a rising trend means degraded hf propagation; values of 4 and up may mean auroral conditions on the vhf bands.

Solar Activity and Geomagnetic Field

Solar activity, as the term is used in the final portion of the WWV bulletins, relates to fast-changing conditions that affect propagation adversely. Activity is given as very low, low, moderate, high or very high.

Geomagnetic field conditions are termed quiet, unsettled or active. These relate roughly to K-indices of 0 to 1 for quiet, 1 to 3 for unsettled, and 4 or higher for active. These three words rather aptly describe the propagation conditions associated with them and with the K-indices. Often any K value of 4 or higher will be associated with a "geomagnetic storm," usually described as "minor" or major." Either is bad news for the amateur interested in high-latitude hf communication. "Major" may include a total or near-total interruption of all communication on the lower hf bands. It is likely to be accompanied by auroral conditions on the vhf bands, at least in the higher latitudes.

K values are used to derive the A-index for the whole UTC day. The manner in which this is done is rather involved. — *Carl Bixby, W1TKG, and James Morris, K1UJ*

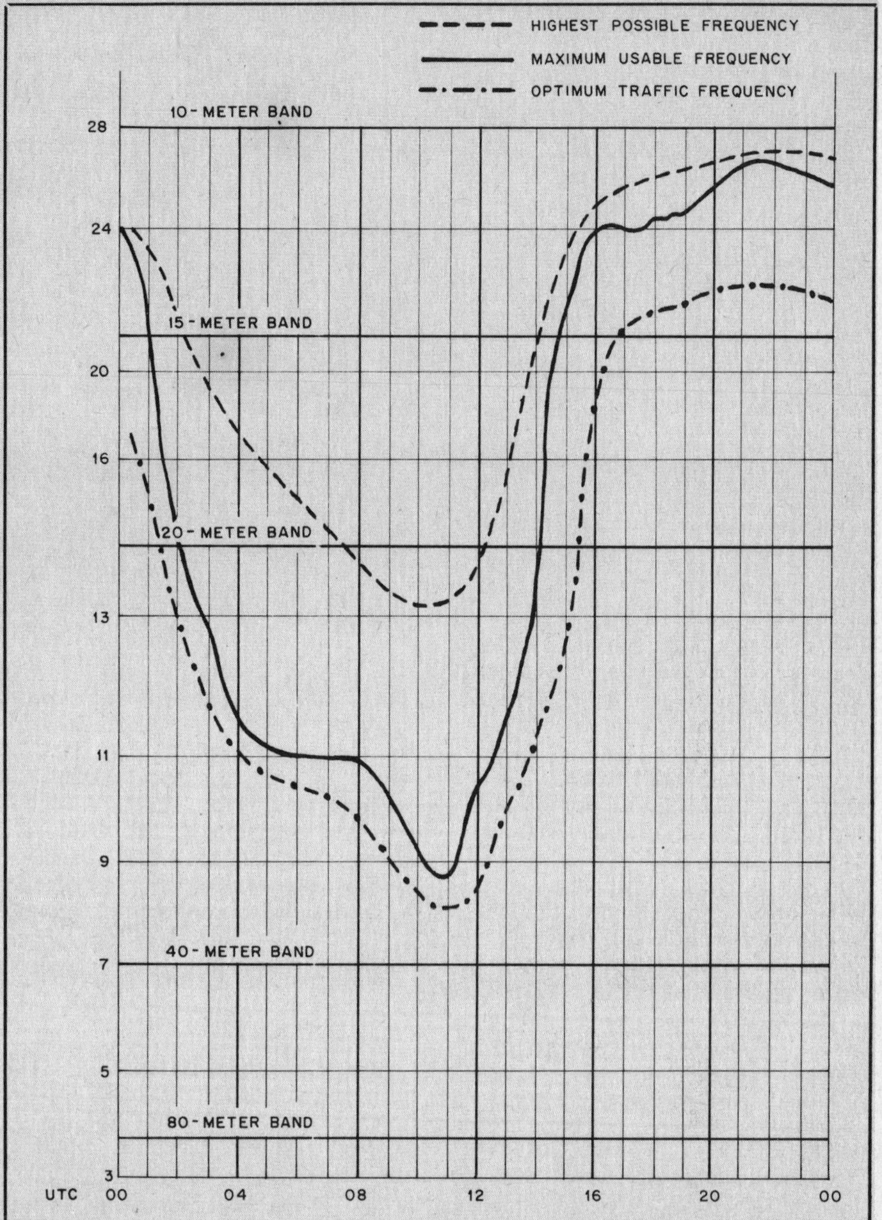

Fig. 1 — The chart predicting one month's propagation conditions for the East to West Coast circuit is typical of the 30 charts published monthly in *QST*. On 10 percent of the days, the highest amateur band that is open will be just below the highest possible frequency (hpf) curve. On 50 percent of the days it will be just below the maximum usable frequency (muf) curve. And on 90 percent of the days the highest band that is open will be just below the optimum traffic frequency (fot) curve. The horizontal axis shows Universal Coordinated Time (UTC); the vertical axis, frequency in MHz.

Table 2

Some time and frequency stations useful for propagation monitoring.

Call	*Frequency (kHz)*	*Location*
WWV	2500; 5000; 10,000; 15,000	Ft. Collins, CO
WWVH	Same as WWV	Kekaha, Kauai, HI
CHU	3330; 7335; 14,670	Ottawa, ON, Canada
RID	5004; 10,004; 15,004	Irkutsk, USSR*
RWM	4996; 9996; 14,996	Novosibirsk, USSR*
ZUO	2500; 5000	Pretoria, South Africa
VNG	7500	Lyndhurst, Australia
BPV	5000; 10,000; 15,000	Shanghai, China
JJY	2500; 5000; 10,000; 15,000	Tokyo, Japan
LOL	5000; 10,000; 15,000	Buenos Aires, Argentina

*Call, from international table, may not check with actual reception. Locations and frequencies appear to be as given.

Table 3

3.5-MHz Phone Allocations Throughout the World

Area	*Frequency (MHz)*
Africa	3.600-3.800
Asia (most)	3.600-3.900
Australia	3.535-3.700
Canada	3.725-4.000
Europe (except USSR)	3.600-3.800
Japan	3.500-3.575[1]
Pacific	3.600-3.900
South America	3.600-4.000
USA and possessions	3.775-4.000
USSR	3.600-3.650

[1]Japan also has a "DX band" between 3.793 and 3.802 MHz.

IARU Region 1 has officially recommended that 3.790-3.800 MHz be used only for intercontinental contacts.

There are several exceptions to the above table; local license-class restrictions are not accounted for.

The low bands (80 and 40 meters) are pretty productive hunting grounds for the DXer willing to take the time. Some of the low-band QSLs from DX worked on 80 and 40 in recent months.

during the hours of darkness. Again as on 160, long-haul DX paths between two given areas are enhanced when sunset at one end of the DX path corresponds with sunrise in the area at the other end. This is called *shadow edge* or *grayline* DXing.

It's hard to imagine that the daytime short-hop ragchewing and net operation that seems to be the mainstay of 80 meters gives way to a world of some pretty-heavy-duty DXing by night. Several 80-meter DXers have worked from 200 to well over 300 countries. If they can do it, so can you, with a little time and effort. On 80, an Advanced or Extra Class license can be the difference between success and failure in your DXing efforts, although a respectable country total can be amassed with one of the other license classes.

Eighty-meter WAC (Worked All Continents) and racking up 100 countries for your Five-Band DXCC are well within the grasp of any operator with even a modest station/antenna setup and a little bit of operating savvy. Antennas for 80 can range from simple dipoles and trapped verticals to elaborate delta loops and quad configurations. The rig should have split-frequency capabilities if you are interested in working 80-meter phone DX. To work some of the countries that can't transmit in the U.S. phone band, an external receiver or external VFO would definitely work to your advantage.

The Extra Class portion of the phone band (3.775-3.800 MHz) is the hotbed of DX activity, but many DX stations will listen up in frequency announcing what frequency (above 3800) they are listening on. The knowledgeable 80-meter DXer will listen well below 3800 for a DX station who may wish to work the U.S. using split-frequency operation. On cw, the majority of DX stations work in the 3.500- to 3.525-MHz range. Table 3 shows the 80-meter phone frequency allocations from around the world.

For those interested in serious DXing on 80 meters, there is a definitive work published on the subject. The book *80 Meter DXing* by John Devoldere, ON4UN, is published by Communications Technology, Inc., Greenville, NH 03048, and is available from the Ham Radio Bookstore. This book is probably the most complete work on the subject of 80-meter DXing and is written by an operator with 300 countries worked on 80!

40 Meters (7.0-7.3 MHz)

Forty, like the other low bands, 160 and 80 meters, is basically a night-time DX band. Again, grayline-DXing methods prove most fruitful.

On 40, though, the DXer faces a new enemy. Propagation and band characteristics are not the only obstacles to overcome. Forty meters is (like 160 and 80) a *shared* band; broadcast stations have assigned frequencies on 40 also. Thus, as the band conditions begin to change and become more favorable for working DX, it seems as if every broadcast station in existence is trying to claim a piece of 40 meters as its own. The bc stations use several kHz of bandwidth and legally run tens or hundreds of kilowatts of power. Strangely enough, it's possible to tell when 40 is open to certain areas by the strength of a received bc signal.

At 7 MHz, rotatable/steerable antenna arrays start to become more manageable in size and some of the simpler wire and vertical antennas that work well on 80 meters do a creditable job on 40. By running near or at full legal transmitter power to an antenna array, one can put out quite a respectable signal on 40 meters. The steerable antennas make it possible to explore different signal paths utilized on the higher bands, such as the long-path signal. Long path is a phenomenon of propagation where a signal from a given area of the world will peak if you steer your antenna so that it is facing approximately 180° away from the azimuthal heading (beam heading, or short-path beam heading) as determined from a great-circle projection centered on your QTH. (See the example of a great circle map and list of beam headings elsewhere in this chapter.) For example, if you usually point your antenna at 120° to work a station in Africa, a long-path opening would cause the same African station's signal to peak up when you rotated your antenna to a bearing of about 300°. It's worth checking for long-path signals, especially when signals seem to be down early in the morning or early evening.

As they do on most of the other bands, the cw DXers like to stay in the lower 25 kHz of the 40-meter band. But since there is such strong competition for space on 7 MHz, DX stations can regularly be found on cw much higher in the band — as high as 7.100 MHz and sometimes higher.

Working DX on 40 phone will quickly bring home the need to have the ability to work split frequency. The U.S. phone band extends from 7.150-7.300 MHz, and South and Central Americans as well as Africans can be found and worked here *transceive* (on their own frequency, not split). European and some Asian stations *must* be worked split-frequency on 40. The Europeans can be found between 7.080 and 7.100 MHz a good portion of the time, listening for U.S. types.

When the band is quiet and not crowded it is not necessary to have a "killer" signal to easily work DX. Many QRPers (operators who run 10 or fewer

JA9BE and his antenna farm. Mori has one of the big signals on the low bands, 80 and 40 meters. Among others, there is a 2-element Yagi on 40 meters atop that tower. You can see Mori's 5-Band DXCC plaque in the shack, but JA9BE has recently added a 5-Band Worked All States plaque.

watts to their antennas) are quite successful in achieving respectable country totals on this band. QRPers on 40, as on most of the other bands, tend to be cw operators and gather at or near 7.060 MHz. Forty meters, for all its quirks, is quite an interesting band and has a lot to offer for a serious or casual DXer.

20 Meters (14.000-14.350 MHz)

The main DX band: When working DX is mentioned, most hams immediately think of good old reliable 20 meters. During the peak of the solar cycle, you can work DX stations on this band 24 hours a day. Every serious DXer has good 20-meter capabilities, because sooner or later most of the juicy DX will show up there. Whether sunspot maximum or sunspot minimum, you can depend on 20 meters to open up, however briefly, to some DX location every day.

For U.S. operators, the common band divisions exist. DX can be worked, usually, from 14.000 to 14.100 on cw, and 14.200 to 14.350 MHz on phone. The QRP enthusiast finds a home around 14.060, while the RTTY buff and the SSTV op can work their share, too.

DX operators find a haven in the DX phone segment on 20, below 14.200 MHz where the DX station can go to escape the U.S. DX hordes and work us split frequency or other foreign stations transceive. Again, split-frequency capability is important.

Full-sized beam and quad arrays are quite manageable at 14 MHz, but the ever-popular trapped triband beam antennas offer good performance despite reduced size.

On 20 meters, as on 40, early-morning and early-evening band openings may hold some surprises if you were to swing your antenna in the long-path direction. Long path may give way to multipath (signal coming to your antenna over more than one path) signals with that echo-chamber sound. ***Skew path*** signals, which are propagated when neither operator has his antenna in the long-path or short-path direction and the signals are being propagated in some roundabout route, also become a possibility. A nighttime signal that is coming over the North Pole is quite easy to identify by the fluttery quality of the signal.

Twenty meters, being the mainstay of the DXer, has a very forgiving nature. There's room for practically everyone who wants to work DX and, given the volume of DX stations that show up, anyone who can radiate any sort of signal is able to work DX successfully.

15 Meters (21.000-21.450 MHz)

More prone to the whims and variations of the solar cycle than 20 meters, 15 meters can be depended upon to open up to one area of the world or another. When 15's hot it's really hot, but when it's not it's still at least lukewarm. Fifteen is a nifty halfway point between the always-dependable 20 meters and the somewhat-more-finicky 10-meter band. When 10 closes, a short trip down to 15 meters might pay off with signals booming in from all over the world.

Cw DXers on 15 are usually found from the bottom of the band, at 21.0 up to 21.1 MHz. The Novice segment, from 21.1 to 21.2 MHz, provides plenty of DX action for the Novice/Technician licensees.

On phone, split-frequency operation is used only by DXpeditions (trips to a foreign country arranged specifically to put it on the air for DXers to work) to the less-common DXCC Countries and other more-sought-after DX stations, to cut down on QRM (interference) from the pile-ups that they generate. Even in these cases, the split used is usually only on the order of a few kHz. In the really rare operation (SV/Mount Athos or FO0/Clipperton Island, for example)

Tim Chen, BV2B, is the only source of DX activity on Taiwan, as he is the only amateur. Tim spends a lot of time on phone on 20 meters, as BV2B and likes to hand out contest cw QSOs as BV2A (his licensed cw call). If you want to see a first-class pile-up, wait until word gets out that Tim is on a given frequency.

Father Moran, 9N1MM, the only licensed amateur in Nepal. He is very active on 15- and 20-meter phone, giving out that rare country to needy DXers.

where overwhelming numbers of eager DXers try to get through at the same time, the split used might be 20 or 30 kHz or more. Of course, the bottom of the U.S. phone band is 21.250 MHz, and a weary DX station can take refuge there from the multitudes of anxious U.S. DXers.

During the times of high solar activity, as at the time of this writing when solar cycle 21 is reaching its peak, 15 meters will give you all the DX action you can handle.

10 Meters (28.0-29.7 MHz)

It's been said that when 10 meters is open, you can work the world by loading a few watts of power into a "wet noodle." Maybe it's not that simple, but you'd be surprised at the astounding results that a modest station can produce when 10 is alive.

Ten-meter beam/quad antennas, even full-sized arrays, are relatively small and easy to rotate with conventional TV antenna rotators, and can be mounted on relatively small TV towers and masts. Apartment dwellers and folks with small yards take heart! Vertical and dipole antennas are more than up to the job of snagging that DX station when the band is jumping.

When conditions are right, 10 meters will open up nicely, providing reliable paths to various areas of the world until

9M2PV, West Malaysia, is QRV quite often for those needing 9M2, especially on 10 meters at the appropriate times. Andy likes to chat with the DX gang and has an incredibly quick QSL turn around time.

well after nightfall. Don't forget to check for the long-path openings, especially when 10 is just starting to open up early in the morning.

Cw DX on 10 can be found from the bottom band edge at 28.0 to the upper end of the Novice portion at 28.2 MHz. Phone band starts at 28.5 MHz to the upper limit of 10 at 29.7 MHz. Be sure to be wary of the Amateur Satellite (OSCAR and RS) downlinks located between 29.4 and 29.55 MHz.

Ten-meter operators have at their disposal, a foolproof tool to judge whether or not the band is open — beacons. Beacons are stations, erected and maintained by individuals and groups who are interested in propagation studies. These beacons transmit continuously, allowing the operator to determine if the band is open and to where it is open. A list of some 10-meter beacons appears in Table 4.

Ten meters is a band that can be full of surprises. A CQ on this band, when it appears to be dead, can produce some surprisingly good DX contacts. It sometimes turns out that everyone was just tuning a "dead" band looking for a DX signal to pounce on.

New Hf Bands (10, 18 and 24 MHz)

The 1979 World Administrative Radio Conference, where delegates of every interested country in the world met to decide how best to "divvy up" the radio spectrum, produced some pretty exciting news for hf DXers. New amateur allocations have been made at 10, 18 and 24 MHz. Although amateur occupation of these bands is not scheduled to begin until 1982 and after, we can speculate on the significance of these new allocations.

The DXer will be able to more-closely follow the muf (maximum usable frequency). How many times have you heard 20 meters hopping, and tuned down to 15 meters only to find the relative silence of a "dead" band? Ten MHz will be an excellent check point between 7 and 14 MHz, just as 18 MHz will be between 21 and 14 MHz, or 24 MHz, the intermediate stop between 10 and 15 meters. It's a lot easier to follow the muf up the bands as they open up in 3- or 4-MHz steps rather than in leaps of 7 MHz at a clip.

DXing on the Vhf Bands

Entering the vhf spectrum in search of DX requires equipment and operating techniques that are somewhat different than those required for DXing on the lower bands. Detailed information on vhf operating, equipment and propagation patterns are found in the vhf section of this book and in appropriate chapters of the ARRL *Radio Amateur's Handbook* and the ARRL *VHF Manual*.

6 Meters (50-54 MHz)

We'd be remiss if we didn't mention 6 meters as a DX band. In times of peak solar activity when the muf climbs above 54 MHz, you'd be hard put to tell 6 meters from 10 meters by the call signs being heard. Europe, Japan and Oceania are all being heard and worked by mainland-U.S. stations under optimum conditions.

Six meters is a band that can't decide if it is a hf (high frequency) or a vhf (very high frequency) band. It is capable of exhibiting the best in propagation conditions of both types of bands and passes as an acceptable transitional band between the hf and vhf bands. The WAC Award is definitely attainable on 6.

Other Vhf Bands (144-220-432 MHz)

Some truly remarkable DXing has been done on these bands. Paths of 1500 or so miles on 144 MHz are not all that uncommon. Such exotic vehicles of propagation as meteor scatter (reflecting signals from the ionized trail of meteors entering the ionosphere), tropospheric ducting (a path created by unsettled weather between two opposite frontal systems) and aurora are utilized.

EME (earth-moon-earth) is the technique that can turn these vhf bands into DX bands. In EME, signals are bounced (reflected, if you will) off the moon from one station on earth to another. Great distances can be covered in this mode. Station equipment and antenna arrays for moonbounce become quite elaborate. See the vhf/uhf chapter for more definitive information.

Satellite DXing

DXer purists will argue that working DX through one of the amateur satellites is like working DX through a great spaceborne repeater. The amateur satellites include: OSCAR (Orbiting Spacecraft Carrying Amateur Radio), RS (Russian Radiosport satellite) and the new Phase III, which will allow (after its launch in 1980) easy access to a new "DX band" that would be capable of supporting trans-continental/trans-hemispheral communications for many hours a day.

Several DX-oriented awards for satellite work already exist. Again, as on vhf work, satellite DXing requires a whole new set of skills. Satellite operation is throughly discussed elsewhere in this book.

Where is All the DX Hiding?

A lot of good information on DXing, including "where to find 'em," is as close as your fellow DXer. The two things that a DXer likes to talk about are, of course, DX and himself — in that order. A "big gun" (serious) DXer is a gold mine of information on the haunts and habits of DX

Table 4

International Beacon Project on 10 Meters*

Freq. (kHz)	*Station*	*Location*	*Remarks*
28,185	OA4VHF	Lima, Peru	1
28,200	Common		
28,202.5	9J2BB	Lusaka, Zambia	0500-0600GMT 1500-1600GMT
28,205	DL0IGI	Mt. Predigtstuhl, Germany	28,200 H . . 00-05 H . . 30-35
28,207.5	N4RD	Englewood, FL, USA	
28,210	3B8MS	Signal Mt, Mauritius	
28,212.5	ZD9GI	Gough Island S. Atlantic	3
28,215	GB3SX	Crowborough, England	
28,217.5	VK2WI	Sydney, Australia	1
28,220	5B4CY	Limassol, Cyprus	
28,222.5	YU	Yugoslavia	1
28,227.5	FX3	France	2
28,230	ZL2MHF	Mt. Climie, New Zealand	
28,232.5	VP8	Falkland Is.	2
28,235	VP9BA	Southampton, Bermuda	
28,237.5	LA	Norway	1
28,240	PY1CK	Rio de Janeiro, Brazil	4
28,242.5	ZS	South Africa	1
28,245	A9XC	Hamala, Bahrain	
28,247.5	EA20IZ	Spain	Unofficial
28,250	W		1
28,255	W6	CA, USA	1
28,260	VK5	Australia	1
28,265	VK6	Australia	1
28,270	VK8	Australia	1

*This project is sponsored and encouraged by IARU Region 1 and organized by Alan Taylor, G3DME.

Notes:
1) In planning stage
2) Under construction
3) Expected to be operational soon
4) Being rebuilt after period of unserviceability

operators. How do you find a DXer to get to the source of all this dandy information? Check the monthly DXCC endorsement/Honor Roll listings in *QST*. You'll more than likely note the calls of a couple of locals, perhaps someone in your own Amateur Radio club. It's fairly easy to get a dedicated DXer talking about his favorite subject, then just sit back and soak up these choice tidbits of knowledge that are the fruits of years of serious DXing.

The best way to find out where that DX is hiding is to spend some time on the bands. Get to know the band openings, some of the regular DXers on the band, and the general feel for band conditions — specifically which areas of the world are heard at your QTH and when. The general gossip and information from other DXers is known as the DXer's grapevine. The grapevine is usually very accurate when it comes to news of upcoming DXpeditions and the whereabouts at any particular time of some of the more sought-after DX stations.

A given DX station's operating habits may, at first, seem less than predictable. Not so. More than a few of the operators of the rarer DX stations seem to be creatures of habit who favor certain portions of particular bands of frequencies. Information about a DX station's habits is available from various sources.

Is it Hard to Work 9-V Land?

The lists that follow contain degree-of-difficulty rankings for countries (by prefix) on the ARRL DXCC country list. The prefixes are listed by groups numbered from 1 to 5.

Group 1 contains countries that have healthy and active populations of amateurs and can be worked almost any time you get on the air.

Group 2 countries are countries on the air almost as frequently and in the same quantity as those in group 1. It shouldn't be too hard to work all countries in groups 1 and 2 in a single contest weekend.

Group 3 countries for the most part have resident amateur populations. Propagation conditions play a large part in whether you can work these countries. It takes a little looking to find all of these countries.

Group 4 countries might have a small resident amateur population, but a DXpedition and very good propagation are a great help in making these countries easy to work.

Group 5 If you don't have a contact from any of these countries, you'll have to wait for a major DXpedition or a major change in their political situation. Good luck.

Tnx to Honor Roll DXer Ken Bolin, W1NG, for these evaluations.

Group 1

CT DJ DM EA F G GW HA HB I JA K/W/N/A KV/KP2 LZ OE OH OK ON OZ PA PY SM SP VE/VO YU YV ZS1-6

Group 2

3D2 4U(ITU) 4U(UN) 4X 5T 5Z 6W 6Y 8P 8R 9A(MI) 9H 9J 9L 9Y C6 CE CM/CO CN CT2 CX EA8 EI EL FC FG FM FP FY GD GI GJ GM GU HC HH HI HK HP HR IS J3/VP2G J6/VP2L J7/VP2D KH6 KL7 KP4 LA LU LX OA OX PJ(NA) SV(GR) TI UA(EUR) UB UC2 UP2 UQ2 UR2 VK VP2M VP9 VQ9 XE YN YO YS ZE ZF ZP

Group 3

3B8 3D6 3V 5B 5H 5N 5U 5W 7P 7X 8Z4 9G 9K 9M2 9N 9Q 9V A2 A4X A9X AP C3 C5 CEØA CT3 D4 DU EA6 FO(POL) FR(RE) FS HBØ HKØ(SA) HM HS HZ/7Z J2 JT JY KC4 KG4 KG6/KH2 KG6R KS6/KH8 KX OD OHØ OY PJ(SM) S7 TF TG TJ TU UA2 UA9-Ø UD6 UF6 UL7 UM8 UO5 VK9(NI) VP1 VP2A VP2E VP2K VP2S VP2V VP5 VP8(FI) VR1(BPI) VR1(G&O) VR3 VR8/T2 VS6 VU YB YJ YK ZB ZD8 ZK1(S) ZS3

Group 4

3A 3B6 3B9 3C 3CØ 4S 5V 6O 8Q 9M8 9U 9X A3 A5 A6X A7X BV C2 CEØZ CR3/J5 CR9 D2 D6 EA9 EP FB8X FB8Z FH FK FR(G) FR(J) FR(T) FW H4 HC8 HKØ(BN) HKØ(MI) HV JD(MT) JD(OG) JW JX KB/KH1 KC4/KP1 KC6(ECI) KC6(WCI) KH7 KJ/KH3 KM/KH4 KP3/HKØ KP4(DI) KP6/KH5 KW/KH9 OJØ P2 PYØ(FN) PYØ(TR) S2 SSV(CR) ST STØ SU SV(DOD) SV(MA) TA TI9 TN TR TY UA1(FJL) UG6 UH8 UI8 UJ8 VE1(SI) VE1(SPI) VK(LH) VKØ(MAC) VK9(CI) VK9(COI) VP8(SGA) VP8(SOR) VP8(SSA) VP8(SSH) VR6 VS5 XF4 XT YI YVØ ZD7 ZD9 ZK1(N) ZK2 ZL(A&C) ZL(C) ZL(K) ZM7 ZS2

Group 5

1S 3X 3Y 4W 5A 5R 5X 7O 7Q BY C9 CEØX ET FB8W FO(CI) JD(OKI) KP6/KH5K PYØ(P&P) S9/CR5 TL TZ VKØ(HI) VK9(WI) VS9K VU7(A&N) VU7(LAC) XU XV XW XZ YA ZA

On-The-Air Bulletins

On-the-air bulletins with news of interest to the DXer are fairly common. Many local radio clubs utilize their 2-meter or 220-MHz repeaters to alert DX-oriented members to up-to-the-minute happenings in the DX world. Imagine, if you will, how much more efficient it would be to have four or five operators sitting at the rig, scanning the bands in search of DX to work, than to try to cover the happenings on five bands all by yourself. This is effectively what a group of DXers, having access to a repeater, accomplish. When one of the ops spots a choice piece of DX, he can simply access the repeater and give the information to his cohorts or anyone else monitoring the repeater. A simple "9N1MM on 21.350" (of course giving call signs to comply with FCC regulations) would be enough to put everyone else on the right track. You'd probably be surprised at how much support you could muster from local DXers for starting a DX-spotting net or DX-information bulletin service on your local repeater. After all, "the more ears, the more you hears."

Bulletins are broadcast as a service to the worldwide DXing community, too. Three of many are:

1) ARRL DX Bulletins — broadcast from station W1AW on Friday (UTC) on cw, ssb and RTTY. See *QST* for W1AW's bulletin schedule.

2) The Northern California DX Association DX Bulletin — sent on cw at 0200Z (UTC) on each Monday at 14.003 MHz.

3) The WØRJU DX Bulletin — sent at 0100Z on cw Tuesday at 7.002 MHz and at 1200Z on Monday at 14.002 MHz.

These bulletins contain timely information on happenings in the DX world.

Published Information

All the major U.S. Amateur Radio periodicals have a section devoted to news for the DXer, containing news of upcoming DXpeditions, QSL managers and associated happenings. Since these publications are published monthly and have fairly long lead times for publication, their value for disseminating really current news is limited. The great value of the DX information contained in these publications is found in following up an operation with correct QSL routes, news of what will occur (DXpeditions planned months in advance) in future months, as well as the ever-interesting narratives from those lucky operators just returned from successful DXpeditions.

Several individuals and DX-oriented clubs across the U.S. and Canada publish frequent newsletters (most weekly or bi-weekly) or bulletins to keep their subscribers abreast of happenings in the DX scene. A s.a.s.e. to any of the bulletins listed below will bring subscription/membership rates and frequently, a sample copy of the DX bulletin for your inspection.

The DX Bulletin
306 Vernon Ave.
Vernon, CT 06066

The Long Island DX Bulletin
P. O. Box 173
Huntington, NY 11743

Long Skip
c/o The Canadian DX Association
P. O. Box 717 — Station "Q"
Toronto, ON M4T 2N7
Canada

QRZ DX
P. O. Box 494
Howe, TX 75059

Because of the frequency of publication of these bulletins, their news is quite timely and can be of great assistance to the DXer. For example, a business-sized s.a.s.e. to the *Long Island DX Bulletin*

International Third-Party Traffic — Proceed with Caution

Every once in a while, DX stations may ask you to handle a third-party message to a friend or relative in the states. This is all right as long as we have an official Third-Party Traffic Agreement with that particular country and the traffic is noncommercial and of a personal unimportant nature. During an emergency, our State Department will often work out a special temporary agreement with the country involved, but in normal times, *never* run traffic without first making sure it is legally permitted.

U.S. Amateurs May Handle Third-Party Traffic With:

Argentina	Honduras
Bolivia	Israel
Brazil	Jamaica
Canada	Jordan
Chile	Liberia
Colombia	Mexico
Costa Rica	Nicaragua
Cuba	Panama
Dominican Republic	Paraguay
Ecuador	Peru
El Salvador	Trinidad and Tobago
Ghana	Uruguay
Guatemala	Venezuela
Guyana	4U1ITU, Geneva
Haiti	

Canadian Amateurs May Handle Third-Party Traffic With:

Bolivia	Jamaica
Chile	Israel
Colombia	Mexico
Costa Rica	Nicaragua
Dominican Republic	Peru
El Salvador	Trinidad and Tobago
Guatemala	United States
Guyana	Uruguay
Honduras	Venezuela

garnered a sample copy a few days later. That particular issue contained the welcome news that P29CC in Papua-New Guinea "with his new beam is dynamite near 14.265 MHz around 1200 UTC. QSL via W2NHZ." The day after the bulletin was received, P29CC was found and worked (and QSL was received in short order) at 1241 UTC at 14.264 MHz. QED.

Many local radio club newsletters carry DX information as a service to the DXers on the membership list. One or two active DXers simply reporting what DX they have heard and worked since the last issue of the club paper could easily constitute a small column for the newsletter, and in so doing perform a service for other club members who happen to be DXers, too. If your club paper has no DX news at this time, why not approach the editor? Chances are, he'll be more than happy to cooperate. It's that much less he has to "fill in" to put out a decent-looking newsletter!

DX Nets

DX nets are a haven for the rare and semi-rare DX stations who cannot or will not cope with the frustration of trying to control a pile-up. The DX net provides a regularly scheduled time and place for a DX station to be able, through a well-controlled frequency situation, to work a good many of the DXers that need him for a "new country."

Below, we'll list some of the more-well-known DX nets. The average DXer can fill a lot of the holes in his "needed country" list with a few check ins to these nets. Beware! Be prepared to follow the instructions of the net control station. Those who don't play the game by the rules in the net situation *do not* get to work the choice DX that frequently checks in.

A sampling of DX Nets follows:

Arabian Nights DX Net
14.250 MHz, 0500 UTC Friday only

Forty-Meter DX Net
7.182 MHz, 0500 UTC Sunday only

P29JS DX Net
14.265 MHz, 0700 UTC daily

Pacific International DX Net
14.265 MHz, 0530 UTC Tuesday and Friday

W7PHO Family Hour (20-meter version)
14.225 MHz, 1400 UTC daily

Arabian Christian Net
14.305 MHz, 1700 daily

DX to DX Net
21.280 MHz, 1730 UTC Monday, Wednesday and Friday only

Africana Net
21.355 MHz, 1800 UTC daily

10 Meter DX Net
28.510 MHz, 2100 UTC Monday, Wednesday and Friday only

W7PHO Family Hour (15-meter version)
21.345 MHz, 0000 UTC daily

There are other DX Nets. A little listening on the bands will tell you when and where they meet. Those DX nets listed above are known producers of some very choice DX.

Alternatives

After the "tips" on where the elusive rare DX stations can be found do not pan out, it really does not hurt to scan the bands a while. It just might be your lucky day. 8Q7 might be calling CQ over and over with no takers. Fat chance? Stranger things have happened.

When all else fails, why not try a short CQ or two. You never can tell when that new country might be just a CQ away.

There are indeed "tricks of the trade" that take some of the guess work out of chasing DX.

How to Work DX

Now that you have spotted the DX station that you want to work, and you're all tuned up and ready to call him, the rest is "duck soup," right? Maybe and maybe not. In working the DX station, as in finding him, variables come into play that can spell the difference between success and failure.

Station Equipment

The most valuable pieces of equipment that the DXer has at his disposal are his two ears and the "gray matter" that lives between them. Being able to listen and apply a few principles based on common sense cannot be overly stressed as desirable attributes for the successful DXer.

As one might expect, in most cases the quality of the station used plays a large part in the ability to work a given DX station in a given set of circumstances. After all, the more elaborate the station hardware, the more rf one can generate in the desired direction, the louder will be the signal at the DX station's receiver.

We are all victims of our circumstances. It's the lucky DXer who has unlimited resources to build his station with. Therefore, compromise is necessary. If you don't have the room to erect beam antennas, you can buy or build a linear amplifier — thereby increasing the quality of your signal through watts rather than aluminum. It will come naturally to optimize your station layout/equipment through on-the-air experience. You will also learn when it is best to use a vertically polarized antenna vs one that is horizontally polarized, higher vs lower antenna, when to use that speech clipper and when not to, etc. And, to a degree, any station that will radiate a signal is capable of working DX. We won't dwell on the station layout/design (that subject is covered quite nicely elsewhere) but we will recommend quite strongly that, all else aside, when you consider buying a rig for serious DXing, that the rig be capable of split-frequency operation. It makes no difference if the gear is a separate transmitter/receiver, transceiver with external remote VFO, or a transceiver with provisions for using a separate receiver. To be useful in the split-frequency situation, the rig should be capable of transmitting and receiving at least 50 kHz apart.

The *ARRL Antenna Book* provides many ideas and construction hints for those DXers interested in upgrading their antenna systems.

Operating Techniques

Surprisingly enough, success in working DX also is determined by the operating ability of the person at the microphone or telegraph key. A successful operating technique can, in many situations, seemingly add as many decibels to your signal as another element in you antenna array.

The Pile-up

We'll assume that if we find one of the

more-common DX stations calling CQ or QRZ, not many stations will respond and it will be an easy matter to zero beat (get on the same frequency as the DX station), call and work that station. If the DX station happens to be a very rare catch, literally hundreds or thousands of eager DXers will be calling at once. Join the callers, and you're in that meatgrinder known as the *DX pile-up*. The pile-up, also known as the "free-for-all," causes all those stations to pile up on or near the DX station's frequency, calling (or shrieking) to catch the DX station's attention, creating unbelievable interference on the frequency. In this situation, as might be expected, the stations with the strongest signals are the first to be worked. The operators with the lesser signals must, if they have the patience, call for great lengths of time before getting an answer — if they ever do. Even in this, the worst of cases, there are tricks an operator can employ to increase the chances of success.

Listen. Listen. Listen. That's the only way you'll be able to hear that DX station to figure out how to work him. W6ISQ offers some Pile-up Operating Tips (phone and cw):

1) Don't jump right into a pile-up; listen for five or six contacts. That way you can figure out how the DX station is handling the pile-up.

2) Keep your calls short. Don't call the DX after he has gone back to someone else. Very tacky. Delays the whole operation.

3) Don't send the DX station's call more than once, if at all. Who else would you be calling in the pile-up? Better you should be signing your call than his.

4) If the DX station is working "tail-enders" (folks who call immediately at the end of his last QSO), try it. If he is not working tail-enders, do not disrupt the proceedings by trying it.

5) Try using different phonetics on phone. Keep your calling time short, but be sure to sign both call signs when working the DX stations. Those are the FCC regulations.

6) If the DX station is working only stations in an area other than your own, perhaps conditions are not yet quite right for you. Call, but don't go nuts. Stay loose and hope you get propagation before the DX goes QRT (signs clear).

Again, *listen* and follow the DX station's instructions to the letter. Being courteous and fair to all the stations on frequency by employing good operating procedures should be every bit as important to the DXer as logging the DX station.

Split-Frequency Operation

The DX operator has the option, if he has the proper equipment and operating skills, to help reduce the pile-up and the ensuing QRM. In operating split frequency, the DX station directs stations to call him on a frequency other than his own. The DX op may specify a single frequency for callers or, if the pile-up becomes really big, the DX station may specify a frequency segment where he will listen for all callers. Example TLØBQ is transmitting on 14.195 MHz and says he will listen for U.S. DXers from 14.210 through 14.250 MHz. The theory behind split operation is that with stations calling off the DX station's frequency, there will be no QRM to the DX station's transmissions, making it easier for the eager masses of DXers to hear, call and ultimately work the DX.

Here are a few more tips from W6ISQ on Pile-up Split-Frequency Operation (phone and cw).

1) Be sure your transmitter dial is calibrated. The DX station will direct you to call on (at times) a specific frequency and if your calibration is off, even slightly, you will not be heard.

2) If possible, use two receivers or a receiver/transceiver set up so that you may listen to the DX station, and monitor the frequency that the DX operator is listening to. This will help you to locate the station being worked and to determine the DX station's pattern of working stations.

3) Consciously or not, most DX operators form a pattern in how they work a pileup. Listen, carefully. See if you can figure out how he is working the pile-up.

a) Does he move up or down a set number of kHz after each QSO to work the next station? If so, try to keep in step by changing your transmitting frequency by the same increment each time you call.

b) Does he tune to the edges of the segment he has specified? If so, position yourself at one of these edges and do your calling from there.

c) Does the DX station work "hit or miss" among the callers? Maybe he is looking for signals in clear spots. Therefore, position yourself on one of these "clear frequencies" and remain there as long as it remains relatively clear.

d) We'll add that if the DX station is working "hit or miss," a good suggestion would be to tune the specified listening segment and copy the calls of a few of the other stations who are calling the DX, making sure to note or chart their exact frequencies. When the DX station works one of these stations you are using as "band markers," you can immediately dial your transmitter to that frequency and increase your chances of bagging the DX (see Fig. 2).

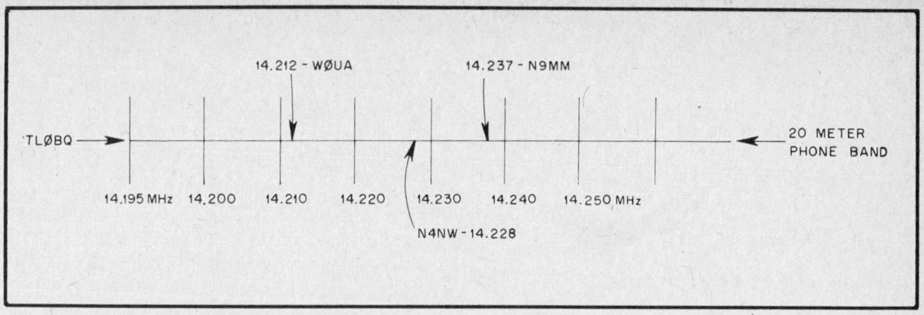

Fig. 2 — We are trying to work TLØBQ, Central African Empire, who is transmitting on 14.195 MHz. TLØBQ says he is listening for stations to call him from 14.210 to 14.240 MHz. We listen a bit and find: WØUA calling at 14.212, N4NW calling at 14.228, and N9MM calling at 14.237. We write these calls and frequencies down. Then when TLØBQ works one of the stations that we are using for a benchmark, say N9MM at 14.237, we can tune our transmitter quickly to this frequency and have a good shot at working TLØBQ.

Always check the frequency that you intend to use to call the DX station to see if another station has prior claim to that frequency. The fact that you are chasing a rare DX station does not allow you to drop all the rules of common courtesy and run roughshod over the rest of the amateur population that is trying to use the band for other purposes.

Selective Calling

Another method of controlling the size of the pile-up is selective calling. In selective calling, the DX station sets up certain guidelines for the calling stations to follow. Usually, stations are invited to call by U.S./Canadian call areas. For example, AP5HQ, transmitting on 14.190 MHz, says: "Listening for stations from the sixth U.S. call district. W6s only, please."

This situation may not be ideal, but is surely a lot better, especially for stations with the weaker signals, to compete on a local level with several hundred stations rather than on a national or a worldwide level with tens of thousands of calling stations.

The Controlled Frequency or DX Nets and Lists

We'll discuss two areas that come under the "controlled frequency" heading: the DX list and the DX net.

The use of DX nets and lists to run up one's DXCC country totals is a controversial subject. One faction of the DX community claims that DX stations worked via nets and lists should not count for DXCC credit. It's like shooting fish in a barrel, they say. On the other hand, a DX

net or list, properly used, can be a very effective and expedient tool for the DXing community as a whole.

The MC (master/mistress of ceremonies, also known as host, list taker, etc.) who runs a tight ship can put a large number of stations through to a given DX station in a relatively short time. The MC can also help transcend language barriers.

Simply put, a DX net is a regular on-the-air gathering of a group of individuals interested in working DX and in helping others to increase their country totals. The net should have a regular MC/host who will exercise control over the proceedings of the net, check stations into the net in an orderly manner, then ultimately get the DX stations in contact with those amateurs who wish to work the DX. The MC should have an adequate signal and a basic sense of good operating procedures.

Those things said of the MC can also be said of the operator who takes it upon himself to run a list. The major difference between a net and a list is that a net is a *regularly* scheduled event, whereas a list might be taken on the spur of the moment to cover a particular DXpedition, or to introduce a non-English-speaking DX station to the art of working a DX pile-up of English-speaking amateurs. The list operation might prove useful to the DX operator until he becomes proficient enough to be able to carry on a fundamental QSO. The list operation is very useful in the case where the needed DX station is capable of putting out only a modest signal. In this instance, a disciplined MC with a big signal could be most effective in keeping order on the DX station's frequency. If you've had occasion to listen to some of the shenanigans going on in the pile-ups, you can appreciate what an asset a good MC can be. By the same token that the MC is obligated to make things run in the most efficient manner possible, those who avail themselves of the list's/net's services are obligated, if by nothing other than common courtesy, to play by the rules.

When using the list/net, there is one overriding commandment to abide by: *Follow the instructions of the MC to the letter.* The more cooperation that the MC receives, the faster things move on frequency and the more stations can work the DX on frequency. Common courtesy, common sense and simple cooperation make net/list operations better for all concerned.

Anyone who has sat in on a net/list operation, expecially one that hosts a rare DX station has heard the bickering and the obnoxious behavior of those who wish to work the rare station. Then, there are the "frequency policemen" who take it upon themselves to sit on the DX station's frequency shouting orders and making pithy comments, in the process causing more QRM than they prevent. We haven't forgotten but are reserving a special place on the lid (very poor operator) list for that demented individual, the intentional QRMer. Lists/nets have a whole gamut of obstacles to overcome. Orderly, courteous operating on the part of everyone involved is the key to successful list/net operation.

Using Contests to Add Countries

Amateur Radio DX contests, which usually involve some of the rare ones are sponsored by several organizations (ARRL, *CQ* Magazine, The Japan Amateur Radio League and so on). Although the format of these contests favors the "free-for-all" pile-up situation, weaker stations should take heart. Contest operation functions at a much more rapid pace than the usual QSO format. Thus, in the course of a weekend of contesting (depending on the rules for that particular contest), a DX station can work many thousands of stations! Even though he may not care much for contest operation, the DXer should try contests as a way to increase his country totals.

Modes

A few suggestions covering each mode follow:

Cw — In a pile-up, call the DX at the periphery of the mass of calling stations. Don't be afraid to call the DX station between a few hertz to a couple of hundred hertz off his zero-beat frequency.

Try to match your sending speed to that of the DX you are calling. In trying to pare down the pile-up, he is likely to be sending at his maximum speed and will be able to copy your call and report, at least. If everyone in the pile-up is sending at about the same speed, try varying your code speed up or down. It might catch the DX station's attention.

Listen. Try to find out which part of the pile-up the DX station is working. If he says "u 3" or "d 10," he means he is going to listen for callers up 3 kHz in frequency or down 10 kHz. If the DX station specifies that he is listening elsewhere, do *not* under any circumstances, call him on his frequency. Aside from QRMing the DX station, making it difficult for other calling stations to hear the DX station, you might, if the DX op is transmitting on a frequency below the privileges granted by your license, very well be out of band. The serious DXer is just as concerned with operating within the guidelines set by his licensing authority as with "bagging that new one."

A memory keyer is a definite bonus in those really big pile-ups. Just load your call and push the buttons. Keeps both hands free for tuning.

Phone — Again, keep your calls short. In a pile-up, sign your call once only at each calling period. Ignore the poor operators who start calling and never stop. Making comments on frequency during the pile-up only adds to the QRM. The heckler does not deserve the satisfaction of getting a rise out of those DXers who are waiting on frequency. Let all the comments pass, even though you may feel that it would be very satisfying to do otherwise.

When signing your call in a pile-up on phone, use good solid phonetics. Standard phonetics work best, as they are nearly universally understood. "Cute" ones have a way of getting lost, especially to someone who speaks your language only marginally. On phone in the pile-up, for example, "We're One Obnoxious Donkey" would not come through to the DX station as well as "Whiskey One

Low-Power DXing and Station Design for the Low-Power DXer

DXing while running relatively low power can be one of the most exciting and at times most frustrating aspects of Amateur Radio. In this era of kilowatt amplifiers and multi-element monoband antenna systems (including giant 80- and 40-meter arrays), a respectable DXCC total can be accrued while running no more than 100 watts and using simple wire antennas. What may be lacking in signal strength may be more than made up in operating skill. Patience, confidence and experience are three virtues of the low-power DXer. Confidence in your operating abilities and operating skill and know-how come with on-the-air experience. So don't be afraid to pit your low-power signal against the kilowatt stations in those DX pile-ups and DX contests.

Right from the start, it must be understood that the low-power DXer will have to work a little harder than his "kilowatt contemporaries." It will take at least several months of serious low-power DXing to work a respectable number of DXCC countries.

One way to work a lot of DX in a short period is to operate in one of the major DX contests. It is very possible to work 50 or more DXCC countries in a single contest weekend. There are occasions when DXCC can be worked in a single day, if not a few hours, by low-power stations during a contest.

I would like to pass along a few hints that I have learned in my many years of low-power DXing/DX contesting.

Equipment: It is important that you be thoroughly familiar with all phases of the operation of your station equipment. This is true whether your are waiting for the "big DXpedition" to show up or are going to try to increase your DXCC country totals in one of the DX contests. When the operator on the Clipperton Island DXpedition says he is going to work a 50-kHz split on 20-meter phone, you had better know right away how to make your rig do that split or spend a frustrating time listening to other stations working him.

Time spent on preventive maintenance of your station equipment is very wisely spent. What could be more disheartening to a DXer than to hear an operator on the 8Q7AF (Maldives) DXpedition call for his particular call area only to find that a dirty relay in his transceiver prevents proper transmit operation? Check the tubes, if your rig has them, in your transmitter and receiver, and replace any weak ones. Check all the visable discrete components for abnormal discoloration and replace as necessary. Also inspect jumper wires/cables and resolder/replace any that might be making poor connection. Two particularly vulnerable weak points that I have found in my station equipment are rotary switches and relay contacts. These should be cleaned periodically with a contact cleaner and a lint-free cloth. Simply spraying the contacts will not clean them, but will only move the dirt around.

The station should be set up as neatly and compactly as possible. *All* controls should be within arm's reach while you are seated at the operating position. It's hard enough to crack that pile-up or work a DX contest without having to get up from your operating position each time you change bands or antennas.

If you find it necessary to throw more than one switch to go from transmit to receive, and vice versa, it may be wise to consider installing a multipole relay so that only one switch need be thrown.

Break-in on cw (the better to hear what is happening on frequency) and VOX (voice-operated transmit which keeps both hands free to tune the external receiver/VFO) on phone are the best ways to go.

Quick and precise band changes may sometimes make the difference in success or failure in that competition for the rare DXpedition or in the DX contest. An easy way to help effect quick band changes is to mark the control positions on the transmitter for each band/segment. Using these markings, it should only take a few seconds for some final peaking before you're ready to go after changing bands. If a transmatch is used, it should also be similarly marked. Keep in mind that there will probably be different settings on the same band if more than one antenna is used on that band. If you do not desire to permanently mark your rig, try using small triangles, cut from colored tape; a different color for each band/antenna works nicely, and they are easily removed.

Remember to also mark your receiver calibration dial/control, since it is likely that the calibration will change from band to band and in the excitement of that DX pile-up, it's all too easy to find yourself transmitting out of band. The calibration marks are also useful when operating split frequency.

Antenna Systems: Antennas play a very important role for the low-power DXer. If you're running 1000 watts and 50 watts is being wasted in heating your coax or some nearby tree, it's hardly noticed. But if your output power is only 100 or so watts that same 50 watts that's being used to warm up the neighborhood could be better used to add a couple of decibels to your signal at the DX station's end. The low-power DXer, even more than the "kilowatt kid" across town, should continually strive to improve his antenna system for maximum efficiency. Good solid starting points for antenna system design/improvement are to be found in the ARRL *Radio Amateur's Handbook* and *The ARRL Antenna Book*.

It's not really necessary to have a big antenna system to be successful in low-power DXing and DX contesting. In the 1974 ARRL DX Competition, the first year I finished at the top of the low-power heap on *both* phone and cw, my antenna system was a Hy-Gain 18-AVQ five-band trapped vertical, a 40- through 10-meter, multiband dipole running northwest to southeast and a 20-15-10 meter multiband dipole running northeast to southwest. This same basic antenna system also accounted for more than 260 of my total DXCC countries.

Pile-ups: There is no set formula that guarantees success in breaking a pile-up, either on that rare DXpedition or on a rare multiplier in a DX contest, but a few hints follow that should help a little.

Send your call-sign as quickly, but as clearly, as possible. The longer it takes to send your call, the better the chances are that QRM will cover you up. I find that using "hard/curt" phonetics such as "Willie Alpha One Sugar Sugar Hotel" sometimes gets better results than using the "softer" phonetics like "Washington America One Sierra Sierra Hotel." In a DX contest, spend only a few minutes in a particular pile-up, if there are a number going on the band. I usually keep a list of pile-ups, the call of the DX station and the frequency he is working on. That way, I can bounce back and forth between pile-ups and by calling when the pile-up lessens, I can maximize my chances for success in working the DX station. If the DX station at the pile-up is particularly rare, you may naturally want to invest more time in trying to work it.

Be aggressive, but don't go overboard. Listen to how the DX station is working his pile-up. Try to get in step with how the DX station is working the band. If he is listening high in frequency, call him high, and so on.

All major DX contests can be very productive in your efforts to add to your DXCC totals. If you want to get all that's possible from your DX contesting, you should plan to be on the air for at least 50 percent of the contest period. A lot of good DX is going to be missed if you operate only three hours out of a 48-hour contest period. Remember, even if you run really low power to a relatively small antenna system, *do not be afraid to call CQ.* Some DX stations are not very proficient operators when it comes to handling the pile-ups that they generate. These DX stations will spend the entire contest answering the CQs of other stations. There's no reason why your CQ should not be one of those answered. Find a clear frequency near the bottom of the band and start calling CQ. Don't become discouraged if you continue to call CQ and get only a few replies; stick with it and you're likely to get a surprise or two.

Good luck with your low-power DXing. Be sure to lay in a healthy stash of QSLs to send to all those "new ones" that are bound to be worked as a result of your efforts. — *Ken Bolin, W1NG*

QSLs from some of the more recent successful DXpeditions.

Oscar Delta." The ITU-recommended phonetics are:

A — Alpha
B — Bravo
C — Charlie
D — Delta
E — Echo
F — Foxtrot
G — Golf
H — Hotel
I — India
J — Juliette
K — Kilo
L — Lima
M — Mike
N — November
O — Oscar
P — Papa
Q — Quebec
R — Romeo
S — Sierra
T — Tango
U — Uniform
V — Victor
W — Whiskey
X — X-Ray
Y — Yankee
Z — Zulu

Once again, if the DX station is working "split frequency," do *not* transmit on his frequency.

Power

In the DXing community, there are both "full legal power" advocates and QRP (10 watts or less input) enthusiasts, QRPers accept the challenge of trying to break the pile-ups in full competition with the kilowatt folks. Most of us fit somewhere between these two extremes. Operating technique and a superior antenna system can compensate for lack of a linear. Honor Roll member W1NG offers some advice to the low-power DXer elsewhere in this chapter.

Dxpeditions and DXpeditioning

Dxpeditions are trips made by amateurs to put a rare country on the air. The idea is to work as many of the "deserving DXers" (a phrase initiated by Hugh Cassidy, WA6AUD, former editor of the *West Coast DX Bulletin*) as possible. In fact, more than just a few of the "countries" have no amateur population — or no population at all — and would be impossible to work were it not for a DXpedition.

A DXpedition can be that multi-thousand dollar assault on that barren island that hasn't been on the air in 20 years, or a DXpedition can be a few hours of leisurely operating during a vacation to the sunny Caribbean.

A well-known DXpeditioneer was once heard to remark that "to teach a brand-new DXer how proper operating techniques can make like easier for all DXers, he should be made to sit for just a few minutes at the other end of the pile-up." We dare say that being the DX would quickly show the poor operator the error of his ways. A mini DXpedition, incorporated with your overseas vacation, is within the reach of many amateurs.

Even on the most basic level, a DXpedition requires a lot of thought and planning. The most important aspect is to get the license from the authorities in the country where you wish to operate. The ARRL International Services Branch offers invaluable help with licensing information for various DX countries.

With new DXers showing up on the bands every day, no DXpedition, even to the most commonly heard DX location, is for naught. Rarity is a relative term. Everything is "rare" to the beginning DXer.

Want to become a "legend" in DXing lore? Embark on a DXpedition to one of the uninhabited DX locations that hasn't been on the air in several years. Instant fame. DXers who've been around the last few years can probably name most of the operators of the major DXpeditions to such rare locations as FO Clipperton Island, 3Y1 Bouvet Island, 8Z4 Saudi/Iraqui Neutral Zone and 3CØ Annobon.

Who, among the DXers of recent years, has without a twinge of envy, read or heard of the DXploits of super DXpeditioneer, Gus Browning, W4BPD? Or those U.S. ambassadors of goodwill and veterans of many DXpeditions, the Colvins, Lloyd, W6KG and Iris, W6QL, who have operated from more DXCC countries than many amateurs will even work in a lifetime of on-the-air activity? Bill Poellmitz, K1MM, relatively new on the DXpedition circuit, has (with Austin Regal, N4WW) put CEØ, Juan Fernandez, on the air; has been to South and Central America for a little operating; has been on a little junket to Reunion Island (FRØMM) and a foray to Spratly, 1S1DX, with notable DXers Harry, VK2BJL, and Stew, K4SMX. If you want to talk about dedication to DXing, the 1S1DX group drew gunfire from hostile military types while they were trying to find an operating site in the Spratly Island Group. That the Spratly DXpedition even made one QSO, much less their final total of over 13,000, is a tribute to the dedication of the DXer's spirit. Try to name a DXer who isn't grateful to Chuck, ZL1ADI, and Ron, ZL1AMO, for their superb operating from Pitcairn Island as VR6HI. They very efficiently erased a "much-needed one" from many a DXer's want list. And their next DXpedition? It'll come BY and BY. Surprise!

This isn't meant to be a canonization of any specific DXer or group (just you try to tell the DXer who has just worked the latest DXpedition for the country that puts him on the Honor Roll that those involved *aren't* saints!), but if you get to know the players, you get to know a little more about the game — *DXpeditioning*.

When trying to work a DXpedition for that "new one," it's extremely important to follow all the guidelines for efficient and courteous operating. *Remember*, many thousands of eager DXers are also waiting to work a "new one," the DXpedition's time is limited and every time someone on frequency "goofs it up" and wastes a minute or two of the DXpedition operator's time, that's a few QSOs that will never be made.

Every major Amateur Radio publication carries news of the upcoming DXpeditions and, quite often, the details on recently completed DXpeditions by one or more of the participants. Reading these narratives is a good way to learn some of the pitfalls as well as some of the rewards for a successful DXpedition.

Some of the larger DX clubs in the U.S. and Canada sponsor DX conventions, just like your run-of-the-mill "hamvention," but devoted entirely to DXers and DXing. Operators from some of the recent DXpeditions are usually the featured speakers.

Any "hamvention" worth its salt will schedule one or more DX Forums for the DXing crowd. It's a good place to meet fellow DXers, trade "war stories" and keep up on what is happening in the DX world. *QST* and the other major Amateur Radio publications carry schedules of upcoming "hamventions."

It would be worth the price of travel expenses to import one of the participants of a recent DXpedition to be featured speaker at a meeting of your local radio club. Quite a few of the operators from the various DXpeditions are more than glad to speak and show slides or movies of their trip. Such an event would turn a humdrum club meeting into a special event. It's common knowledge in DXing circles who has been where and who is willing to go on the club lecture circuit. Ask around.

You might have heard on the air or read that donations or "help in the form of green stamps" (U.S. dollars) would be appreciated by the ops/QSL manager of a recent DXpedition. A major DXpedition to one of the more-remote DXCC countries sometimes takes a rather healthy chunk out of someone's bankroll and such an effort is worthy of support.

In fact, a dollar or two with your QSL, if nothing else, will help to defray the cost of the QSL cards, no small expense in itself if 15 or 20 thousand QSOs have been made. What better way to "participate" in the DXpedition than to help financially, if possible? The DXer has to decide for himself if a request for aid is legitimate.

A good way to positively support DXpeditioning is to support one of the legitimate DX foundations. Foundations not only support DXpeditions, but are quite helpful in assisting overseas indigenous amateurs with donations of equipment and supplies (QSLs). The two foundations that come immediately to mind are The Northern California DX Foundation and the YASME Foundation. There are others. Many local DX clubs have DXpedition funds, and several of the major DXpeditions would have found it a little harder if not impossible, to make it without assistance from one or more of these foundations/funds. You can be sure that your support will be well used by one of these foundations.

What it boils down to is that with the *proper* support, more DXpeditions will be able to put more "new ones" on the air for us all.

Tools of the DXer

Some of the accessories that you might like to have to make DXing a little easier and more enjoyable are:

1) A 24-hour clock. Get it, set it to UTC (Universal Coordinated Time — also known as GMT) and never touch it again. Uniformity is the name of the game in DXing. All DXers play the game together in UTC.

2) A beam-heading chart. A simple, easy-to-read chart, like the one in Fig. 3, that lists DX prefixes, followed by the short-path Great Circle bearings (beam headings calculated by using your QTH as the center point) in degrees. The Ham Ads in *QST* offer several different sources for beam-heading charts. Stan Horzepa, WA1LOU, 72 Stiles St., Waterbury, CT 06706, and Bill Johnston, N5KR, 1808 Pomona Dr., Las Cruces, NM 88001, offer good versions of the headings charts. Send them a s.a.s.e. for prices and details. In addition, N5KR offers a computer-generated azimuthal equidistant map centered on your QTH (see Fig. 4), a great help for quick beam-heading references.

3) A subscription to one of the previously mentioned DX bulletins. They're not a substitute for being on the air and collecting DX data for yourself, but a nice adjunct and alternate source of DX information. There's no sense missing a "rare one" simply for lack of information.

4) A DX *Callbook*. A must for the dedicated DXer who is anxious enough for his QSLs to QSL directly to the DX station, the *Callbook* is a wealth of other DX-related materials, including postal information. Write *Radio Amateur Callbook*, 925 Sherwood Dr., Lake Bluff, IL 60044, for prices and a list of their other fine printed products.

5) A wall-size world prefix map. A large, color version of the map is available from ARRL. Nice to be able to "see 'em" as well as to "work 'em."

6) A world atlas, for those interested in geography. A nice version is available from the *Radio Amateur Callbook*. The address is listed above.

7) A QSL manager list. No sense in sending your QSL direct or via the bureau if the DX station has a manager who can turn around the card faster and more quickly. *QST* and most of the other Amateur Radio publications have monthly QSL manager listings. You might want to start an index card file on QSL managers using the information from these publications and from tips picked up on the air. The *Callbook* has a fairly up-to-date QSL manager listing in the appendix. A comprehensive QSL manager list, updated weekly, with over 4000 DX station listings, is available for $1.75 (for U.S./VE amateurs) from Jay O'Brien, W6GO, 6606 Fifth St., Rio Linda, CA 95673.

8) A DXCC country list with a system to record countries worked and confirmed (such as the list in the reference section of this book) helps you to keep track of your progress towards DXCC, 5-Band DXCC and WAZ (the *CQ* Magazine Worked All Zones Award).

As you progress up the ranks in the world of DXing, you'll find these "tools of the DXer," as well as other specialized operating aids, invaluable in your efforts.

The ability to work even the more sought-after DX station is not limited all the time by the amount of station hardware available. Proficiency in good operating techniques, which comes with on-the-air practice, can indeed increase one's chances of working a given DX station despite seemingly impossible odds.

Tom Christian, VR6TC, a direct descendant of Fletcher Christian of *The HMS Bounty* fame, is famous for his regularity of operation in handing out one of the truly "rare" DXCC countries, Pitcairn Island, to the eager hordes of DXers. *(AH6H photo).*

One of the big DXpeditions was mounted to the exotic location of Abu Ail. OE6XG/A was ably manned by J28AZ, DJ9ZB, I2FGP, I2CBM and OE6EEG. Four days of superb operating by the members of OE6XG/A took Abu Ail off the most-needed list for a long time to come. I2FGP, the main cw op, is shown above. Giampaolo is no stranger to exotic call signs. Giampaolo has also operated as 6O1FG, J2ØBS and JY8FG.

What To Do After Work DX — QSLs/Awards

It's a great feeling of accomplishment. Several hundred stations were calling that DX station and he came back to *you*.

You exchange signal reports and *brief* pleasantries (in consideration for the many ops on frequency waiting for their chance to work the DX) and enter the contact in your logbook. And even though the FCC does not require that precise QSO records be kept, the conscientious DXer keeps very detailed records of each DX station worked. This information will become invaluable when the time comes to go back through your list to try to get confirmations (which have not yet arrived) for a QSO that is several years old. It's nice to be able to go back and remember your climb up the DXCC listings.

A QSL card to confirm the QSO would

GREAT CIRCLE BEARINGS FROM NEWINGTON CT

CALL	DEG	CALL	DEG	CALL	DEG	CALL	DEG	CALL	DEG	CALL	DEG
A2	103	FG	157	KJ	287	TR	94	VQ9	52	3C	93
A3	259	FH	81	KL	326	TT	78	VR1	280	3D2	265
A4	48	FK	274	KM	301	TU	99	VR3	262	3D6	102
A5	345	FL8	64	KP4	164	TY	92	VR4	291	3V	68
A6	48	FM	156	KP6	272	TZ	92	VR6	230	3X	101
A7	49	FO	244	KS4	195	UA	346	VR7	262	3Y	180
A9	48	FP	59	KS6	265	UA1	10	VS5	349	4K	180
AP	34	FR	81	KV	159	UA2	42	VS6	324	4S	327
BV	345	FS	156	KW	304	UB	44	VS9	317	4U	57
BY	358	FW	273	KX	300	UC	42	VS9K	62	4W	59
C2	290	FY	150	KZ	193	UD	42	VU	331	4X&Z	55
C3	63	G	51	LA&B	34	UF	43	VU7	342	5A	69
C5	102	GC&J	56	LF&G	34	UG	40	W	274	5B	54
C6	187	GD	50	LU	174	UH	35	XE	243	5H	83
C9	92	GI	49	LX	52	UI	31	XF4	247	5N	89
CE	179	GM	46	LZ	52	UJ	27	XP	17	5R	86
CE9	180	GU	55	M1	56	UK	346	XT	91	5T	92
CE0A	212	GW	53	OA	183	UK2	39	XU	357	5U	82
CE0X	187	H4	291	OD	53	UK5	47	XV	359	5V	94
CE0Z	186	HA	51	OE	51	UK6	42	XW	356	5W	267
CM	197	HB	55	OH	31	UK7	25	XZ	348	5X	78
CN	77	HB0	54	OH0	37	UK8	31	YA	35	5Z	76
CO	197	HC	188	OK	49	UK8J	27	YB&C	339	6O	66
CP	170	HC8	205	ON	51	UK8M	337	YI	49	6W	101
CR3	103	HH	181	OR4	180	UK8N	337	YJ	276	6Y	190
CR5	96	HI	176	OX	17	UK8R	27	YK	52	7O	59
CR9	354	HK	180	OY	39	UL	25	YN	204	7P	107
CT	69	HK0	198	OZ	44	UM	336	YO	49	7Q	89
CT2	81	HL&M	344	P2	317	UO	47	YS	212	7X	77
CT3	84	HP	193	PA&D	50	UP	41	YU	53	7Z	53
CX	166	HR	205	PE&I	50	UQ	40	YV	168	8J	180
D2&3	99	HS	353	PJ	166	UR&T	36	ZA	56	8P	155
D4	108	HV	58	PY	137	UW	346	ZB	73	8Q	317
D6	81	HZ	53	PY0	136	UY	51	ZC	54	8R	158
DA-L	48	I&IT	58	PZ	154	VE1	53	ZD7	117	8Z4	50
DM&T	50	IS	62	S2	343	VE2	2	ZD8	118	9A	56
DU	341	J2	64	S7	58	VE3	318	ZD9	139	9G	98
EA	68	J3	158	S9	96	VE4	318	ZE	96	9H	63
EA6	66	JA-R	335	SK-M	36	VE5	313	ZF	200	9J	95
EA8	88	JT	358	SP	45	VE6	310	ZK1	252	9K	49
EA9	72	JW	16	ST	70	VE7	310	ZK2	260	9L	104
EI	52	JX	27	SU	62	VE8	332	ZL	248	9M2	339
EL	104	JY	54	SV	56	VK	297	ZL5	180	9M6	339
EP	43	K	274	TA	50	VK9	211	ZM7	269	9M8	352
ET	68	KA	335	TF	35	VO1	52	ZP	165	9N	339
F	57	KB	277	TG	215	VO2	26	ZS	109	9Q	88
FB8W	117	KC4	185	TI	201	VP1	214	ZS3	113	9U	84
FB8X	121	KC6	311	TI9	203	VP2	156	3A	59	9V	355
FB8Y	180	KG4	186	TJ	89	VP5	178	3B8	78	9X	83
FB8Z	267	KG6	319	TL	83	VP8	172	3B9	81	9Y	158
FC	60	KH	282	TN	92	VP9	142				

Fig. 3 — A simple format for your beam-heading chart is best. It's easy to quickly find the short-path beam heading when you hear the call sign of the DX station you wish to call.

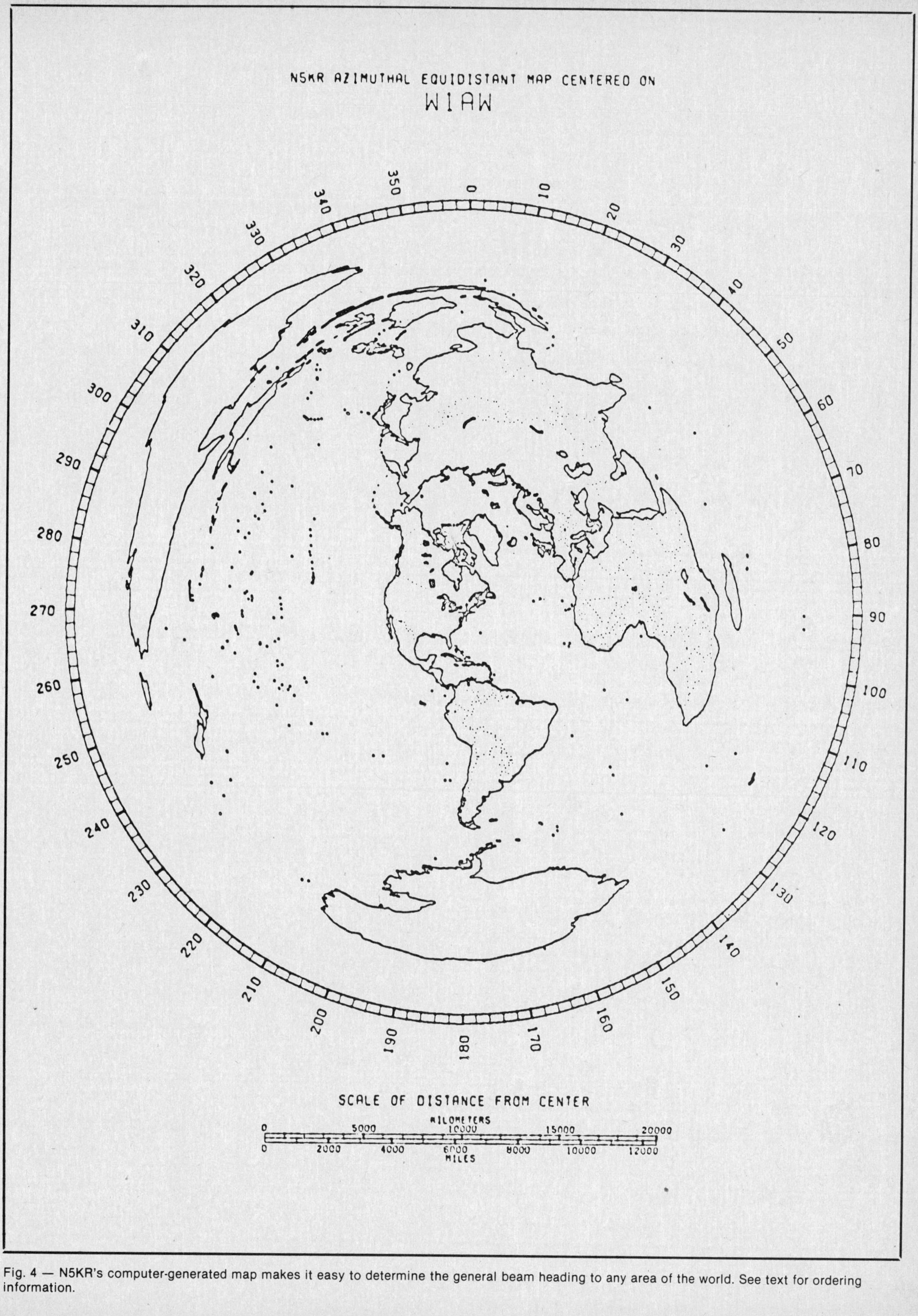

Fig. 4 — N5KR's computer-generated map makes it easy to determine the general beam heading to any area of the world. See text for ordering information.

The ARRL DX QSL Bureau System

Within the U.S. and Canada, the ARRL DX QSL Bureau System is made up of 22 call area bureaus that act as central clearing houses for QSLs arriving from foreign countries. These "incoming" bureaus are staffed by volunteer workers. The service is free and ARRL membership is not required.

Most countries have "outgoing" QSL bureaus that operate in much the same manner as the ARRL-Membership Overseas QSL Service. The member sends his cards to his outgoing bureau, where they are packaged and shipped to the appropriate countries.

A majority of the DX QSLs are shipped directly to the individual incoming bureaus where volunteer workers sort the incoming QSLs by the first letter of the call sign suffix. One individual may be assigned the responsibility of handling from one to three letters of the alphabet.

To claim your QSLs, send a 5- × 7-1/2-inch self-addressed, stamped envelope to the bureau serving your district. Neatly print your call sign in the upper-left corner of the envelope. A preferred way to send envelopes is to affix a 15-cent stamp and clip extra postage to the envelope. Then, if you receive more than 1 oz of cards, they can be sent in the single package.

Some incoming bureaus sell envelopes or postage credits in addition to the normal handling of s.a.s.e.s. They provide the proper envelope and postage upon prepayment of a certain fee. *The exact arrangements can be obtained by sending your inquiry with an s.a.s.e. to your area bureau.* A list of bureaus can be found below.

Good cooperation between the DXer and the bureau is important to ensure a smooth flow of cards. Remember that the people who work in the area bureaus are volunteers. They are providing you with a free but valuable service. With that thought in mind, please pay close attention to the following DOs and DON'Ts.

DOs

Do keep self-addressed 5- × 7-1/2-in. envelopes on file at your bureau, with your call in the upper-left corner, and affix at least one unit of first-class postage.

Do send the bureau enough postage to cover envelopes on file and enough to take care of possible postage-rate increases.

Do respond quickly to any bureau request for envelopes, stamps or money. Unclaimed card backlogs are the bureau's biggest problem.

Do notify the bureau of your new call as you upgrade.

Do include an s.a.s.e. with any information requests to the bureau.

Do notify the bureau *in writing* if you *don't* want your cards.

Do be appreciative of the fine efforts of these volunteers.

DON'Ts

Don't expect DX cards to arrive for several months after the QSO. Overseas delivery is very slow. Many cards coming from overseas bureaus are over a year old.

Don't send your outgoing DX cards to this bureau (see ARRL-Membership Overseas QSL Service).

Don't send envelopes to your "portable" bureau. For example, WA1SQB/2 sends envelopes to the W1 bureau, *not* the W2 bureau.

ARRL DX QSL Bureau System

First Call Area: all calls* — Hampden County Radio Association, Box 216, Forest Park Station, Springfield, MA 01108.

Second Call Area: all calls* — North Jersey DX Assn., P. O. Box 8160, Haledon, NJ 07508.

Third Call Area: all calls* — Leon Lapkiewicz, K3GM, P. O. Box 6238, Philadelphia, PA 19136.

Fourth Call Area: all single-letter prefixes — National Capitol DX Assn., Box DX, Boyce, VA 22620.

Fourth Call Area: all two-letter prefixes — Sterling Park Amateur Radio Club, P. O. Box 599, Sterling Park, VA 22170.

Fifth Call Area: all calls* — ARRL W5 QSL Bureau, Box 1690, Sherman, TX 75090.

Sixth Call Area: all calls* — ARRL Sixth (6th) District DX QSL Bureau, P. O. Box 1460, Sun Valley, CA 91352.

Seventh Call Area: all calls — Willamette Valley DX Club, Inc., P. O. Box 555, Portland, OR 97207.

Eighth Call Area: all calls — Columbus Amateur Radio Assn., Radio Room, 280 E. Broad St., Columbus, OH 43215.

Ninth Call Area: all calls — Northern Illinois DX Assn., Box 519, Elmhurst, IL 60126.

Zero Call Area: all calls* — WØ QSL Bureau, Ak-Sar-Ben Radio Club, P. O. Box 291, Omaha, NE 68101.

Puerto Rico: all calls* — Radio Club de Puerto Rico, P. O. Box 1061, San Juan, PR 00902.

U.S. Virgin Islands: all calls — Graciano Belardo, KV4CF, P. O. Box 572, Christiansted, St. Croix, VI 00820.

Hawaiian Islands: all calls* — John H. Oka, KH6DQ, P. O. Box 101, Aiea, Oahu, HI 96701.

Alaska: all calls* — Alaska QSL Bureau, 4304 Garfield St., Anchorage, AK 99503.

SWL — Leroy Waite, 39 Hannum St., Ballston Spa, NY 12020.

QSL Cards for Canada (VE and VO) may be sent to: CRRL Central QSL Bureau, P. O. Box 663, Halifax, NS B3J 2T3. Or, QSL cards may be sent to the individual bureaus.

VE1* — L. J. Fader, VE1FQ, P. O. Box 663, Halifax, NS B3J 2T3.

VE2 — A. G. Daemen, VE2IJ, 2960 Douglas Ave., Montreal, PQ H3R 2E3.

VE3 — The Ontario Trilliums, P. O. Box 157, Downsview, ON M3M 3A3.

VE4* — W. A. Stunden, VE4BJ, 578 Oxford St., Winnipeg, MB R3M 3J9.

VE5 — A. Lloyd Jones, VE5JI, 2328 Grant Rd., Regina, SK S4S 5E3.

VE6* — G. D. Holeton, VE6AGV, 4003 1st St., N.W., Calgary, AB T2K 0X2.

VE7* — Howard Martin, VE7AFY, No. 45-9960 Wilson Rd., Ruskin, BC V0M 1R0.

VE8* — Al Sturko, VE8NS, P. O. Box 72, Fort Smith, NT X0E 0P0.

VO1, VO2 — CRRL VO QSL Bureau, P. O. Box 6, St. John's NF A1C 5H5.

*These bureaus sell envelopes or postage credits. Send an s.a.s.e. to the bureau for further information.

be nice as a momento of a short-but-pleasant QSO, a decoration for the shack wall or the needed confirmation for a sought-after award. Most awards are issued only after proof, in the form of QSL cards, is submitted to the society that issues the award.

QSLs and QSLing

There is no sure-fire way to guarantee that you will receive that QSL card from the DX station, but there are ways to greatly increase your chances for a sure, if not rapid, QSL return.

The QSL — your card should be attractive (to attract attention, perhaps a step above the common, bland QSL) but straightforward. All necessary QSO information should be on the card: in fact, information on one side of the card will make answering the card a relatively simple matter.

The QSL card should have places for the station worked, band, mode (cw/ssb), signal report and, most importantly, the date and time in UTC (UTC only). If you make a mistake, do not correct it by writing over it on the card; destroy that card and start again. Most societies that issue awards will *not* accept an "altered" card as a valid confirmation.

Aside from using UTC, another convention that you should adopt when sending QSLs to overseas operators is the "Dxer's Date Recording System." Instead of writing the date December 22, 1979 as 12/22/79, write 22/XII/1979. The date is in Arabic numerals, followed by the month in Roman numerals, then the year in Arabic numerals. This, like the use of UTC, is a convention to try to keep every DXer in tune with others across the seas.

QSL Routes

Direct: Most certainly the fastest, but also the most costly, way to QSL is the direct route. That is, sending the QSL

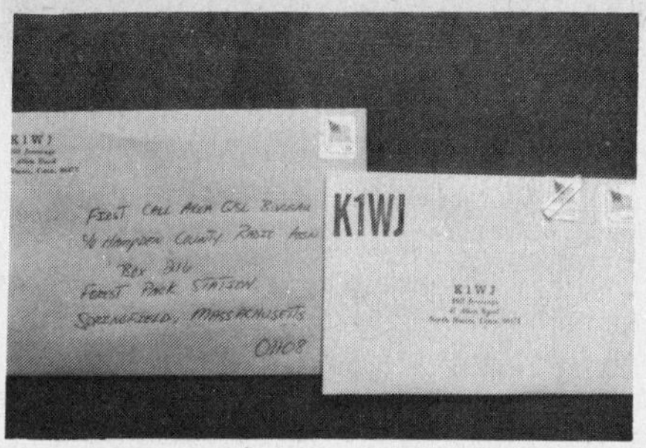

Remember to keep a supply of envelopes on file with your incoming DX QSL manager (see text for address). Affix one unit of first-class postage to your self-addressed, stamped envelope, put your call sign in the upper-left-hand corner of the envelope and clip additional postage to the envelope if you are expecting many cards.

directly to the home of the DX Station. The high cost results from the expense of providing sufficient return postage for the DX station to send his QSL back to you.

The price of an overseas airmail letter is currently 31 cents U.S. (1/2 oz). Most foreign stations will not and should not be expected to QSL direct unless you include a sufficient number of International Reply Coupons (IRCs). You can purchase IRCs from your local post office currently at 42 cents apiece, although some DXers accumulate IRCs they receive and sell them to fellow DXers at a discount. Anywhere from two to five IRCs are required for return airmail service from a given DX country (the *Radio Amateur Callbook* contains complete information on overseas mail rates). Thus a direct QSL card exchange with a country that requires five IRCs for return postage may cost you as much as $3 U.S.

Foreign Stamp Service: A DX stamp service is a good alternative to using IRCs. Several such services exist (check *QST* advertisements). The DX stamp service sells airmail and surface-grade postage stamps from different countries. You only need to purchase the stamp for the country you are sending the QSL to. Affix the stamp to a self-addressed envelope and enclose it with your QSL to the DX station. The stamp usually costs much less than the several IRCs for the same mail service. This would represent about a $2 savings over the worst case (the five-IRC country) involving IRCs. One DX Stamp Service is provided by G. N. Robertson, W2AZX, 83 Roder Pkwy., Ontario, NY 14519; send an s.a.s.e. for details.

Bureaus: Not quite as quick, but much less expensive, are the various QSL bureaus one can use to handle outgoing DX QSLs (your QSL to DX stations). These bureaus advertise their services in the major Amateur Radio publications. The ARRL provides an outgoing QSL Bureau *for members only*. The ARRL Bureau is discussed in a sidebar in this chapter. This bureau provides a fast, inexpensive QSL route for ARRL members.

The services of the ARRL Incoming QSL Bureau are provided *free* for all amateurs throughout the United States and Canada. The ARRL coordinates the bureau, with the actual QSL sorting and mailing done by volunteer individuals and radio clubs throughout W/VE-land. To learn how to claim any QSLs due you from your incoming bureau, see the user's guide, elsewhere in this chapter.

As long as you are in no hurry for your DX QSLs, the bureau is the way to go. Turn-around times of one year or more are not uncommon.

QSL managers: Many active and sought-after DX stations employ QSL managers. A QSL manager is usually an amateur (hopefully with an efficient postal system), who voluntarily takes on QSLing chores for the DX station. A self-addressed, stamped envelope accompanying your QSL card to the QSL manager's address will usually get you a return QSL very promptly.

Think you would like to be a QSL manager? Think you have some talents that might be put to good use as a manager? On an adjacent page, W7OK

ARRL-Membership Overseas QSL Service

This is an "outgoing" service that allows ARRL members to send DX QSL cards to foreign countries at a minimum of cost and effort.

While QSLing direct to foreign amateurs is faster, it is also more tedious. Time spent searching for addresses in the foreign *Callbook*, addressing and stuffing envelopes, and mailing, could be better spent operating DX. And, the cost of IRCs, airmail postage, and envelopes can be prohibitive. An unlimited number of QSLs may be sent for distribution 12 times per year. The fee is just $1 per pound or portion thereof (155 QSL cards average a pound).

The ARRL-Membership Overseas QSL Service operates *only* in an "outgoing" capacity. To receive QSL from DX stations, see the information sheet entitled "The ARRL DX QSL Bureau System."

U.S. amateurs may send SWL reports to foreign short-wave listeners.

Unlicensed (associate) members may send SWL cards to foreign amateurs.

QSL managers: Write for details.

Requirements

1) Presort your DX QSLs alphabetically by call sign prefix (A3, AP, C6, CE F, FG, G, GI, GM, JA, 3A2, etc.).

2) Enclose the address label from the brown wrapper of your current copy of *QST*. This information shows that you are a current ARRL member.

Family members may also use the service by enclosing their QSLs with those of the primary member. Include the appropriate fee with each individual's cards and indicate "family membership." Sightless members who do not receive *QST* should indicate that the QSLs are from a "sightless member." ARRL-affiliated club stations may utilize the service when submitting club QSLs by indicating the club name. Club secretaries should check affiliation papers to ensure that membership is current.

3) Enclose payment in the form of a check, money order or cash.

Sending large amounts of cash through the mail is not suggested. Please do not send stamps.

This is a current list of countries for which QSLs may be forwarded. Keep in mind that QSLs received for countries that do not have bureaus will be held until such time as a distribution point exists. In the interest of economics and efficiency, QSLs cannot be returned.

Afghanistan, Alaska, Algeria, Angola, Antarctica (no KC4A), Antigua, Argentina, Ascension Island, Austral/French Antarctic Lands, Australia, Austria, Azores, Bahama Islands, Bahrain, Barbados, Belgium, Bermuda, Bolivia, Brazil, Bulgaria, Canada, Cape Verde Islands, Cayman Island, Chile, Colombia, Congo Republic, Cook Islands, Costa Rica, Cuba, Cyprus, Czechoslovakia, Denmark, Dominica, Dominican Rep., Ecuador, Egypt, El Salvador, Ethiopia, Falkland Islands, Faroe Islands, Fiji Islands, Finland, France, French Guiana, French Oceania, German Dem. Rep., Fed. Rep. of Germany, Ghana, Gibraltar, Grenada, Great Britain (or British Commonwealth), Greece, Greenland, Guadaloupe, Guam, Guantanamo Bay, Guatemala, Guyana, Haiti, Hawaiian Islands, Honduras, Hong Kong, Hungary, Iceland, India, Indonesia, Iran, Ireland, Israel, Italy, ITU — Geneva, Ivory Coast, Jamaica, Jan Mayen, Japan, Johnston Island, Jordan, Kenya, Korea (no HL), Kuwait, Lesotho, Liberia, Liechtenstein, Luxembourg, Madeira Islands, Malagasy Republic, Malawi, Malaysia, Maldives, Malta, Mariana Islands, Marshall Islands, Mauritius, Mexico, Midway Islands, Monaco, Mongolia, Morocco, Nauru, Netherlands, Netherlands Antilles, New Caledonia, New Zealand, Nicaragua, Nigeria, Northern Ireland, Norway, Oman, Pakistan, Panama, Papua New Guinea, Paraguay, Peru, Philippine Islands, Poland, Portugal, Puerto Rico, Republic of China, Rhodesia, Romania, Rwanda, Samoa (American), San Marino, Scotland, Senegal, Seychelles, Sierra Leone, Singapore, South Africa, Spain, Sri Lanka, St. Helena, St. Lucia, St. Vincent, Suriname, Svalbard, Swaziland, Sweden, Switzerland, Syria, Thailand, Togo, Tonga, Transkei, Trinidad and Tobago, Turkey, Uruguay, USSR, Vatican, Venezuela, Virgin Islands, Wales, Western Samoa, Yugoslavia, Zaire, Zambia.

So You Want To Be a QSL Manager!!

There have been many fine articles directed at new and old-time DXers on the proper ways to use the services of QSL managers — and service is exactly what a QSL manager provides.

Let's determine which DX station will be interested in your efforts. First, there are the well-equipped stations who have the time to work quite a few stateside and other DX stations, but simply do not have the time to spend with QSLs. Second, there are fine DX operators who make many QSOs but simply have no desire to send or receive any, even though they have the time and financial means to do so. Third, many DX operators find that QSLing is a financial burden. We often do not realize the financial sacrifice so many of our DX friends have had to make just to get a transceiver and a good antenna.

What personal qualifications are necessary to make a good QSL manager? A good manager will be a DXer. This will give him or her the insight into, and understanding of, the feelings of those he'll be trying to help. A DXer will also know what is happening on the bands and will probably subscribe to one or more of the published DX bulletins. Time is also very important. If you are pressed for time by your occupation or family, it is best that you do not consider this added demand on your time. And, remember, time spent on QSL manager duties takes that much time away from your own operating activities.

Being a QSL manager is not a job that you start and then abruptly stop. It takes months and years to become well known and to have your call published in the several QSL manager directories. Don't start unless you are prepared to stay with it for a long time.

About the expenses involved: Being a QSL manager will cost you money if you intend to do a good job. Regardless of the s.a.s.e.'s received and an occasional dollar bill from some thoughtful DXer, you will find that you will have to lay out some money. You will always have the cost of mailing cards you have received to the DX station. There is also the cost of printing the QSLs for the DX station who just can't afford the expense. In some cases, you will have to send postage money to the DX station so he can send you the logs. Expenses are probably the main reason why many QSL managers "throw in the towel" after just a few months or a year on the job. This can leave hundreds of DXers waiting for QSL cards that they will never receive, and can also give managers, in general, a bad name. If spending a few extra dollars will be a financial burden to you, don't consider becoming a QSL manager. The post office charges 42 cents for each International Reply Coupon (IRC), but will only give you 20 cents in postage for each IRC you wish to redeem. Try to find local DXers who need the IRC that you receive and might be willing to pay you 25 to 30 cents each. This will help somewhat with expenses.

A good manager must be able to deal with detailed paperwork and be very organized. Now, if you are still interested in becoing a QSL manager, how do you go about finding a DX station who is interested in using your services? You must "sell yourself" and your abilities to the DX operator. After all, you will be his representative to the DX community and he has every right to be able to feel that he can trust you and depend on you.

Type up a description of yourself with a friendly offer to be a QSL manager, stating why you wish to be a manager and why you would make a good job of the required duties for as long as required. When you QSO a DX station who does not have a manager, and for whom you'd like to assume those duties, mail your typed description along with a picture of yourself and your QSL to him. If you have made a good presentation of yourself, in time you will have made some good contacts as well as having made some good DX friends.

Say you got the manager's job. Now you are ready to start. Requests are coming in for QSLs for "your" DX station. The DX station is telling his contacts to QSL to your call and, finally, the log sheets of QSOs that he has sent to you arrive. Get into it!

Probably, no two managers organize their work the same way, but the first thing they all have to do is open the mail. See who the card is for and note the date and call on the back of the envelope and file all requests by station (if you manage for more than one station) and date.

You will need some supplies. Stock up on envelopes and postage stamps. Remember that all envelopes must meet the postal code for size. From the local stationery store, get a glue stick, adhesive tape, a roll of 3-inch gummed paper tape, scissors, a moistener to wet the glue on the envelopes, and a rubber stamp with your name, call and address.

Most DXers will make it easy for you to send them their QSLs. They will have their call sign and all QSO information printed on the same side of their QSL. They will also have the correct date and time in UTC. And most importantly, they will enclose an s.a.s.e. with their QSL request. The DXers will also keep in mind that it may take several months for you to get the logs from the DX station for verification.

Overseas stations should send you a self-addressed envelope with two IRCs for airmail return or one IRC for seamail return. Some will only send you an IRC or two without the self-addressed envelope. For these, take you scissors and cut out the return address on their envelope and, using your glue stick, attach that name and return address to one of your stock of envelopes. Never rewrite the names and addresses unless you absolutely have to. About 25 percent of the overseas stations will QSL you via the ARRL QSL Bureau for your call area. Be sure to inform your district QSL Bureau of the calls of the stations you manage. Keep plenty of proper-size envelopes and postage at your bureau. Return overseas cards received via the bureau through the ARRL outgoing QSL Bureau. This will save you the expense of sending direct to the overseas bureaus. Be sure to answer all the SWL (shortwave listener) QSLs received. These SWL cards mean a lot to prospective amateurs in many foreign countries.

About 10 percent of the stateside stations will fail to send you an s.a.s.e. or even a postage stamp. For these, you must pay the postage yourself or else return the DX QSL via the bureau and hope the requesting station has an envelope at his or her district QSL bureau. Otherwise, he'll never get that QSL.

Carefully check the requests for cards with your DX station's logs. Very often the time and or date on the QSL is off considerably, so make every effort to find the QSO in the logs. If you can't find the QSO, you can do either of two things: send the card back to the station making the request with the notation "not in log, please check time and date," or you can send the card direct to the DX station for him to decide whether or not to QSL. The best thing to do is to return the card to the station making the request. This is the most disappointing thing that you, as manager, will have to do. Operators requesting QSLs from your DX station really believe they have made the contact, but in the QRM of the pile-up, it is easy to be mistaken. This happens to all DXers at some time or another. Don't just disregard the request; let the DXer know that he is not in the log.

Carefully fill out all QSLs. Be sure that the date, year, frequency and mode are correct. Use a ball-point or felt-tipped pen. Sign your name or initials and your call sign, then stamp the back of the QSL with your rubber stamp to add authenticity.

Some DX stations that you manage for will tell you they do not want any cards received for them. It's best to keep these cards, so store them away in a dry place in case the DX station changes his mind at some future time. For those DX stations who want their incoming cards sent to them, wrap the cards in heavy wrapping paper (grocery bags work fine) seal securely with gummed wrapping tape in packets that weigh less than 2 pounds (approximately 310 QSLs). Send them via seamail with the required customs declaration stating that the contents are "used radio postcards" with no monetary value. This is the most economical way to send the cards to your DX station. Airmail costs over twice as much.

Keep up with your QSLing. Send out the QSLs every day if you have the logs for confirmation. If you get far behind, the job will seem more like drudgery than a pleasant chore. If necessary, enlist the aid of your XYL/OM or an amateur friend or two to help out. Remember, some anxious DXer is waiting for the mail every day.

In spite of all the work, it will never cease to be rewarding when you tune in on your DX friends and hear them say to QSL via your call sign, or most of all, when you open an envelope and find a note from an appreciative DXer who simply says "thanks, OM."

Now, get with it, do a good job, and good luck! — *Don Brickey, W7OK* tells what it takes to handle the manager's chores.

Awards

For the majority of DXers, having one or more of the various DX-oriented awards to shoot for provides the impetus to spend long — and very often some very strange — hours, headphones clamped tightly to ears, waiting for that "new one" to show up on the bands. And what could be more rewarding to the DXer than to see the fruits of his DXing labors as he advances up the DXCC listings in competition with his contemporaries?

The left photo shows what was sent to the QSL manager of 3Y1VC (Bouvet Is.). Note the self-addressed envelope and the IRCs for return postage. On the right: 3Y1VC's QSL was received fairly promptly.

DXCC and All Its Variations

When DX awards are mentioned, DXCC is the first thing that comes to mind for most DXers. DXCC, the DX Century Club, administered by ARRL (send an s.a.s.e. to ARRL for details) for U.S./VE ARRL members and all DX stations throughout the world, is the king of all DXing awards. With a modern starting date of November 15, 1945, the DXCC countries list and country criteria provide the basis for a majority of all the DX awards issued throughout the world. An annotated countries list appears in the reference section of this book.

To qualify for the initial DXCC award it is necessary to work 100 countries on the DXCC countries list, receive confirmations in the form of QSLs, and submit those 100 QSLs, along with the necessary paperwork, to ARRL for verification. Endorsements are issued for further achievement levels as you work new countries and receive the QSLs. With your initial DXCC submission, and each subsequent endorsement, your call and endorsement level are published in the DXCC section of *QST*. DXCC is available for mixed modes (both phone and cw) and exclusively phone or exclusively cw.

Honor Roll

Just as DXCC is the epitome of DX Awards, Honor Roll is the high point of the DXCC program. To qualify, you must have confirmations from DXCC countries worked that total within 10 of the current maximum number of countries on the list. Each DXCC country must have been worked by someone or have been proven possible to work. For example if there are 319 "current" countries on the DXCC countries list and at least one DXer has all 319 worked and confirmed, then everyone who has 309 or more current countries confirmed and credited to his DXCC is on the Honor Roll. Since the DXCC Honor Roll is a competitive listing from the beginnings of DXCC, and some countries are only on the air when a major DXpedition is mounted, it may take even the most avid DXer 10 or more years to reach the Honor Roll. All Honor Rollers are listed in *QST* twice a year. Honor Roll listings apply to a) mixed, b) phone and c) cw DXCC awards only.

Specialty DXCC Awards

160-Meter DXCC: Work and confirm at least 100 DXCC countries on 160 meters. Only a handful of 160 DXCCs have been issued, attesting to the difficulty of attaining this award. 160 DXCC is a one-shot award; it is not endorsable.

RTTY DXCC: Work and confirm 100 DXCC countries via teleprinter. RTTY DXCC is also not endorsable.

Satellite DXCC: Work and confirm 100 DXCC countries through amateur satellites (OSCARs and RS), with a starting date of March 1965. Non-endorsable.

Five-Band DXCC: Starting date of January 1, 1970. Work and confirm at least 100 countries on each of five amateur bands. All DXCC rules apply. A handsome plaque is awarded to all successful applicants.

5BDXCC was created by the ARRL Board of Directors to promote proficiency in operating on different amateur bands. Quite a boon to DXing on the low bands (80 and 40 meters), 5BDXCC is a good test of the real operating abilities of the DXer. It is within the reach of those who are willing to work at it. 5BDXCC is also unendorsable.

Other Awards

WAC (Worked All Continents), sponsored by the IARU (International Amateur Radio Union) is awarded to those who work and confirm all the continents. Endorsements are available for: all phone (ssb), working all continents on each of five bands, all on 3.5 MHz, and all on 50 MHz. An s.a.s.e. to IARU/ARRL hq. will bring an application and rules form.

WAZ (Worked All Zones), sponsored by *CQ* magazine, is available for working and confirming all 40 of the CQ zones throughout the world. Band/mode endorsements are available. It takes a little bit of effort to meet these requirements and qualify for this attractive award. Send an s.a.s.e. to *CQ* magazine, 76 North Bdwy., Hicksville, NY 11801, for rules and application forms.

CQ DX Awards. *CQ* magazine sponsors an awards program of the endorsable type (as DXCC), based on DXCC countries. Honor Roll listings here start at 275 DXCC countries confirmed. Issued for both cw and ssb. Send an s.a.s.e. to *CQ* Magazine for details.

Most countries throughout the world have Amateur Radio societies that issue DX-oriented awards. From time to time the requirements for these awards are listed in *QST*. A comprehensive collection of the rules for DX awards from around the world can be found in *Amateur Radio Awards*, by C. R. Emary, published by the Radio Society of Great Britain, 35 Doughty St., London, England WC1N 2AE. It is available in the U.S. from the Ham Radio Bookstore.

Whether you choose to work DX for the fun of it, or strive to complete the requirements for earning a DX award, you will find DXing one of the more challenging and enjoyable aspects of Amateur Radio.

Chapter 7

Contests

Contests. What do they mean to you? To some, the sheer enjoyment of making a large volume of new contacts in a weekend is the thing. Included in that group of contacts might be a new DXCC country, or a new state, or maybe even a new friend. Others view contests as a chance to up their code speed, or test a new antenna or check propagation. The serious participant gets butterflies as he quickly relives the anticipation of the clock flipping over to 0000 UTC and the instantaneous onslaught of signals over the bands. Eastern European countries view contesting as a sport, with the winners regarded as highly as their top Olympic athletes.

Regardless of the outlook taken, contests provide a unique opportunity to blend much of what can be learned in other chapters of this manual into a single operation. The operator in a November Sweepstakes (SS), for example, would benefit from knowledge about traffic handling, since each contest exchange can be viewed as a piece of traffic. Knowledge of DX prefixes, propagation and beam headings would help him to find that elusive Alaskan contact to complete a "clean sweep" (of all 74 multipliers available). Once the contest is over, our SS operator might be hard at work typing his log into a computer terminal to allow quick, foolproof location of duplicate contacts. N5JB, writing in the October 1979 *North Texas Contest Club News* advocates the position of contesters:

"Have you ever had some ragchewer give you a hard time about being a proclaimed contester? You know; the old bit about all the contesters messing up the bands on assorted weekends. How did you handle the situation? Did you present our position effectively? Well, just in case you run into such a situation again, remember that the best defense is a good offense. Here are a few points for you to pass on to the ragchewer for him to ponder and maybe a few for you to ponder too:

"Ask him what he's done lately to improve *his* communications skills and his station engineering to improve his ability to communicate.

"A good contester always strives to improve his skills. He may often miss a good band opening or seldom be found on the air during non-contest periods because he is making station or antenna improvements. For some reason, dedicated contesters seem imbued with a 'bigger and better next time' syndrome.

"Ask him if he has any idea what bands to operate at a certain time to be able to communicate with a desired location.

"Remind him that a good contester develops a keen sense of propagation and usually has the facilities to work each band effectively. I remember during the Nicaraguan earthquake crisis that a number of contesters handled emergency traffic on 40 because the ragchewer types didn't have antennas sufficient for decent communications. Maybe you remember a similar instance.

"Remind him, too, that 80% of the true-blue contesters have their Extra Class licenses, and that more than 90% have, at least, the Advanced. That fact certainly speaks well of the technical proficiency of contesters as a group. Can he say the same about *his* club?

"The ragchewer types also need be advised of a few other facts of life:

"Band occupancy is the name of the game when it comes to justifying our claims to the frequencies we've been allocated. That's one of the principal reasons why the ARRL promotes contests the way it does.

"Many of the Soviet bloc countries require evidence of successful participation in a certain number of contests before a licensee can advance his class. If the ragchewer types think incentive licensing here is tough, that ought to make them pause and reflect. It also helps explain why there is such heavy participation of Soviet bloc countries in all of the DX contests.

"Radio contesting is also one of the least expensive ways to gain personal recognition nationally and internationally (at least among our fraternity), and it can be done without buying a Lotus or Ferrari and entering the Grand Prix, and without ever leaving home. There is virtually no other activity I know of where an individual can accomplish this. In some countries, the recognition goes beyond the fraternity and considerably more esteem is

Traffic-Handling in Contests

Roughly speaking, each contest exchange can be viewed as a piece of traffic. Whether the exchange is 599 PA or 198 B K3LR 72 WPA, a definite format is followed. We might compare this to a formal message in which a standard format is required. Neither the contest contact nor the formal piece of traffic is complete until each and every part is received and confirmed.

When the ARRL November Sweepstakes began, it was as much a traffic-handling exercise as a contest. Although there have been numerous other rule changes since this contest's birth, the exchange remains the same; therefore, the original outlook is certainly still valid.

Perhaps the best example of a contest with traffic-handling as an integral part is the DARC-sponsored Worked-All-Europe contest. In addition to the usual contest exchanges, the rules incorporate the unique concept of contest QTCs. Non-European stations are permitted to send traffic similar to a book of messages (see chapter 4) to European stations. The traffic contains information about contacts that the non-European has already made during the contest. A more-detailed explanation of the QTC concept may be found in the contest rules.

publicly attached to the demonstrated talent than is the case here in the U.S.

"Contesting also promotes a degree of comaraderie that goes beyond the casual relationships established in DX contacts because of the obvious mutual interest. It thereby serves the purpose of international friendship that is meant to be a part of Amateur Radio.

"Finally, contesting is an outlet for those of us who are competitive by nature. That is why many who are successful in other walks of life gravitate to contesting. The activity is taken up by the sportsmen in our hobby."

New Horizons

Contests also broaden horizons. One need only read a few "soapbox" comments in *QST* appearing with the contest results to understand the reasoning here. It might be a vhf operator who finds normally unreachable states (or countries!) suddenly quite workable, a Novice working JH1WIX around the world, or an experienced operator finally breaking the 60-QSO-per-hour barrier. In each case, new horizons are reached.

The veteran operator may get as much satisfaction out of his achievement as the rookie Novice. Perhaps this best exemplifies the beauty of contests. Since there is no ultimate achievement for a given contest, there is always room for improvement. When a ham completes WAC (Worked All Continents) or WAS (Worked All States), that's it. There simply are no more continents or states left to work. However, contesters continually shatter previous records. The sky's the limit!

Factors in Contesting

Advancing technology, greater ham population, favorable propagation, better preparation — all are definite factors in the rapid growth in scores. This growth is occuring at *all* levels, not just at the top. Plain and simply put, it's just getting easier to make contacts.

Of the four factors cited in the previous paragraph, three are basically constant for all participants.

Technology is advancing for all of us — any contester has available a wider variety of better equipment. How to pick the arsenal of equipment best suited to your needs is a matter of personal preference. More on this later.

A greater amateur population means heavier contest participation. The number of active stations, at least in the major contests, is absolutely staggering. K1TN did a study of the Sweepstakes logs received for the 1975 event and found almost 10,000 different call signs in the 2500 logs received! It is interesting that the highest number of QSOs made by any station that year was 1742 — a small percentage of the total number of stations active during the contest period, indeed!

The third constant factor, propagation, is a sore spot to some. Each DX contest ends with the East Coasters complaining about the long strings of Japanese stations worked by West Coast stations, while the Westerners counter with envy over the large number of European country multipliers unavailable to them but easily workable from the shores of the Atlantic!

This involves competition on a national scale, however. To overcome these extreme differences in propagation, contest sponsors award the top stations in each geographical area, whether it be ARRL section, state or call area. This reduces the area of competition down to a size in which propagation differences are, for the most part, nullified. Therefore, propagation can be viewed as a constant.

Preparation, the fourth factor cited, is the factor that varies considerably among contest participants. Preparation can mean the difference between a successful contest operation and a frustrating weekend.

Preparation

By no means does a contest begin with the first contact. The object of contesting is to contact as many stations as possible within the prespecified time frame. But to allow the operator to concentrate wholly toward this end, a number of things can be done beforehand to eliminate wasting time during the contest period.

A station that is convenient to use is one important consideration. If changing bands involves unscrewing one PL-259 and finding the correct coax for the right band and screwing that connector onto the antenna jack, 30 seconds of carrier to peak the settings on the transmitter, receiver, amplifier and Transmatch, and grabbing a copy of the *License Manual* to refresh one's memory on what segment of that particular band he or she is allowed to transmit on, then that operator is *not* prepared.

A coax switch from the local radio store, masking tape with appropriate markings for each band's setting, and familiarity with privileges allowed under whatever class of license held, would save this operator a couple of minutes *every* time a band change is desired.

How's a couple of minutes going to hurt that much, you ask? For the sake of illustration, let's say that our unprepared operator wants to make a big splash in a 48-hour DX contest. He likes to change bands every 15 minutes because he always feels like he is missing an important contact on another band. Four band changes an hour, at two minutes per band change, for 48 hours, is . . . 384. That's six-and-a-half hours! He could've had a comfortable sleep in that time.

Any amateur station is normally capable of making contest contacts — in fact, a number of QRP operators pride themselves on the fine results possible with just a few watts of power. On the other extreme, stations with more power using kinky long wires or indoor antennas are capable of similar results. K1PR likes to tell the story of an EA station who called him during the ARRL DX Contest on 10-meter ssb. Seems the EA was loading his window frame for an antenna! Some contest sponsors offer special awards to QRP stations producing respectable results.

Major Contests

What constitutes a major contest? By definition, major means "greater in number."

Based on the number of logs received from participants (which gives a good relative* indication of overall participation), the major domestic contests are the ARRL Field Day and November Sweepstakes. The major worldwide contests are the ARRL DX Contest and *CQ* Magazine's World-Wide DX and WPX contests. We should also mention the top vhf contests offered by ARRL, which include the January SS, two vhf QSO parties and a uhf contest.

Of course, each contester has his own favorite contest(s), which become his "major" contests. The contests cited above might then be viewed loosely as the most-popular candidates, with the number of logs representing the tally of votes. Rules for all ARRL contests appear in appropriate issues of *QST* prior to the event.

*The percentage of participants actually submitting logs is not constant for all contests.

Receiving Effectively

This brings us to the subject of receivers. Contests attracting a large number of participants create a wide spread of signals across the band. It is mandatory in this type of situation to have a receiver that effectively eliminates the interference from adjacent frequencies. This is especially important during cw contests. Most commercial manufacturers supply as "standard equipment" a crystal filter for ssb only. These filters work pretty well on ssb, although some operators prefer a somewhat narrower passband. Problems arise when one tries to use this same filter during a cw contest, however. The bandwidth of a cw signal is much smaller; therefore, many more cw signals can be heard in the filter's passband than ssb signals. As a result, many manufacturers offer an optional cw filter, usually with a 500-Hz passband. A few companies offer even narrower filters, but all except the most fussy operators fare quite nicely without these extras.

Armed with these two basic filters, one can do fairly well with almost any receiver. It would be misleading to give the impression that all receivers, whether they are part of a transceiver package or not, are essentially equivalent. Most definitely, some receivers provide very good performance for the investment, whereas others requiring a heavier investment are of rather poor quality. A quick informal survey of experienced operators, contesters or not, will weed out the bad

A 120-footer at WA1MAO/K1PR. The tower features a 40-meter beam and stacked arrays on 20, 15 and 10. (*K1TO photo*)

prospects and highlight the good. Which receiver is the "creme de la creme" is a matter of personal preference. One particular model might excel in one area and do poorly in another. A look at the top stations in contesting today will yield almost as many different receivers as stations.

Here is where psychology plays an integral part in the discussion of receivers. If an operator feels that his particular receiver is the best, then he'll be completely comfortable with it. Another operator placed in front of that very same receiver might have a completely different outlook, constantly contemplating the shortcomings of the receiver. Some operators like lots of knobs to play with; others couldn't be bothered.

As far as a transmitter goes, anything with a VFO (variable frequency oscillator) and a clean signal is adequate. A clean, crisp signal, especially important on voice, is a most definite plus. Often, if two stations are calling another station — one mushy and loud, the other clean and crisp but weaker — the weaker station will be answered first.

Power Play

Obviously, any station not using the legal limit of power may be at a disadvantage signalwise. Because of certain characteristics of ssb and cw contests, however, a lower-power cw station will do relatively better than a comparable station on ssb.

To bear this out, let's look at the "High Claimed Scores" for ARRL's 1979 SS: The top low-power ssb score is only *half* the top high-power score and over 500 contacts below the Top Ten. To reach the Top Ten, this station would have needed 44% more contacts. On cw, the highest claimed low-power total has almost 75% as many contacts as the highest claimed high-power entrant. He needed to add only 210 more QSOs (or 22%) to reach the Top Ten. These figures seem to reveal that, indeed, low-power stations score *relatively* better on cw than on ssb.

Thus we see that depriving an operator of his amplifier causes a much greater drop in his ability to make ssb contacts than cw contacts. This comparison can be extended to all weak-signal contacts. Amateurs generally agree that many weak-signal contacts are made using cw when the contact might be impossible on ssb. VHFers who attempt contacts via ionospheric scatter often use cw since the signals are extremely weak. QRP enthusiasts find that their best results come from using cw. DX lovers often find one marginal path available on cw and useless on ssb.

Antennas

You have to hear 'em before you can work 'em. This means having an efficient antenna system. The most important part of an amateur's station is the antenna system. Different antennas work (or don't!) for different people for different reasons.

Finding the selection of antennas for your backyard that works best for you with your operating habits takes a long period of experimentation. Sure, a four-element 40-meter beam is nice to have, but what height above ground is optimal? If domestic contesting is your bag, is that high-gain, low-angle-of-radiation antenna really necessary? Would an auxiliary single-quad loop or simple half-wavelength dipole meet your needs for closer-in, higher-angle signals? Or would a fixed, switchable, two-element wire array be more economical and provide similar results? How about a system of phased verticals (if the lawn mower avoids chewing up the ground radial system!)? What about a quarter-wavelength sloper facing south to supplement the beam?

Questions like these face many amateurs. Whereas very little can be accomplished as far as modifying the gear in the shack, a few innovative ideas for antennas may gain a few easy decibels (or S units!). Perhaps the most important consideration in choosing antennas is your intended audience. A low inverted-vee antenna might work well in a domestic contest. But attempting to use that same antenna in a DX contest would probably net only a handful of contacts and a lot of frustration. If you are gearing up for the WAE (Worked All Europe) or AA (All Asian) or even a state QSO party (other than your own), your intended audience is all in one direction. A multielement, highly directive array would work quite well here.

But let's just for a second put ourselves in the shoes of a W9 or WØ operator who has just decided to operate the November SS. What is his intended audience? The whole of the United States and Canada, right? He needs to work stations on both coasts. Imagine his frustration if a weak W1 were to call in while his big directive array was pointed west. It might be a Vermont station, so he turns the beam 180 degrees. Thirty seconds later, the contact begins — assuming that the frequency is not overwhelmed by other anxious, contact-hungry contesters in the meantime! And if the W1 turned out to be portable 5 . . . That switchable array would obviously have been more convenient here.

Another consideration might be the level of interference generated from one particular direction. An East Coast SS operator might design a 40-meter antenna with a high front-to-back ratio to attenuate the European broadcast stations as much as possible. A West Coaster working Japanese stations in a DX contest wants a good front-to-side ratio on his antenna to eliminate the interference from other stateside stations.

The same type of reasoning applies when choosing between a dipole and a vertical antenna. Put up one of each for any band sometime and get comparative signal reports from stations at various distances. Contests thus provide an excellent opportunity for antenna testing. If it's the contest that you put the antenna up for and the antenna turns out not to work well, comparative signal reports beforehand would have been quite beneficial. In testing antennas for major competitions, many contesters make a few quick contacts in a smaller contest the weekend before, netting some quick information about relative signal strengths between the two antennas.

Some contesters use separate antennas for transmit and receive. The most common examples are "beverage" antennas — long and low-lying lengths of wire terminated with a resistor to ground — which are used on 1.8 and 3.5 MHz to enhance weak-signal reception of low-angle signals. Since these antennas have such a limited application and require a large backyard, they really aren't practical for most contest stations. In any case, it is rare for the first antenna system selected and erected to work to one's complete satisfaction.

Station Arrangement

Once the station has been put together, the contest operator must become familiar with its use. By all means, if you are more comfortable and familiar with your own gear, you will fare considerably better using it as opposed to any other piece of equipment, even if that equipment is "better." Try going over to a buddy's house and operating his new fancy radio and you'll see what that means. Complete familiarity with one's station has permitted operators using much smaller antennas to emerge victorious over operators with much better equipment.

Preparation also includes smaller things such as reviewing the list of things to fix (written down after the previous contest) and remedying them, keeping rotor boxes within reach, securing the keyer paddle to the operating table, labeling the positions on the coax switch, preloading the memory keyer, synchronizing the clock to a time station such as WWV or CHU, prenumbering log sheets, getting the VOX settings just right, setting up a tape recorder to complement the logs, gathering fuses and spare parts (or even spare equipment), sharpening a batch of pencils, clearly labeling dupe sheets for each band to prevent confusion, preparing propagation data (i.e., what band is open to where at what time) and so on.

Another very important aspect of preparation is knowing the rules *thoroughly*. This means knowing what the object of the contest is — whom to contact and what counts as a multiplier. Is the contest on one mode only, one band only? Can stations contact one another on each band, or just once regardless of where? What are the suggested frequencies for contest activity? When does it start and end? Are there any breaks in the contest period? How many hours of operation are allowed for single- and multi-operator stations? What information is required for a complete exchange?

Digging deeper, this means contacting a station down the street in the CQ WW DX contests on 160 meters to gain *two* additional multipliers (country and zone), even if no other contacts are made on the band. It means realizing that in the WAE contest, TF stations are prime, but OX stations won't count for credit. It means that taking a 10-minute break during SS does not count as off time. It means an extra three hours of operating time for Field Day groups that set up after the "starting gun."

The point is that each set of rules contains its own set of subleties. A thorough knowledge of the rules allows an operator to better plan an operation and avoid a mid-contest search for information to confirm a particular rule.

Most of the major ham periodicals include a contest calendar, including rules for each contest. Further clarification may be obtained from the sponsors. A good opportunity to get clarification of any rules that might appear unclear can be done at the same time you write for log sheets, dupe sheets and summary forms.

Operating a Contest

The basic object of any contest is to contact as many stations as possible within a preset time frame. To make more contacts than others means being more time-efficient than they are. This section, then, is devoted to the theme of increasing an operator's efficiency.

Listening during a contest to find efficient operators may be the best approach in getting valuable hints. To carry this idea a bit further, consider setting up a tape recorder to record stations that appear to be efficient operators. For a prospective apartment-dwelling contester who does not even have his own station, taking a tape recorder to a friend's station for a few hours on contest weekends is a good way to practice and improve.

Many Eastern European countries require their prospective amateurs to first obtain an SWL (Shortwave Listener) permit. Their intent is to allow the amateur to learn about on-the-air operating procedures before being granted transmitting privileges.

A number of contests have special categories for SWL stations, and the awards given to the top SWLs resemble those given to the top amateur entries. By having this separate SWL category, the SWL can feel as much a part of the contest as do the on-the-air participants. See chapter 14 for a discourse on SWLing.

It might bear mentioning here that vhf contests have a special appeal to apartment dwellers because of the widespread interest in hilltop expeditions. Such expeditions are impossible for hf contesters, especially to the extent that most vhfers go. For hf contests, the apartment dweller's best approach would be joining a crew of operators at a multioperator station that is already set up at someone's home. By participating in the construction and operation of such a station, an apartment dweller is an integral part of the overall operation.

Practice Makes . . . High Scores

Practice is vitally important in the field of contesting. Very rarely is it possible for a beginner to compete with more experienced operators. At first, this may discourage the newcomer, but every top operator started the same way. One active ham has admitted that in his first contest effort he made just 25 contacts in 26 hours of operation! Through many years of practice and improvement, this contester now thinks little of making 100 contacts in a single hour. Most top-scoring operators started with modest stations, which forced them to improve their operating skills to increase their contest scores. The best advice to newcomers is: "Don't be discouraged!"

Practice in contesting can take many different forms. Learning by doing is a proven method in education — the same principle applies perfectly to contesting. Listening to tapes of experienced operators is one way of practicing. Sending sample exchanges back and forth using code oscillators is also effective. While driving in a car, a phone operator might voice the exchanges that will occur during ssb contests.

But there is no substitute for actually sitting down in front of a radio and making real contacts during an actual contest. Since there is a contest of some type nearly every weekend of the year, this approach is quite easy. Contests with a high level of activity are better training exercises than contests with few participants, but each contest regardless of size yields some experience in becoming an efficient operator.

K8MR led a group of Mad River Radio Club members to a second-place finish in the 1979 ARRL DX Contest, using the call FG7AR/FS7.

Getting the Message Across

To be an efficient operator for an entire contest means being efficient during each and every minute of operation. Each contact must be made quickly so that the next can be made. To make each contact as quickly as possible means *exchanging* information as quickly as possible.

This does not mean cranking the keyer speed up to 50 wpm and sending the exchange in two seconds, nor does it mean speaking as fast as possible. The other operator must be able to receive your transmission clearly. Sending cw at 50 wpm or speaking too fast won't accomplish this very well. Use of understandable phonetics is vital on ssb. "*B*loomin *Z*ipper *F*lipper" might catch a stateside station's attention, but forget trying to make the DX operators understand! Use of the standard phonetic alphabet or well-known geographical names is best. For BZF, "Bravo Zulu Foxtrot" offers better clarity.

So we see that making contacts efficiently means transmitting the exchange quickly *and* clearly. Each contest operator has limitations here. Perhaps a few illustrations will explain this point.

First of all, send the exchange once only. If the other station needs a repeat, he will ask for it. But if he receives the information completely the first time, a repetition will be annoying and worthless. Valuable time is wasted.

The speed of the transmission, then, is the important vehicle for clear reception. What speed is optimum for each contact is best judged by what speed the other station is going. If a station answers your 30-wpm CQ at 10 wpm, it would probably be fruitless to send your exchange at 30 wpm. That station answered your CQ at that speed only because that speed is comfortable to him. By turning the keyer speed down closer to 10 wpm, this contact can be completed on the first try.

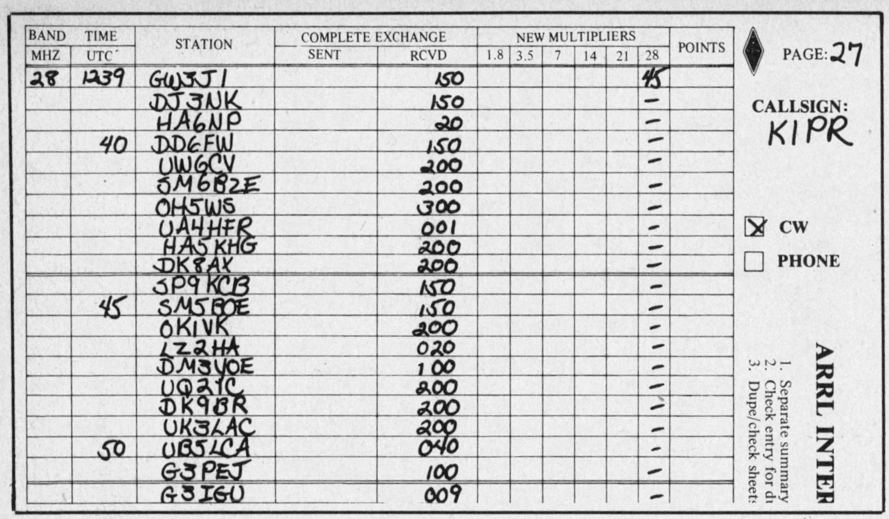

BAND MHZ	TIME UTC	STATION	COMPLETE EXCHANGE SENT	COMPLETE EXCHANGE RCVD	NEW MULTIPLIERS 1.8	3.5	7	14	21	28	POINTS
28	1239	GW3JI		150						45	
		DJ3NK		150						–	
		HA6NP		20						–	
	40	DD6FW		150						–	
		UW6CV		200						–	
		SM6BZE		200						–	
		OH5WS		300						–	
		UA4HFR		001						–	
		HA5KHG		200						–	
		DK8AX		200						–	
		SP9KCB		150						–	
	45	SM5BOE		150						–	
		OK1VK		200						–	
		LZ2HA		020						–	
		DM3VOE		100						–	
		UQ2IC		200						–	
		DK9BR		200						–	
		UK3LAC		200						–	
	50	UB5LCA		040						–	
		G3PEJ		100						–	
		G3IGU		009						–	

PAGE: 27

CALLSIGN: K1PR

☒ CW

☐ PHONE

ARRL INTER

1. Separate summary
2. Check entry for d
3. Dupe/check sheets

Fig. 1 — A page out of the log of K1PR's cw DX Contest effort.

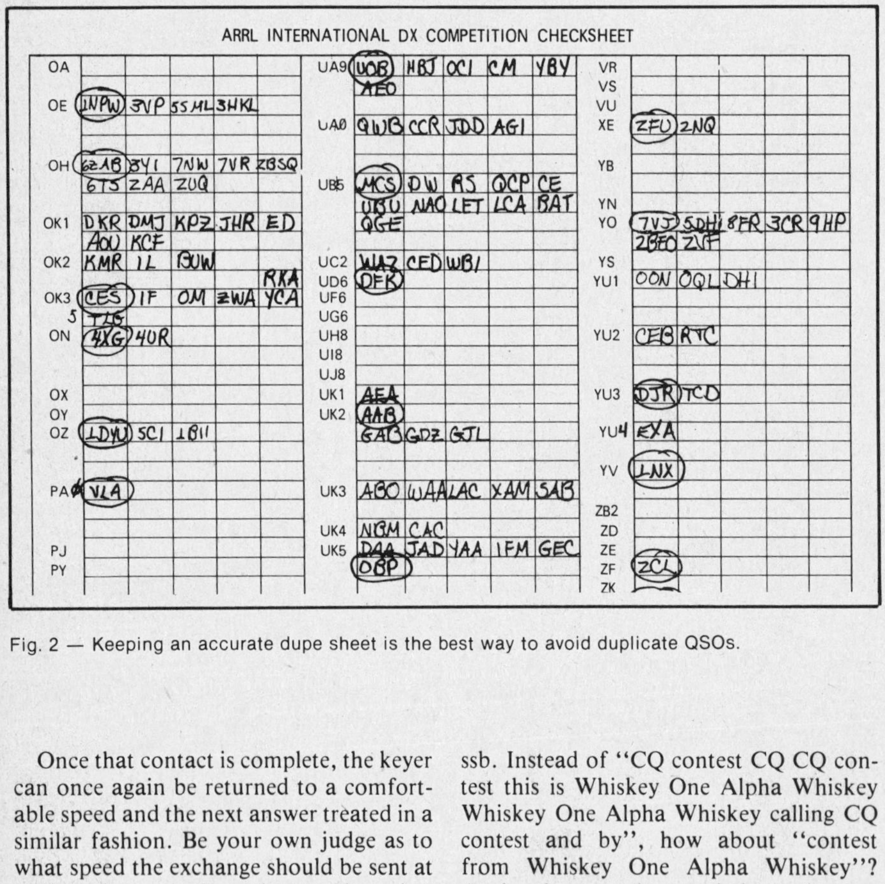

ARRL INTERNATIONAL DX COMPETITION CHECKSHEET

OA

OE INPW 3VP 55 HL 3HKL

OH 62AB 3Y1 7NW 7VR ZBSQ 6TS ZAA ZUQ

OK1 DKR DMJ KPZ JHR ED AOU KCF

OK2 KMR IL BUW

OK3 CES IF OM ZWA YCA RKA 5 TJG

ON 4XG 4OR

OX

OY

OZ LDYU SC1 1B11

PA VLA

PJ

PY

UA9 UOB HBJ OCI CM YBY AEO

UA0 QWB CCR JDD AGI

UB5 MCS DW RS OCP CE UBU NAO LET LCA BAT QGE

UC2 WAZ CED WB1

UD6 DFK

UF6

UG6

UH8

UI8

UJ8

UK1 AEA

UK2 AAB GAB GDZ GJL

UK3 ABO WAA LAC XAM SAB

UK4 NBM CAC

UK5 DAA JAD YAA IFM GEC OBP

VR

VS

VU

XE ZFU 2NQ

YB

YN

YO 7VJ 5DH 8FR 3CR 9HP 2BEO 2VF

YS

YU1 OON OQL DHI

YU2 CEB RTC

YU3 DJR TCD

YU4 EXA

YV LNX

ZB2

ZD

ZE

ZF ZCL

ZK

Fig. 2 — Keeping an accurate dupe sheet is the best way to avoid duplicate QSOs.

Once that contact is complete, the keyer can once again be returned to a comfortable speed and the next answer treated in a similar fashion. Be your own judge as to what speed the exchange should be sent at to be clear to the other station, but remember always to send the exchange only once!

The Exchange

Shorter transmissions take up less time and might convey the same information. Instead of sending CQ TEST CQ TEST DE W1AW W1AW TEST K, wouldn't a simple TEST W1AW convey exactly the same information? In fact, TEST W1AW could be sent three times, listening for an answer each time, in the time consumed by the first example! Notice that CQ DE and K can be inferred from the context of the transmission.

The same approach can be taken on ssb. Instead of "CQ contest CQ CQ contest this is Whiskey One Alpha Whiskey Whiskey One Alpha Whiskey calling CQ contest and by", how about "contest from Whiskey One Alpha Whiskey"? Again, the second example is much more efficient than the first, but it conveys the same information.

By using similar reasoning, the unnecessary extras can be eliminated from the exchange with another station and the subsequent confirmation leading to the next contact. Here are two possible complete contacts in a cw contest — note the brevity and completeness of the second:

CQ TEST CQ TEST DE W1AW W1AW TEST K
W1AW W1AW DE VE3GOL VE3GOL $\overline{AR}$
VE3GOL UR 599 CONN 599 CONN DE W1AW BK

W1AW DE VE3GOL R UR 599 ONT 599 ONT 73 BK
73 CUL CQ TEST . . .

TEST W1AW
VE3GOL
VE3GOL 5NN CT W1AW
W1AW R 5NN ONT VE3GOL
R TEST W1AW

If you were answering W1AW at the same time that VE3GOL was, would you even bother to wait for the exchange to be completed in the first example? Either way, you would probably curse under your breath at the operator's inefficiency.

When a repeat of some part of the exchange is required, the same policy of sending that part only once applies. The operator requiring the fill will appreciate your efficiency and the contact remains as short as need be. Here's an example of an efficient fill in the second example above:

. . . 5NN CT
STATE?
CT
W1AW R . . .

Notice that in each example, each station identifies once and each report is acknowledged by a simple R.

Duping the Log

Operating a contest also includes effective and efficient logging and duping procedures. Each contester has his own view on what sequence works best, but the general approach is the same.

Use of a pencil with a good eraser is practically mandatory for clean logs and dupe sheets. Remember that the contest log checker must look over your log when it is submitted and a clean log is appreciated much more than sloppy one. You will never initially be able to copy every piece of information perfectly, so erasing is fairly frequent.

In contests where more than a few contacts are to be made, keeping a cross-check (dupe) sheet of each station worked is the most convenient way of avoiding duplicate contacts (which don't count). For most contests, this means keeping a separate dupe sheet for each band; for others, only one sheet is needed. Many contest sponsors provide dupe sheets for the asking (ARRL's form CD-77 and CD-175, for example), although you may use a different format if that system works best for you. For domestic U.S. contests, the usual procedure is to group each call area together and further break down the call area by suffixes. Whether you choose to enter the entire call sign (advisable for at least the one- and two-letter suffixed calls, since each suffix might have a number of different possible prefixes), the suffix only, or the entire call sign, is a matter of preference.

For DX contests, the duping procedure is to break the world down by prefixes and enter the suffix only. Since the prefix is already listed, there is usually no need to enter the full call sign.

But the call sign must first be entered into the log — in other words, you have to work 'em before you can dupe 'em! Unlike the regular station log, in which the complete QSO information is entered, a number of shortcuts may be employed in a contest log.

Efficient Logging

Since the date will change only once or twice during an entire contest, there is no need to enter it for each QSO. For entering the time of a QSO when a number of contacts are made in an hour, the general approach is to enter only the minute of each QSO, leaving off the hour except for the first QSO in each hour. When the required exchange includes a QSO number, these should be pre-entered *before* the contest begins. The other parts of the transmitted exchange usually remain the same and need only be entered at the top of each page, with arrows pointing downward to show that they remained the same for each QSO. The most important parts of each QSO are the call sign of the station worked — *never* take shortcuts in logging calls — and the exchange received. Usually the exchange received should be logged completely. One exception would be in the case of repetitive signal reports. If everyone gives you a 599 report, the same procedure as for transmitted exchange applies. Any variation must be logged, though. As far as keeping track of new multipliers on the log sheet goes, don't bother. Leave it until off periods are taken or after the contest is over. Same applies to the number of QSO points gained for each QSO.

Every attempt should be made to get each bit of information correct. If it means asking for the call sign a number of times to be sure, do it! Any miscopied information may result in the removal of credit for that contact from your score by the contest log checkers. Too many errors may cause a log to be disqualified.

Examples of efficient logging and duping procedures using standard forms are illustrated in Figs. 1 through 4.

Logging on only one side of each sheet permits an operator to scan previous pages more easily when necessary, and provides a good psychological feeling of filling more pages, making the log seem thicker and bigger. Using one side of a dupe sheet allows the entire sheet to be visible at once and scanned quicker as well.

BAND 55 CW

	A	B	C	D	E	F	G	H	I	J	K	L	M	N	O	P	Q	R	S	T	U	V	W	X	Y	Z	KH6
1	AW ABJA ACZ	BW BIH		DER	EYI	FLM	GD GQ GOR		IK	JW	KI		MEM		OD	PR		RT RX	STO	TO TN TKZ	ABLU UAK	VTM W	WEF WB WA	XA XM	YEC YY	ZZ ZM ZDW	XX

	A	B	C	D	E	F	G	H	I	J	K	L	M	N	O	P	Q	R	S	T	U	V	W	X	Y	Z	KL7
2	AWH AE AEZA	BJG	CC CQ	DW	ECA ABZE	FGY	GTW	HAE	IC	JCX JAS	KOK	KCZL		NBM NT		POG PE PV		RQ RNT	SQ SX	TCQ	UVB	VFT	WJL WSX	XL	YY AAZY	ZQ	AF D

Fig. 3 — Dupe sheets, such as ARRL's CD-77, are also mandatory in stateside contests.

(do not write above this line)

ARRL INTERNATIONAL DX CONTEST

CW Third FULL weekend in February – Phone First FULL weekend in March

CALL USED N6RZ   CW ☒ PHONE ☐   ARRL SECTION (P. 8 QST) SCV

Separate logs and summaries must be submitted for each mode of operation.

Single operator

☐ All band
☐ Single band(circle one) 160-80-40-20-15-10
☐ QRP (10w or less)

Multioperator

☒ Single transmitter
☐ Multi-transmitter

Multiops list ALL operators:
N6RZ WB6SHD AA6T
WB6CBJ WA6OCV

	Valid QSOs	Countries
160	2	2
80	40	30
40	312	35
20	427	61
15	825	70
10	233	44
Total	1839	242

1839 (Valid QSOs) x 3 = 5517 (QSO Points) x 242 Countries =

1,335,114 Claimed Score.

DC POWER INPUT – CHECK ANY USED: ☐ 10 watts or less ☒ 11-200 ☒ Over 200

YES ☒ AFFILIATED CLUB PARTICIPATION NO ☐

ARRL AFFILIATED CLUB NAME NCCC

(Note: Club secretary must submit list of eligibles in order for club aggregate to be listed. 3 or more entries must be submitted.

"I have observed all competition rules as well as all regulations established for amateur radio in my country. My report is correct and true to the best of my knowledge. I agree to be bound by the decisions of the ARRL Awards Committee."

Date 15 MAR 80   Signature Your Signature   Call N6RZ

Print or type:

NAME: DAVID S. ROWLEY CALL: N6RZ
ADDRESS: 171 McGIVERN WAY
SANTA CRUZ CA 95060

1. Check log for duplicate QSOs.
2. Copy ALL QSO info carefully. A penalty is assessed for incorrectly copied QSO info.
3. Observe mailing deadline
4. Dupe/check sheet must accompany all entries of 500 or more QSOs

CD-53(R1179) Printed in U.S.A.

Fig. 4 — Your contest activity culminates in filling out the summary sheet, which at a glance chronicles your achievement.

Multiplier Lists

Instead of keeping track of multipliers on the log sheets, most contesters use some other piece of paper. During DX contests, a scan of the dupe sheets shows quite accurately (assuming the operator has kept an accurate sheet) which country multipliers have been worked. The best times to work on the multiplier list are during off periods, since on-time should be used exclusively for the purpose of making contacts.

In contests such as the CQ Worldwide, where a special zone multiplier is available, extra check sheets are necessary. One simple method is to use a CQ WW log sheet, which has 40 contacts per page, and number each line to represent the 40 zones. Then let each column represent a different band and simply check the appropriate "box" when a new band zone is worked.

Another unique type of multiplier is the prefix multiplier in the CQ WPX (prefix) contests. Many different approaches toward a convenient check sheet have been taken — again, whatever works for you is best.

Here's the setup at N4UM/C6A, which provided many VHF Sweepstakes participants with a welcome Bahamas contact during the 1979 event.

On the other end of the spectrum, a few contests do not include a multiplier at all. The final score is based solely on the number of contacts made. The classic example is Field Day; others include the ARRL Midnight Specials.

Keeping track of which multipliers have been worked and which have not is essential so an operator may reach a decision about whether to stay in a pile-up to work a new one, or turn the beam toward the Pacific in search of KH6, VK, ZL and other rarer stations, or go to 80 meters looking for a Vermont or Wyoming station.

QSOs vs. Multipliers

The most valuable consideration regarding multipliers is knowing how much a new multiplier is worth in terms of contacts. By dividing the number of contacts anticipated by the number of multipliers expected yields a QSO/multiplier (Q/m) ratio which represents the number of contacts that each multiplier is worth.

A typical Q/m ratio for a DX contest is lower than the corresponding ratio for a domestic contest, since more multipliers are available. The actual figures depend on a number of factors.

For example, a West Coast DX contest "big gun" works vast numbers of Japanese stations and relatively few European stations (who provide many more country multipliers), so that station's Q/m ratio is usually substantially higher than for an East Coast station who works vast numbers of European stations and multipliers and a correspondingly lower number of JA stations. A smaller station might not work the high number of contacts that the big gun would, so his ratio would be even lower. In the November SS, the multipliers come quickly at the beginning but are few and far between later on. Therefore, a station with 300 contacts might have 60 section multipliers and a Q/m ratio of 5:1, whereas a station with 1000 contacts and 70 multipliers finds that additional multipliers are worth almost three times as many contacts.

Knowledge of what Q/m ratio is expected, based on past performances, of either one's own or a similarly equipped station in previous runnings of the contest, is valuable since it allows an operator to know how long to stay in a pileup to work a new multiplier. This knowledge, combined with a good feel of how quickly contacts are being made, can give an operator a good estimate of how much time to spend pursuing new multipliers.

A few examples illustrate this point nicely. If the contact rate is very low at the present time (say 10 QSOs per hour) and the anticipated Q/m ratio approaches 10:1, the operator should continue calling the station representing a new multiplier for as long as it takes to work him. If the QSO rate is very high, then emphasis should be placed on making contacts — the multipliers will take care of themselves. During high QSO-rate periods, calling CQ is usually the best method of working stations quickly. The added problem of finding a frequency to resume CQing on is enough to cause most operators to shy away from getting into pileup situations at these times.

Here's another key point: If this new multiplier is common enough that it will probably be contacted later on in the contest, then less time should be spent trying to work it than if the station represents a rare multiplier that might not show up again for the rest of the weekend. This is especially important in the earlier stages of a contest, when the exclusive emphasis should rest on generating lots of contacts. In other words, don't spend 10 minutes calling a Connecticut station during SS just because it represents a new multiplier at that time. The chances are extremely high that another Connecticut station, much easier to work, will be found later in the contest. On the other hand, VE8 represents a very difficult multiplier, so you should try as aggressively as possible to work it whenever found. An operator, then, can figure out the maximum time to allot to calling a new multiplier based on the three factors — Q/m ratio, present QSO rate and commonality.

Instead of calling one multiplier at a time, many operators will hop from pileup to pileup (an extra VFO is an excellent asset here) until each station is finally worked. Keeping track of call signs and their frequencies is often done on scratch paper, although some contesters even have standardized forms for this.

The actual subject of cracking the pileups is treated in depth in the chapter on DXing. Timing and relative frequency to the station being called are the two most important points when attempting to break the piles.

Outlining Objectives

Decisions on whether to hunt for multipliers or call CQ and run (work strings of) stations are often necessary during a contest. Therefore, a good practice is to have one or more possible approaches outlined *before* the contest, so that crucial errors in emphasizing the wrong thing too much are not made during the contest itself. Having a preset outline ensures that you will not forget about points that might be overlooked or completely forgotten in the heat of battle. These outlines should be based on many different factors, including station capabilities, propagation, intended audience, extent of portable participation, past performances, sleep requirements and class of license.

Such an outline might be multifaceted, with one list by time-period suggesting what band to be on working who at what time, another list by country and best times to work each on each band, and another list of bands with possible openings on each.

A small station might outline the following before Sweepstakes weekend: Start on a band where the level of interference is low and other small signal stations are workable. An East Coaster might start on 40 or 80 meters; a West Coaster on 20. These small-signal stations produce the bulk of one's contact total. There are only so many big guns to work — and they will be active during the entire contest. By Sunday, they will be worked more easily, since they will have "worked out" the upper-crust of signals and will be looking for weaker stations to boost their totals. Working other small-signal stations at the start allows the small-signal station to build up contacts while not fighting the intense interference on other bands. When the lower bands begin to fill with stations, then the small-signal station is more apt to be successful on the higher bands. By Sunday, he will be much more successful at generating contacts on the higher-activity bands, and by 2100 UTC, 24-hours after the start of the contest, he might be found on 10 or 15 meters instead of 20 or 80 meters.

Bigger-signal stations are most successful on the bands with the highest level of activity, where they can use their signal to maximum advantage. By the same token, the big signals tend to congregate toward the lower edges of each band, while the smaller signals are better off avoiding these frequencies and staying higher in the band. A band such as 10 meters, where lots of space is available, has appeal to both levels of stations, whereas other, less-open, bands are tougher for the small-signal station to use.

Picking the Right Band

Staying away from high-activity bands is especially important for the small-signal station during ssb contests, where fewer signals can fit in a band segment. Good cw operators with smaller stations can often do well on high-activity bands by outfinessing the bigger stations. This type of operating is close to impossible in ssb contests, however. This point was brought out in an earlier section where low- and high-power SS scores were compared and analyzed. Many smaller-signal stations make as many or more contacts during a cw contest than ssb counterparts because the operating strategies are different for each, and brute force is less of a factor on cw. Big guns, however, score considerably higher during ssb contests since they are always capable of holding a frequency and CQing.

Propagation factors figure greatly in making a precontest outline. Charts found in *QST*, combined with up-to-the-minute WWV reports and a bit of listening to the bands beforehand can give one a very good idea of just what to expect during the contest period.

Poor conditions tend to exaggerate the difference between the big and small guns. While the big guns will make lots of contacts under any conditions, often the smaller guns are unable to work each other and tend to work only big guns. During good conditions, which occur most frequently during periods of high sunspot activity, station size becomes less of a factor and operating skill is the deciding factor. At these times, high-activity bands are quite productive for smaller stations, who can work plenty of other small-signal stations.

It is impossible during these periods to cover every opening with a single-transmitter station, but knowing which band is most productive at what time is still essential. During sunspot lows, most openings can be covered. It is during these periods that a knowledge of propagation is most helpful, for missing a single opening could mean a big scoring loss. Glancing through previous years' logs and looking at logs of other stations in your area can provide good insight on where to be when.

A few other operating hints warrant mention. At the beginning of a contest, if you are hunting and calling stations, don't wait to hear the complete call sign if only the tail end is heard. The chances are very good that you have not worked him yet, and even if you have, he'll remember or notice you on his dupe sheet. You cannot rely on the other station to keep the dupe sheet, however, especially if it is a DX station. When you do hear the call sign completely, be sure to check the dupe sheets before answering. Nothing is more aggravating to a contester than a large number of duplicates. Duplicate contacts are a complete waste of time for both stations involved.

JA7UCS cooks up a winning combination for his group during the DX Contest.

Your License

Another important consideration when preparing an outline is the class of license held. A General class licensee is permitted to operate cw on all but the bottom 25 kHz of each band, except on 10 meters where full privileges are granted. Generals are quite limited in the range of ssb frequencies available, however, with 10 meters again the exception. Upgrading to an Advanced class license allows much greater freedom in the ssb portions of each band, with only small segments of 15 and 80 meters being reserved exclusively for Extra Class licensees. The incentive for upgrading from General to Advanced lies solely in the additional spectrum space available for ssb. For an operator whose main or exclusive interest is ssb, obtaining an Advanced Class license is often enough.

For cw operators, upgrading to Extra Class is almost imperative. This is especially true for DX contest operators, since foreign activity levels are quite high down in those lower 25 kHz (especially on 40 and 80 meters, where the majority of cw DX stations are found). Cw contesters oriented mainly toward domestic contest operating can get by without the bottom 25 kHz, since very little activity occurs in that portion. The only domestic contacts possible in that portion are between two Extra Class licensees, and since Extras represent a smaller percentage of the total U.S. amateur population, they tend to stay above that portion, where many more stations are workable. The vast majority of DX stations are permitted use of the bottom 25 kHz and are worked just as easily in the lower 25 kHz span as in any other part of the band.

Beginner's Luck

A few cw contests have enough activity to make participation by Novices and Technicians fruitful. ARRL's 10-Meter Contest and November SS, along with the major DX contests, usually attract enough stations such that activity extends more than 100 kHz above the lower band edge, spilling over into the Novice/Technician segments. The 10-Meter Contest provides bonus points for contacts with Novices and Techs to encourage activity in that portion of the band.

The contests highlight for Novice/Technician cw operators is definitely the ARRL Novice Roundup (NR), held annually in early February. In this one, the Novices and Techs are the "celebrities." In other words, upper class licensees and DX stations can work only Novices and Techs. Operators are free to pick the most convenient operating times for them during a nine-day period. For many potential contesters who are otherwise occupied on weekends, this contest provides a great opportunity for some midweek contesting.

The NR provides a perfect opportunity for the budding contester to improve operating skills by working and observing the techniques of more-experienced upper-class licensees. Most Novices and Technicians have probably worked mostly other Novices and Techs during their casual operating periods and therefore learn only from each other. By bringing the upper-class licensees together with these less-experienced operators, more can be learned in the nine days of NR than in the rest of a new operator's brief career!

Another hint worth mentioning is the idea of setting a quota before the contest. Often, if a preset quota is reached, the operator will feel good about his performance. Again, don't be discouraged if you don't make your quota the first time — just remember to set a more realistic goal the next time!

Population Centers

One more point to remember is that the highest volume of contacts come from areas with a large ham population. These areas usually correspond to large population areas. This is why a West Coast operator in SS spends his time on the high bands. Since these bands are open coast to coast, and since the greatest percentage of U.S. hams on are on the East Coast, this is where the volume of contacts is to be found. For the same reason, an East Coast SS operator tends toward the low

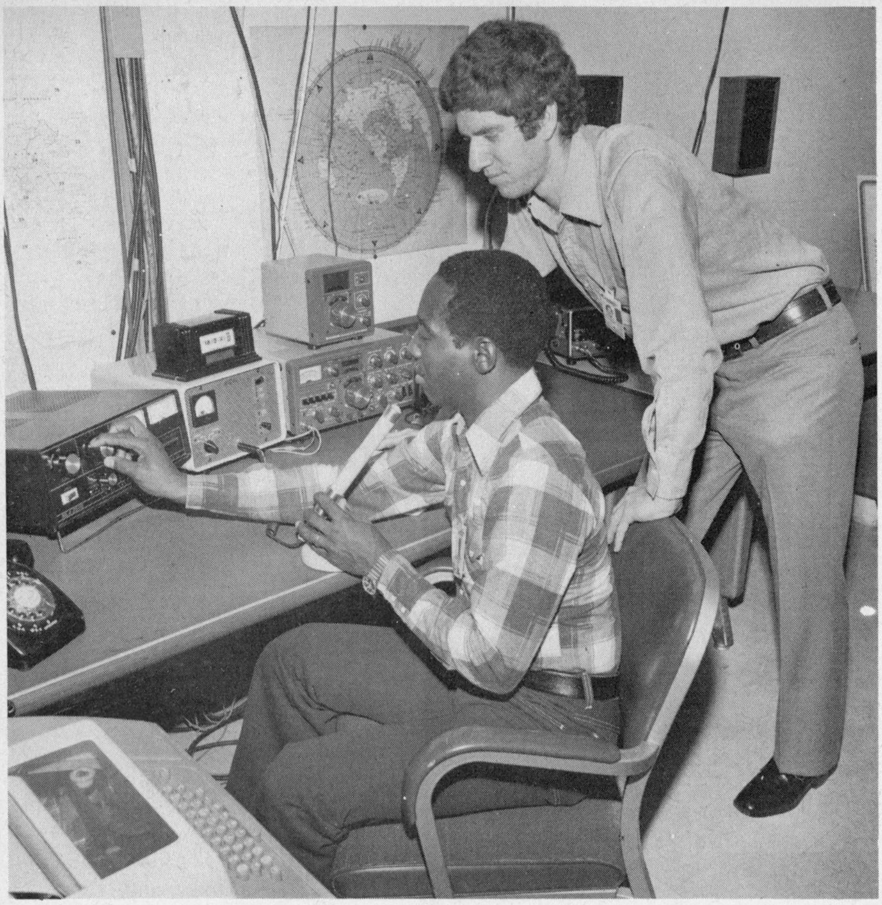

The W6VLD club regularly submits a multioperator entry in contests. Club President WA6EUM (seated) and WA6DPQ check out the station prior to the start of the ARRL 10-Meter Contest.

bands to work other East Coast stations. Japan has a greater ham population than any other country in the world, so the wise DX contester spends as much time working JAs as conditions will permit.

It is usually very difficult to wrangle a contest exchange out of a noncontest participant. If a kind and generous approach is taken, the noncontester will feel helpful, but if numerous stations harass him no end, then his view of contesting is apt to drop another notch. This situation often arises when the noncontester represents a multiplier difficult to find elsewhere.

If the noncontester answers your CQ, it shows that he is interested in your operation. A patient but brief explanation of what is going on will give him a more favorable outlook on contesters in general, but if he receives a terse response, then you are unnecessarily giving contesters a bad name.

Contest Forums

One good source of helpful information on contest operating is the contest forum held at many hamfests and conventions. The best-known example is at the Dayton Hamvention, where the contest forum draws hundreds each year. Most ARRL division conventions include a good contest forum, as well. Forums bring hams together who share a specific interest within the hobby, so attending a contest confab might yield interesting new acquaintances. Who knows, you might even find that that fellow around the corner with the tribander on his roof that you've been meaning to drop in on "one of these days" turns out to have a budding interest in contests, but didn't know whom, to talk to about it!

Contest forums often feature presentations by prominent contesters, slide shows of contest DXpeditions, talks concerning one specific aspect — maybe microprocessors or station design — and many other interesting topics. In any case, they are great learning experiences for the veteran and newcomer alike.

Multioperator Activity

Operation of a multioperator station is quite similar to that of a single-operator station, except that a few extra roles might be created to take advantage of the wealth of manpower. Operating an entire contest period is difficult for many single operators because of restrictions on stamina, time or just plain motivation. The gaps created by these problems can be filled nicely by inviting a few other operators over to join the operation.

Field Day

Perhaps the classic example of multioperator stations occurs during Field Day. Since the concept of emergency preparedness is an integral part of FD, many amateurs who are not particularly contest-oriented will find some motivation in this aspect. Multioperator stations participating in FD are most often set up at locations not usually used for amateur operations, so people with more of an inclination toward station setup and tower work would get as much out of a FD weekend as the operators. Operations of this type are rarely heard in other major hf contests, although they are quite prevalent in vhf contests in the form of hilltop operations. Field Day has such a widespread variety of categories that a group of QRP operators with a battery and a small transceiver can compete as enthusiastically in their category as can a contest group that sets up transmitters and antennas for every band and mode in their category.

Many clubs participate in FD each year, but find no particular appeal for any other contest activities. If one questions the members of such a club as to why FD is so appealing, they most often respond with comments about how much fun it is, how it gives them a unique opportunity to learn both set-up and operating procedures from the more-experienced operators of the group, and the camaraderie — the feeling of being able to contribute in some way toward the success of the operation. These reasons are precisely why multioperator groups in *any* contest can be so successful.

Organization of a multioperator group is absolutely essential for a smooth, successful operation. Having a number of operators together for one operation provides an opportunity for a newcomer to practice his skills under the supervision of more-experienced operators; hams with a mediocre (or no) station get to use better equipment and antennas; relief is available whenever needed so that a fresh operator is always operating; paperwork which may be too cumbersome for a single operator — whether it be duping, keeping track of multipliers, or filling out "rate sheets" — is accomplished more easily; openings on other bands which might otherwise go unnoticed can be found and taken advantage of; an extra set of ears can pull out the call sign of a weak station; and so on.

After the Contest

Once the contest has ended, by no means is the mission completed. A number of contests have been won and lost long after the equipment has been shut off. How so?

The answer lies in the paperwork.

Turning in a log to the contest sponsor is by no means a matter of filling out and signing a summary sheet. The logs must first be reduped and checked for accuracy, each new multiplier numbered, operating time calculated, sloppy log entries cleaned up, etc.

Contest log checkers are quite harsh on sloppily submitted logs; in fact, an overabundance of errors in the log may result in the log being completely disqualified. That's why the importance of being accurate in copying every part of every exchange correctly was stressed earlier. Any miscopied information may result in the removal of credit for that contact, and in many cases, an additional penalty is assessed. The policy of most contest sponsors is to disqualify entries containing errors (not including penalty) in excess of 2%. This means that having one erroneous contact on each log page may spell disqualification.

Contacts may be removed for any number of reasons. As mentioned, any miscopied portion of an exchange is one reason. A lack of any part of the received exchange means that the contact is incomplete and not eligible for credit. Years ago, some sponsors allowed partial credit for partially completed contacts, but this policy no longer holds. Operation for more than the allotted time means removal of all of the contacts made after the allowed time has elapsed. This means that a station can not operate SS for 30 hours straight and take the best 24 hours!

How is an erroneous contact detected? It's simple. Remember that most of the stations you work during a contest also send logs in to the sponsor. Therefore, by cross-checking the other station's log, the log checker can easily discover whether the contact was indeed made and if the exchange was copied correctly. Even if a log is not received from the other station, a letter could be written to that station requesting log data.

Just as in any other form of competition, there are entrants who will attempt to bend or break the rules. It would not be fair to an honest entrant to lose in the final results to a dishonest one.

Many contests have been won and lost in the paperwork. Unfortunately, a few truly honest competitors have been disqualified simply because they didn't dupe their logs after the contest and turned out to have an unacceptable number of duplicates. Every single contact in those logs may have even been copied perfectly! The problem is that the log checkers had no idea whether these entrants intended to leave the contacts in the log, thereby padding the score, or simply neglected the paperwork.

Since the deadline for submitting logs is usually many weeks after the contest, there is absolutely no excuse for turning in a log with duplicate contacts unremoved. If it means double- and triple-checking an exceptionally large log, do it! When removing duplicates from your log, just draw a light line through the contact information and put a "0" in the points column. Never erase duplicates! (We have no intention of scaring anyone away from submitting a log. What we are stressing is the necessity for accuracy.)

W5MYA and K5FN were two of the operators at K5RW/5 during Field Day. (*KA5Q photo*)

This nice location was the scene of the K6OR/6 Field Day activity.

On the positive side, reduping a log may reveal that a contact originally thought to be a duplicate is indeed a valid contact after all. New multipliers may be found as well. In both cases, the checking revealed scoring boosts that would have otherwise gone unnoticed. Log checkers often find the same thing and probably do as much increasing of scores as decreasing!

The Results

By submitting a log, you can see from the results exactly how well you fared against others in your area. Even if you know of others in your area who claim a higher score after the contest, they may not submit their logs at all (or not turn out to have the score they claimed), making you the award winner! By keeping track of your results, you watch your progress as you gain experience.

When submitting a log to the sponsor, it is wise to keep a copy for yourself. Responding to QSLs, comparing with other entrants and making further statistical analysis are easier if a copy is handy.

Although recopying a log may make it look neater to the log checkers, they always will request a copy of the original log. The best idea may be to keep the recopied version and submit the original. The same policy applies to computerized logs. Although typing one's log into a computer terminal will provide foolproof duping, it may also mean losing contacts because of typing errors — be careful! Log checkers have good reason for requesting the original log, since erroneous typing errors may be checked against the original log and prevent removal of those contacts. As stressed earlier, use of log sheets, dupe sheets and summary sheets provided by the sponsor makes it easier for you to remember all crucial log and summary data and makes it easier for the log checker to check your entry.

Results of contests sponsored by the ARRL and *CQ* Magazine may be found in subsequent issues of their respective publications (*QST* for ARRL contest results). For results of contests sponsored by other groups, it is usually a good idea to include a business-sized s.a.s.e. with the submission of the log.

One additional piece of paperwork that should be done immediately after the contest is a list of problems that occurred during the contest and operating ideas you might have picked up and want to remember for the next one. Remember, reviewing this list and remedying the problems are a part of preparation for the next contest.

Club Participation

A large number of contest sponsors promote participation by club groups. Most sponsors do not bother to delineate between clubs formed exclusively for the purpose of submitting large contest aggregates and general-interest neighborhood groups. The ARRL, however, uses a three-tier club competition system allowing for such a delineation. In this system, distinctions are made between the local clubs, who submit 10 or fewer entries; the medium-sized clubs, who may submit up to 50 entries; and the unlimited clubs, who may submit as many entries as can be mustered. An entry means a log submission on *either* mode for multimode contests.

Membership in any club that submits contest entries can be an invaluable asset to the newcomer and veteran alike. Knowing that one's club will be submitting a group score for a contest may provide that little extra motivation to operate a contest when the inspiration to compete on an individual level is not there. Imagine if your club were to lose in their level by a margin smaller than the score you could have produced by getting on for even a little while.

Even if only one member of a local amateur club is interested in contesting, other members might wish to learn about contesting from him or her. In return, the contester might appreciate being enlightened by another club member on some other aspect of the hobby with

Left to right, K1TO, K1XA, K3OX and W1GNC, smile after a record-breaking effort as W1GNC/PJ2 in the 1978 IARU Radiosport Championship.

which he is unfamiliar. Activities of this type provide a wonderful opportunity for newcomers to sample many different aspects of the hobby.

For those hams who start out in this way and choose to concentrate a bit more heavily on contesting, consideration might be given to joining a club whose activities are oriented in that direction. There are a surprising number of contest clubs throughout the U.S. A good place to look for contest clubs in your area is in contest results in the major ham periodicals. The list of participating clubs should represent most of the clubs actively submitting contest scores. If you happen to be in a sparsely populated area, writing to a few hams in your area whose calls are listed in the contest results may provide the spark necessary to actually begin a small contest group.

Clubs provide a perfect opportunity for newcomers to find other contesters to join for multioperator groups or even just to find a better station to "single-op." If the club has a club station for use by its members, then it might be used for a multioperator effort during contests. This idea is carried to the hilt in Eastern European nations, where club stations are used for serious efforts in almost every contest. All UK prefixes in the USSR represent club stations (all other prefixes are home stations), and are usually more in evidence than other U stations.

Contests Not Yet Mentioned

In the course of our discussion so far, we have mentioned a number of specific contests: the ARRL-sponsored November SS, DX Contest, Field Day, 10-meter Contest, Novice Roundup, VHF contests and Midnight Specials; *CQ* Magazine's WW DX and WPX contests; DARC's Worked All Europe; JARL's All-Asian Contest; and the *National Contest Journal*-sponsored North American Sprint.

The ARRL also sponsors three CD (Communications Department) parties each year. The April version is open to all ARRL members, while the other two are open only to SCM appointees. CD parties include only the better operators, since obtaining an appointment in the first place requires demonstration of certain operating proficiency. For this reason and the fact that no awards are given, CD parties are good tune-ups for the major contests. Through its headquarters society (ARRL), IARU sponsors the Radiosport Championship in July.

CQ Magazine and the ARRL both sponsor wintertime (in the Northern Hemisphere) 160-meter contests. Activity on the "Top Band" is unparalleled, so many operators make these two weekends their exclusive extent of operation on the band. Propagation on 160 closely parallels that found on 80 meters, although signals are considerably weaker. With LORAN leaving the band by December 31, 1982, and many more countries now allowing their amateurs to operate on the band, activity will undoubtedly show a marked increase in the years to come, making the 160-meter contests even more prominent in the yearly calendar.

Working DX on 160 is different than on the other amateur bands, in that most DX stations transmit in the segment between 1825 and 1830 kHz and listen lower in the band. Knowledge of frequency allocations for each country is important on 160, just as it is on 80 meters, since easily worked stations may go unnoticed otherwise. For example, Soviet stations transmit above 1850 kHz and below 3650 kHz.

The North American Sprint (NA) has an interesting QSY rule, preventing loud stations from plopping down on one frequency and having everyone answer them. This feature emphasizes operator skill, and since the contest is only four hours long and exclusively cw, good operators can have the satisfaction of outfinessing less-skilled operators using bigger stations. One other rule unique to this contest is the team concept, with 10 members allowed on a team and no distance requirements.

The NA seems to have started a trend toward shorter contests, which deemphasize the stamina aspect. ARRL is now trying a series of single-band, two-hour, sprint-style contests with the establishment of the Midnight Specials.

And finally, contesting, which has caught the fancy of thousands of amateurs over the years, is just (according to W1HDQ) "normal hamming speeded up!"

Chapter 8

Awards Chasing

DXCC, Worked All Zones, Worked All Continents, the 600 Club, and so on, which are covered in other chapters, aren't by any means the only awards available to hams. In fact, there are a great many certificates available out there. And awards chasing can make even the most mundane QSOs significant. This section will discuss some of the major awards programs.

Some Basic Considerations

Regardless of the award you are after, there are a few points to keep in mind. First and most important, *always* include an s.a.s.e. (self-addressed stamped envelope) when requesting the application form for the award you are interested in. Know and follow all the rules of the particular award. Always use the standard form and make sure your application is neat and legible. If QSLs are required with the application, send them first-class registered mail. Include sufficient postage for their return the same way. Check the cards carefully before mailing them. Make sure each card contains your call and the other substantiating information. Don't send cards with markovers or erasures. And most of all, be patient! Your registry number can come in handy should there be a problem and tracing becomes necessary.

ARRL Awards

Code Proficiency — You don't even have to be a ham to earn this one! But you do have to copy one of the W1AW (or W6OWP) qualifying runs. (The current W1AW schedule is available from ARRL headquarters.) Five minutes worth of text is transmitted at each speed, 10-15-20-25-30-35 wpm. Twice yearly, W1AW transmits 40-wpm runs. Just copy 1 minute solid to qualify. Your copy can be written, printed or typed. Underline the minute you believe you copied perfectly and send this text to ARRL hq. along with your name, call (if licensed) and full mailing address. Include the date of the code run you copied.

In the processing, your copy is checked directly against the official tape of the transmission. You'll be advised if you passed or failed. If the news is good, you'll soon receive either your initial certificate or an appropriate endorsement sticker.

Check the "Contest Corral" section of *QST* for the next qualifying run. There is no membership requirement or any kind of fee for this one, but a large s.a.s.e. is required with your copy. Passing the 20 and 25 run might be just the reassurance needed for that Extra Class code exam.

Rag Chewers Club — The first operating award a new ham aims for is often the RCC. This one is designed to encourage friendly contacts, of more substance than the contest-like RST/name/QTH. RCC has just one requirement: Chew the rag over the air for at least a solid half hour. If you want to obtain the certificate, report the QSO to ARRL hq. and you'll soon be issued the distinctive blue award. If you want to nominate someone else for membership, send the nomination to him or her, not to ARRL. If this person wants the award, he or she can send the nomination to hq. This way, no one gets an unwanted certificate, and confirmed ragchewers can still nominate those they think are qualified.

Amateurs who work/confirm Africa, Asia, Europe, North America, Oceania and South America, on five bands, can qualify for this attractive award. See chapter six.

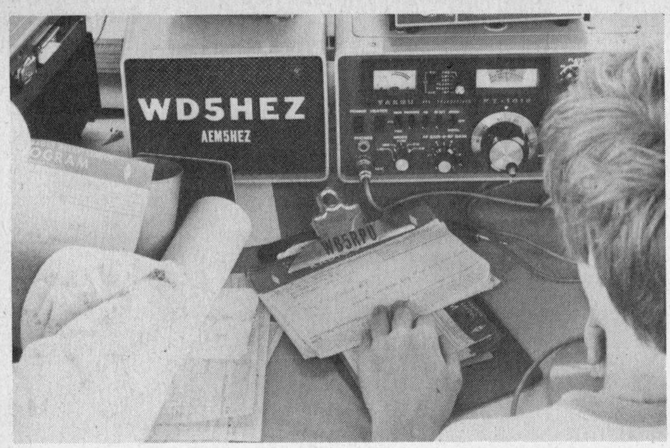

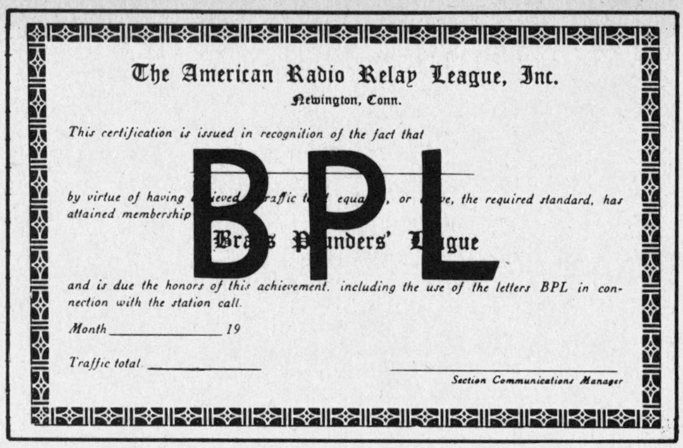
The American Radio Relay League, Inc.
Newington, Conn.
This certification is issued in recognition of the fact that

BPL

by virtue of having achieved traffic totals equal to, or above, the required standard, has attained membership in the

Brass Pounders' League

and is due the honors of this achievement, including the use of the letters BPL in connection with the station call.

Month ______________ 19

Traffic total ______________ ______________ Section Communications Manager

Active hams who handle large amounts of message traffic are eligible for the Brass Pounders League. See text. (*WB5RPU pnoto*)

Worked All States — This one isn't as easy as you may think. You'll spend many hours looking for Delaware or Nevada but it'll all pay off when you work and confirm 'em all. WAS is available to all amateurs. In addition to a "regular" WAS award, separately numbered awards are available in the following categories: 50 MHz, SSTV, OSCAR Satellite, RTTY, 2-letter 75-Meter Extra Class SSB, 144 MHz and 432 MHz. A plaque is offered to the first 10 amateurs earning the 432-MHz award. Additional endorsements to the basic award are offered for SSB, Cw, QRP, Novice and 160 Meters.

Contacts may be made over any period of years. The confirmations must show your call and indicate that two-way communications took place. Contacts with Alaska prior to January 3, 1959, or with Hawaii prior to August 21, 1959, cannot be counted. (These are, of course, their effective dates of statehood.) QSLs for contact with the District of Columbia count for Maryland.

Contacts made through repeater devices or any other power relay method cannot be used for WAS confirmations. The only exception to this is the specially endorsed OSCAR WAS (for working *all* 50 states via the amateur satellite).

ARRL Operating Aid CD-217 details the rules and fees and includes a convenient application form. An additional handy sheet is the WAS "map," suitable for posting on the shack wall. Color it in as you work (or confirm) the states to make a visual record of your progress. If you do a lot of multiband operation, you might wish to post one map for each band.

Five-Band WAS — This award was designed to foster more uniform activity throughout the bands, encourage the development of better antennas, and generally offer a challenge to both old- and new-timers. In addition to the basic WAS rules, 5BWAS carries a "start from scratch" date of January 1, 1970. The rules also require that applicants in the U.S. and possessions, Puerto Rico and Canada, be full members of ARRL. Special fees apply.

Unlike WAS, the 5BWAS is a one-time-only award. No band or mode endorsements are available. The award itself is in the form of a plaque with suitable personalization.

Brass Pounders League — The field of traffic handling is as specialized as almost any other within Amateur Radio. The fun and sport of it have captured the imagination and occupied the spare time of thousands of hams over the years. Based on a system of counting monthly traffic, BPL is open to all amateurs in the United States and possessions, Puerto Rico and Canada who report to their SCM a message total of 500 or a sum of origination and delivery points of 100 or more for any calendar month. All messages must be handled on amateur frequencies within 48 hours of receipt in standard ARRL form. If you meet the requirements, your SCM (Section Communications Manager, page 8 *QST*) will issue you one of the distinctive 3-1/2 × 5-1/2 inch card awards. Additionally, a special engraved medallion is presented to each individual amateur who makes BPL for the third time. For complete details on message-counting procedures, see the traffic-handling chapter.

Public Service Honor Roll — Those amateurs who report to their SCM 60-or-more total points in the published Public Service Honor Roll criteria (for a month) are listed under the PSHR heading in *QST*. Novices/Techs qualify with 40-or-more points. Those amateurs who are listed for 12 consecutive months or 18 out of 24 are in addition eligible for a handsome certificate. See *QST*'s "Public Service" column for more details.

Public Service Awards — ARRL issues two kinds of awards for specific public service activities. The Emergency Service Commendation is given in recognition of an amateur's contribution during a communications emergency. The Public Service Commendation is awarded to amateurs involved in a non-emergency situation or event. In both cases, certificates go *only* to those amateurs who were outstanding and meritorious participants. These types of awards are not worked or applied for — they are awarded upon recommendation by a League Official or some other responsible person.

Old-Timers' Club — In recognition of current-day hams who held an amateur license 20-or-more years ago (lapses permitted), a suitable award is available — the OTC Award. If you qualify under these terms as an "Old-Timer," you'll find the necessary paper work pretty easy. Drop a note to Hq. with the date of your first amateur license and your call then and now. ARRL would like to have a very brief history of what you've been doing in ham radio during that period.

If you're eligible, you'll soon receive your membership certificate. No charge for this one but please include an s.a.s.e.

A-1 Operator Club — Membership in this elite group attests to unusual competence and performance in the many facets of Amateur Radio operation — keying, modulation, procedure, copying ability, judgement and courtesy. Particularly noteworthy about this group is the absence of a "list" to aid those desiring the award. You must be recommended for the certification independently by two operators who already belong. In fact, if you ask to join, you may actually "fail" the test on a courtesy-judgement basis. This honor is truly unsolicited and earned — earned through the continuous observance of the very highest operating standards.

Satellite DX Achievement Award — This multicolor certificate is offered for demonstrated proficiency using the OSCAR communications satellites. Contacts count if made on or after December 15, 1972. (Note: only one award per applicant station.) The cards themselves must confirm two-way communications via the OSCAR satellites, plus the other standard QSL information. You need to accumulate a total of 1000 points to qualify for the award. Each contact with a new station counts 10 points, with a new coun-

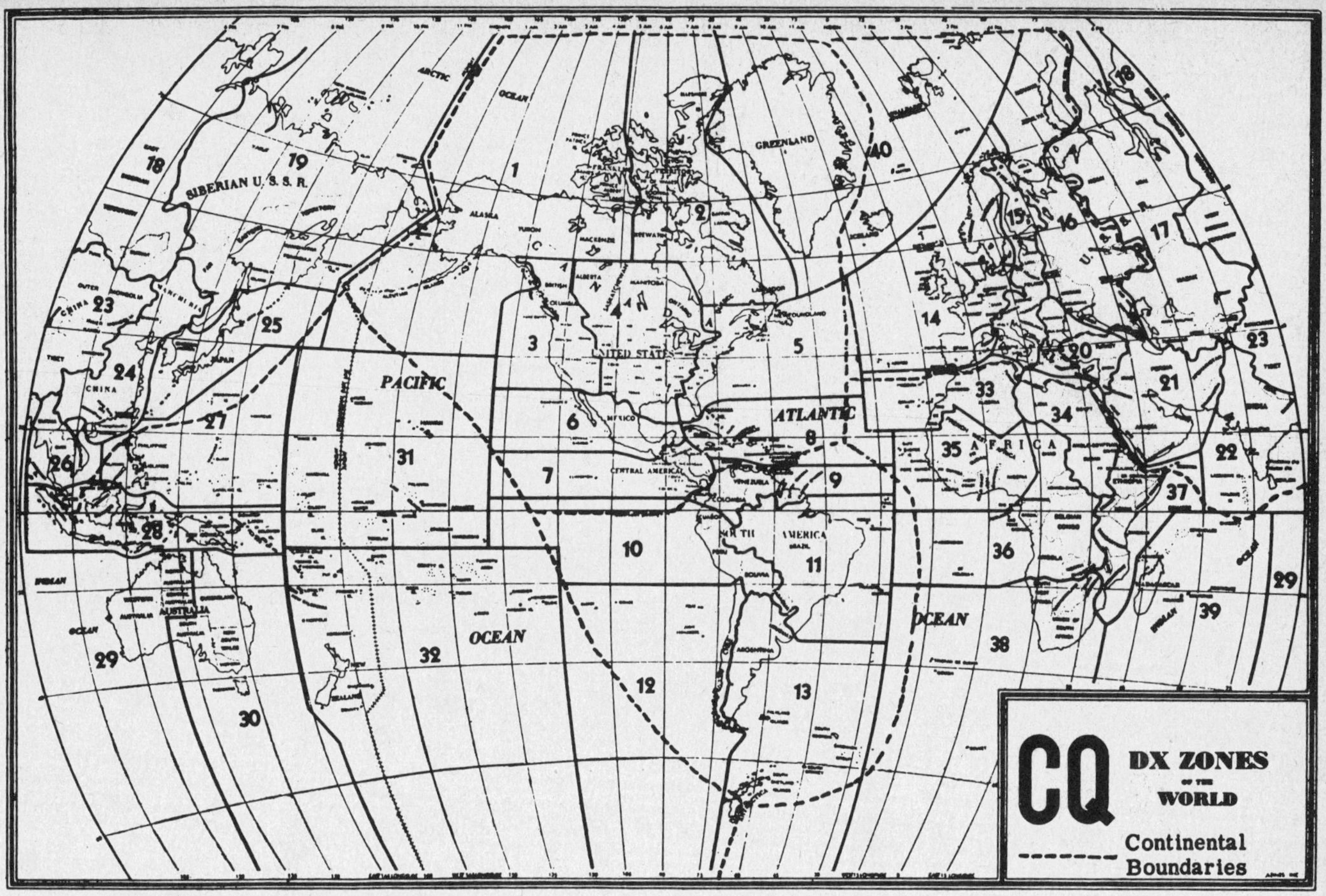

The Worked All Zones certificate is one of the most coveted awards in DX circles. For more information, see chapter six.

try 50 points, and with a new continent 250 points.

Non-ARRL Awards

Besides the WAZ and the USA County Award (explained elsewhere), *CQ* Magazine sponsors the Worked All Prefixes (WPX) award. Certificates and endorsements are offered for working various prefixes. You can earn various forms of this award — mixed mode for 400 prefixes, the cw and phone award for 300 prefix confirmations each. Rules and the official application are available from *CQ* Magazine.

YLRL — The YLRL (Young Ladies Radio League) issues a number of popular attractive certificates. Details can be obtained from Blanche L. Randles, K1IZS, 73 Shed Hill Rd., Stoddard, NH 03464. Please include an s.a.s.e.

Ten-Ten — Amateurs new to 10 meters are often (understandably) bewildered by continual requests for a "ten-ten number." Ten-Ten International, of Southern California, now is a worldwide organization with approximately 20,000 members, promoting activity on 10 meters. To become a member, you must work 10 Ten-Ten members and submit the log data to the appropriate call-area vice president. If everything checks out, you will receive your basic certificate and your own Ten-Ten membership serial number, which you can exchange with other amateurs. By exchanging numbers, you will qualify for an incredible number of 10-meter awards as well as help to bring newcomers into the fold. If wallpaper is your "thing," a Ten-Ten number is highly recommended. Contact Bill Risher, Jr., WB6OMH, 10542 Loch Avon Dr., Whittier, CA 90606 for the address of your call-area coordinator.

SMIRK — Similarly, the Six-Meter International Radio Club exists to promote activity on that band. Membership requires working six SMIRK members to qualify for your own SMIRK number. For more information, contact SMIRK at 7158 Stone Fence Dr., San Antonio, TX 78227.

Specialty Nets

Amateurs can qualify for awards certificates from contest QSOs, casual contacts or schedules. But there is another way — by participating in "specialty nets." These nets serve to increase your chance of getting QSLs, as (like DX nets) their primary purpose is awards.

For example, of all the major operating achievement awards available to the radio amateur, the WAS award is one of the most popular and least difficult to achieve. But surprising as it may seem, Alaska and Hawaii may not be the last states needed to complete WAS; some amateurs have difficulty nailing down the last half dozen or so states. If you're having trouble with WAS, you'll find the 3905 Century Club net a great help.

The 3905 net began in 1976 as the Bicentennial WAS Net, to aid those amateurs working towards the ARRL's Bicentennial WAS award, which was issued for working all states during calendar year 1976. The net was so successful, it has continued operation to provide a meeting place for all those seeking the WAS award.

As the name implies, the net meets daily at 0000 UTC on 3,905 kHz. Although not as formal as traffic nets, the 3905 net does follow a fixed operating procedure. Each check-in is assigned a serial number for that session, with check-ins taken at start and periodically throughout the session. After being called by the Net Control Station in the order of check-in, each station is allowed to make two calls to other stations in the net.

The best way to operate on the 3905 net is make a list of those stations checking in immediately after you, noting their states. Then when your turn comes, you will be reasonably certain of making a good contact. If you call stations that have checked in much earlier or much later than you, there's a good chance they won't be listening. The reason is simple. There are over 100 check-ins for each session, and it can take hours before a station gets his turn. Many operators simply guess when their turn will come and move off until then.

Bill, WA2DPU, achieved "Worked All States" on 6 meters.

The Five-Band WAS plaque awaits those who work all states on five of the amateur bands. Remember, when requesting application forms for awards, always include an s.a.s.e.

Although most users of the 3905 net are trying to pick up some hard-to-get states to complete their mixed-band WAS, there are some amateurs working toward an all-75-meter ssb WAS worked exclusively on 3905. The net also simplifies working the 75/80 meter portion of Five-Band WAS. The biggest benefit of the net, however, is that stations active on 3905 are reliable QSLers. More details can be obtained from the NCS when you first check in.

The Geritol Net

Another WAS-oriented net is the Geritol Net, which meets every Friday and Saturday evening at around 0100 UTC on 3,787 kHz. Because of the frequency, it is open only to Amateur Extra Class operators, but with the growing number of Extras, it isn't quite as exclusive as it may seem.

The Geritol Net provides a convenient hunting ground for amateurs working on the WAS endorsed "all 1 × 2 call signs." Actually, many regulars are prefix hunters as well and may be attempting to work WAS for each possible combination of prefixes — Ws, Ks, Ns, AAs, etc. Of course, with so many U.S. prefixes now in existence, it's possible to earn the WPX award issued by *CQ* Magazine with just Geritol Net contacts.

Geritol Net operation is much less organized than the 3905 net. Check-ins are taken at any time, with the newcomer simply signing in at the appropriate time. The NCS keeps a list of check-ins, but more often than not simply asks for the last letter of the call sign of any station needing a contact. When a station is recognized by the NCS, the operator is permitted two calls.

An exception to this general procedure occurs when a station asks to be checked out. In most cases, the NCS will ask for the last letter of the call signs of any stations on frequency needing a contact with the check-out. Again, NCS recognizes callers one at a time until every station so desiring has made a contact.

The basic Geritol Net award is ARRL's WAS. When applying for the WAS, make sure each card clearly shows the operating frequency to four digits — usually 3.787, and the application form (CD-217) indicates the desired endorsement. Once the WAS is issued, you'll be eligible to work towards the Geritol Net's own awards program. Information about the program can be obtained from the awards custodian.

County Hunting

If trying to work all states is a little too tame for you, try county hunting. The idea here is to work all 3075 counties of the U.S., including the four in Alaska and five in Hawaii. What makes this so difficult is that there are many counties no amateur is active in — or if some are active, they aren't active on the bands or modes you operate on. If you're going to work them all, you'll have to work mobile stations passing through these "barren" counties. The best way to do this is to join the 20 Meter Independent County Hunter Net (ICHN), which meets daily on 14.336 MHz. The actual start-up and close-down times vary from day to day, but generally run from around 1300 to 2200 UTC, depending on the season of the year and the phase of the sunspot cycle. (County hunters also have nets in operation on other bands.)

The word 'independent' in the net's title is no accident — this is without doubt one of the most independent operations you're likely to run into. Although most members follow a generally accepted procedure, there is no formal set of operating rules. Each session is run in the manner the NCS wants, and each has his or her own ideas. Still, the net functions well and everyone has a good time. Although the net exists to help amateurs work all counties, it also provides a nationwide group of operators ready and able to render assistance in cases of emergency.

The net opens each day when one of the county hunters on frequency volunteers to act as the NCS. One or two other stations generally volunteer to act as net relay stations, providing coast-to-coast coverage. A call-up of mobiles follows establishment of the NCS and relays. On some days, mobile operators are few and far between and are run as they check in. On other days, there may be several checking in at the same time. In this case, the NCS may run them in the order they check in, or may juggle the order to accommodate special needs. Although most mobile stations are run on the net frequency, it may be necessary to run a second mobile simultaneously off frequency. This, of course, requires a second voluntary NCS.

A run begins when the NCS and relays announce the call sign, county and state of the mobile, and give the mobile operator a check-in time in UTC. After this, the mobile station simply calls QRZ and tries to sort out the calls — much as a DX station would. Some mobile operators prefer to work on the fly using a tape recorder to log the contacts. Others pull off the road and take lists. In either case, only the signal report need be exchanged directly. If propagation is too poor between the stations involved for an exchange of call signs and other information, the NCS or other designated station may relay that data, but not the signal report; that must be an exchange between the stations made without any assistance.

After every station that can make a contact directly has made it, the NCS and relay stations will look for weak stations unable to get through on their own and set up the contact. After every one has been satisfied, or some arbitrarily established time period has elapsed, the mobile station is given a signout time in UTC and the run is completed.

The goal of every county hunter is, of course, to work them all and earn the Cliff Corne achievement award, named after the first radio amateur to accomplish the feat. The award is currently administered by *CQ* Magazine, whose rules and decisions are the final authority. The basic award is issued to any amateur establishing two-way contact with other amateurs in 500 different counties in any combination of states. Seals are issued in 500-county increments, but the 1000 level must represent at least 25 states, the 1500 level 45 states and the 2000 level all 50 states. Each level can be endorsed for single band, single mode, all fixed, or all mobile. Although the initial application must be made on an official record book, subsequent applications can be on plain paper. For more information, contact *CQ* Magazine, 76 North Broadway, Hicksville, NY 11801.

One problem to all awards hunters is getting the required QSL cards back from the stations worked. As with DX stations, mobile stations can have contacts with hundreds of county hunters each day of operation. The cost of QSLing, both in terms of postage and the time needed to fill out cards, is prohibitive. If you want contacts confirmed, it's up to you to send the mobile operator a filled-in reply QSL. These can be home brewed or purchased from several commercial sources. Either way, the reply card must list the date and time each contact was made and clearly show that the card is confirming your contact with the mobile station. If band or mode endorsements are desired, that data must also be included.

Preparing a reply QSL is only part of the job. You'll still have to mail it (along with your QSL, if it is a first contact) to the mobile operator along with return postage. Although many county hunters follow this procedure, most prefer to make use of the QSL bureaus operated by W6CCM and others. These bureaus function in much the same manner as the ARRL's DX bureau. The county hunter sends a batch of reply cards, sorted by call sign call area, along with a 10-cent per-card service fee, to the bureau. There the cards are re-sorted and sent along with those of other county hunters to the individual mobile operators, who then check the cards against their logs. If the contacts are legitimate, they sign them off. The cards are then returned in bureau provided postpaid return envelopes. As the cards arrive back at the bureau, they are again re-sorted for delivery back to the county hunter. The turn-around time varies from a few weeks to a few months, depending on how fast the mobile operator gets through signing them off.

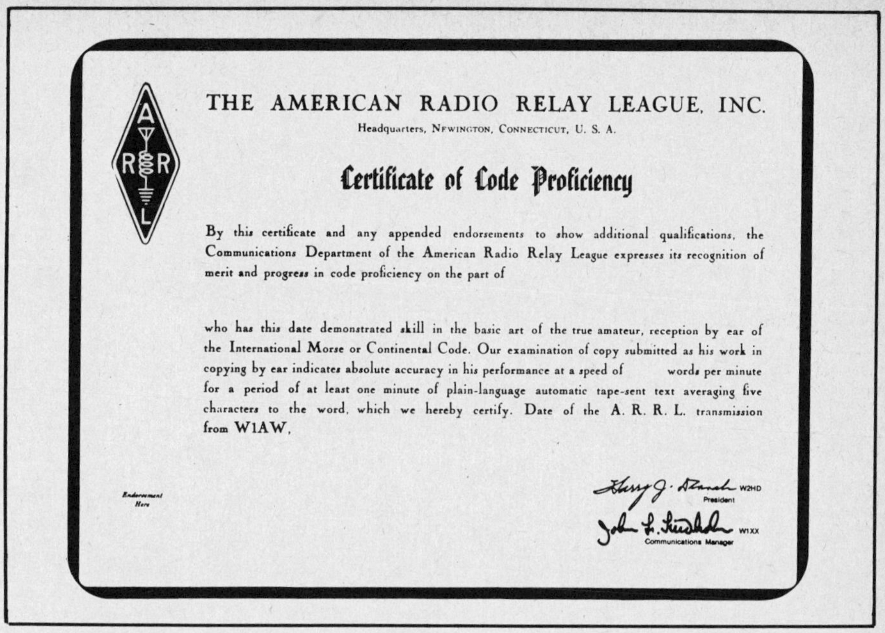
THE AMERICAN RADIO RELAY LEAGUE, INC.
Headquarters, Newington, Connecticut, U. S. A.

Certificate of Code Proficiency

By this certificate and any appended endorsements to show additional qualifications, the Communications Department of the American Radio Relay League expresses its recognition of merit and progress in code proficiency on the part of

who has this date demonstrated skill in the basic art of the true amateur, reception by ear of the International Morse or Continental Code. Our examination of copy submitted as his work in copying by ear indicates absolute accuracy in his performance at a speed of words per minute for a period of at least one minute of plain-language automatic tape-sent text averaging five characters to the word, which we hereby certify. Date of the A. R. R. L. transmission from W1AW.

Endorsement Here

W2HD President

W1XX Communications Manager

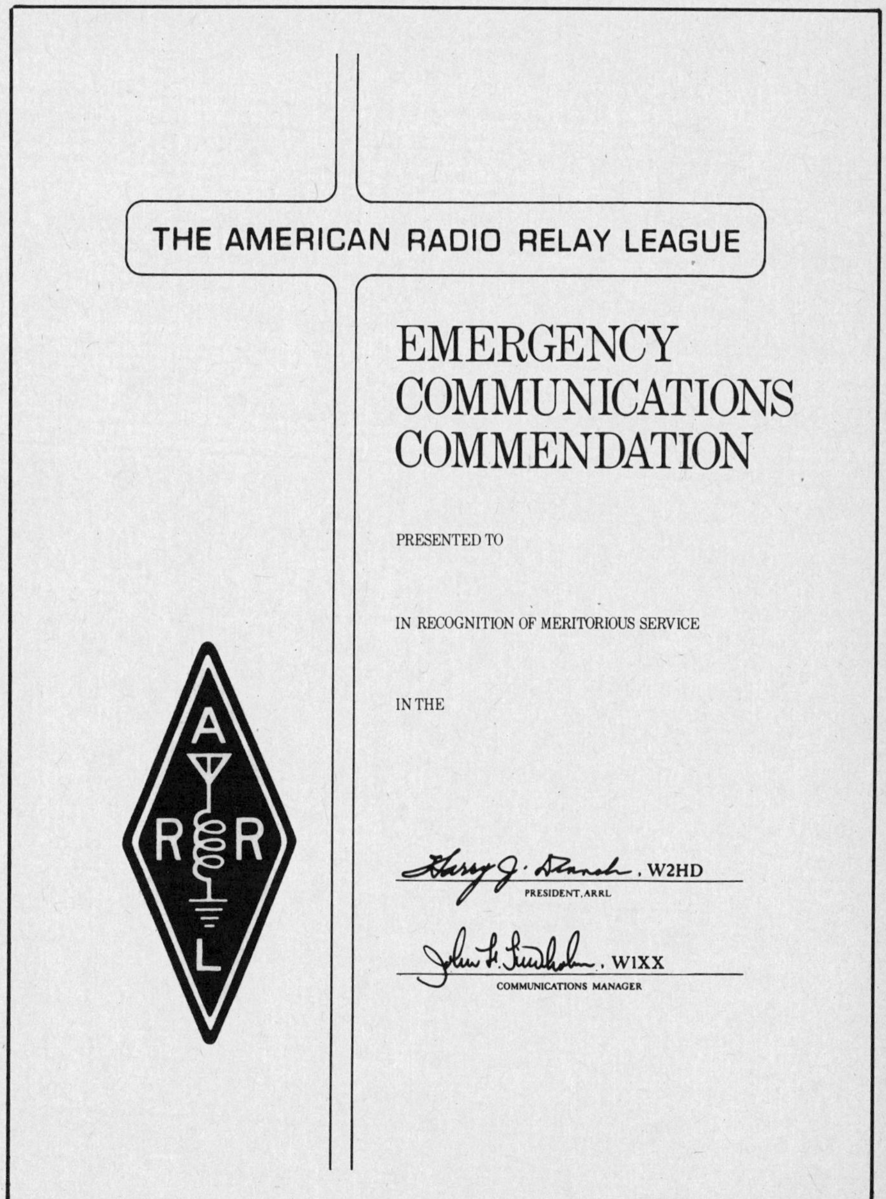
THE AMERICAN RADIO RELAY LEAGUE

EMERGENCY COMMUNICATIONS COMMENDATION

PRESENTED TO

IN RECOGNITION OF MERITORIOUS SERVICE

IN THE

, W2HD
PRESIDENT, ARRL

, W1XX
COMMUNICATIONS MANAGER

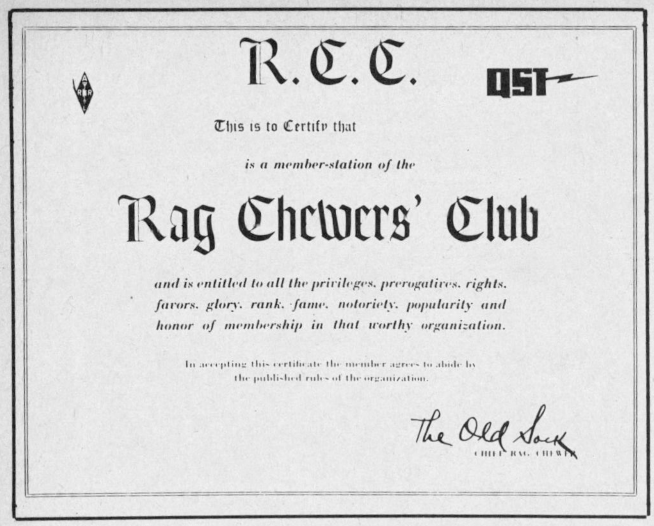

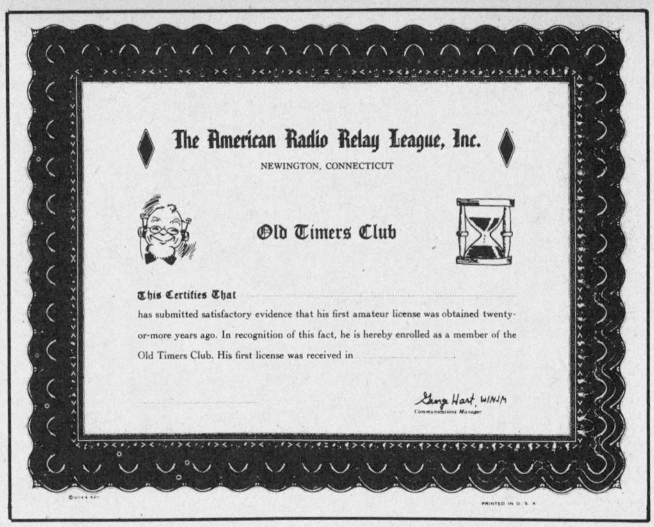

Remember, some of the more active mobilers may run over 100 counties during a single trip.

Although not officially connected with the *CQ* award program, an excellent source of information about county hunting is the Mobile Amateur Radio Awards Club (MARAC). The monthly *MARAC Newsletter* contains the itineraries of planned mobile trips and expeditions, lists of which county hunter needs what to finish a particular state, and the views of its members, most of whom are county hunters. Information about MARAC can be obtained by sending a no. 10, self-addressed envelope, with first-class postage for four ounces affixed, to WØDG, 10310 West 170th Terrace, RR 5, Olathe, KS 66061.

The YL System

As independent in spirit as the county hunters, the YL International Single Sideband System (ISSB) is dedicated to orderly operation. The System meets daily at 14.332 MHz with call-up beginning at 1600 UTC. Operation continues until the band folds. In addition to the daily 20-meter system, there are also 75-, 40-, 15- and 10-meter systems. The 75-meter system meets at 0100 UTC on 3,962 kHz every Thursday. The 40-meter system meets at 0000 UTC on 7,273 kHz every day except Thursday. The 15-meter system meets at 1500 UTC on 21.373 MHz every Thursday, while the 10-meter system meets daily at 1600 UTC on 28.673 MHz. When interest warrants, cw systems are also operated on 3.565, 7.020, 14.080, and 21.080 MHz. An information session providing QSL data and answers to members questions is operated every Monday at 1500 UTC on the 20-meter system.

The word "system" is used rather than net because the operation is systematic rather than a collection of stations operating on the same frequency. The call-up procedure is a good example of this systematic approach. Stations are checked-in each day when the net opens by call area in a 104 736 925 8 sequence. The starting point in the rotation is determined by the last digit of the day's date. For example, call-up begins on the 19th of the month with the 9th district and proceeds with the 2nd, 5th, 8th, and so on. DX is checked-in between each two or three areas, alternating between Europe, Africa and Asia, and the Americas, Pacific and Caribbean. Additional check-ins are taken from each area when operation to that area begins.

The systems are intended to provide emergency communciations, but when no emergency exists it is an ideal place to make a quick contact with an elusive state or DX country. In addition to helping those working on WAS, WAC and DXCC, the ISSB also has a very attractive awards program of its own. Although you don't have to be a member to make occasional use of the system, it is a good idea to join if you plan to make daily use of it. The reason *isn't* that you must pay to use the system, but rather because so many of the regulars are membership-number collectors and will give you a call to make a contact.

Because there are so many members on frequency at any given time, each member is limited to one opportunity to make two calls during each pass through a call area. Stations are called by the NCS in the order of check-in. When called, the station makes the allowed two calls and then returns the system to the NCS. The calls themselves must be brief — generally limited to an exchange of call signs, signal reports, locations and ISSB number. If more information must be passed, the stations should go off frequency to complete the contact. For more information about the Sideband System, its operation and membership, send a stamped, self-addressed envelope to WØUUE, 442 Clover Leaf Dr., Golden Valley, MN 55422.

For Further Information . . .

ARRL's *Operating an Amateur Radio Station* booklet has more information on the League's awards. Many other DX awards of great interest are admirably explained in G5GH's book, *Amateur Radio Awards*, published by the Radio Society of Great Britain. The book is available in the U.S. and Canada through the Ham Radio Bookstore.

Chapter 9

FM and Repeaters

Whether you're walking, mobiling or sitting in the shack, ham radio can be operated everywhere and it's fm repeaters that make it all possible. Fm repeater operating is the most popular mode of communications in Amateur Radio today During the last decade, more gear has been built and purchased for repeater operation than for any other single mode and, perhaps, more than all of the other modes combined. There seems to be no end to this growth.

Today, nearly every active ham, be he a vhfer or a low-band operator, has a transceiver for repeater operation in either his shack or vehicle. Many of the folks who thought of themselves as strictly low-band operators, have found that they can extend their low-band interests up to the vhf frequencies, whether their low-band interest be traffic, RTTY, emergency communications, DXing, contesting or good ole rag-chewing. And there they join the vhfers, who knew where the action was all of the time. So, there is quite a cross-section of hams on our repeaters today.

How We Got Here

Although Amateur Radio vhf repeaters go as far back as the early thirties (on 5 meters), repeaters, as we know them today, began appearing as the result of an FCC directive to the commercial two-way radio service. The FCC required the commercial service to switch from wide-band to narrow-band fm, and in the early sixties, there was a flood of wide-band equipment into the surplus market. Hams began snapping up this surplus equipment and putting it to good use on the ham bands. The modular construction of this equipment lent itself to easy transformation into repeaters. And, *voila,* the first modern-day Amateur Radio repeaters were on the air.

These early repeaters often were off the air more than they were on. Experimentation was the rule of the day, but as repeater circuitry became perfected, more and more repeaters were on the air more and more often. That's when the growth of their popularity really began. As popularity increased, the Amateur Radio equipment manufacturers began designing and selling repeater-oriented gear. This added fuel to the fire; you no longer had to try to convert boat anchors to work in the Amateur Service. Repeater gear could be purchased right off the shelf. The gradual growth curve became an explosion, and the shock waves haven't let up yet!

The Basics: How Repeaters Work

The basic repeater system is not a very complicated collection of equipment. However, there are a lot of misconceptions about how such systems work. Every repeater user should be familiar with the basics in order to promote smoother on-the-air techniques. Users who are unfamiliar with these basics may unintentionally misuse the system and cause interference with others. Furthermore, they may broadcast their misconceptions over the air and lead others astray. An overview of basic repeater operations should help correct the misconceptions.

The basic repeater system consists of five parts (see Fig. 1): receiver, transmitter, carrier-operated relay circuitry (COR) and two antennas, one for transmission and the other for reception. By means of its antenna, the receiver monitors signals on the "input" frequency. When a signal is received, its carrier is detected by the COR, which enables its relay to turn on the transmitter. The transmitter repeats the received signal on the "output" frequency, and this signal is radiated over the air via the transmitter's antenna.

How does your fm transceiver fit into this scheme? Your transceiver set-up has four basic parts (see Fig. 2); receiver, transmitter, transmit/receive switching circuit and antenna. To communicate through the repeater described above, first your transmitter must be tuned to the repeater's receive frequency (the input frequency) and your receiver must be tuned to the repeater's transmit frequency (the

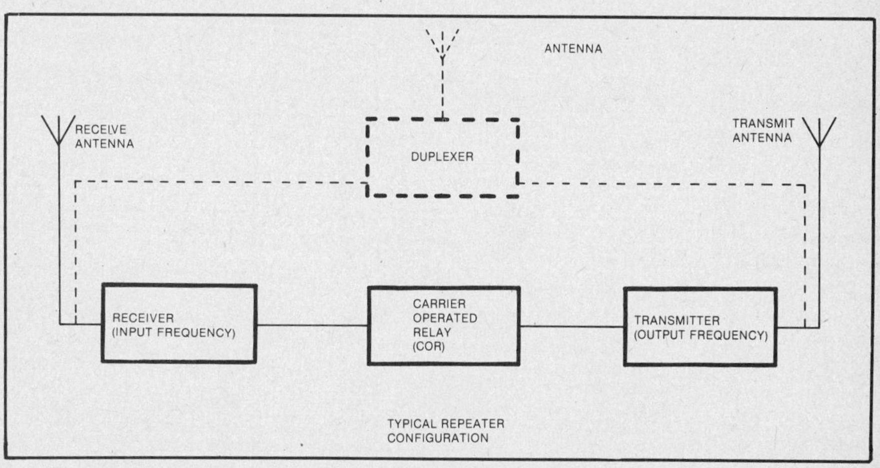

Fig. 1 — A typical basic repeater system with optional duplexer to eliminate dual antennas and desense.

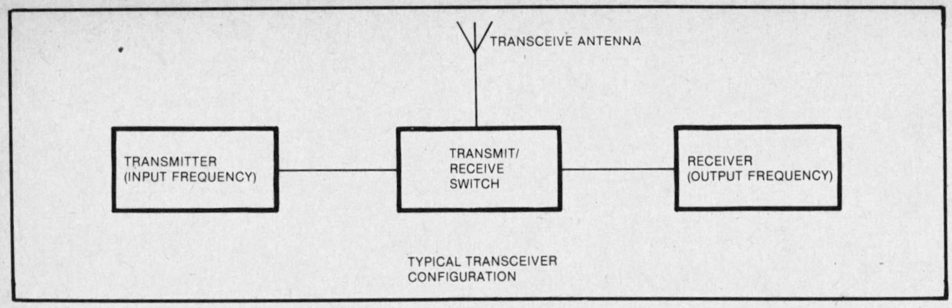

Fig. 2 — The primary parts of a typical fm transceiver for repeater operation.

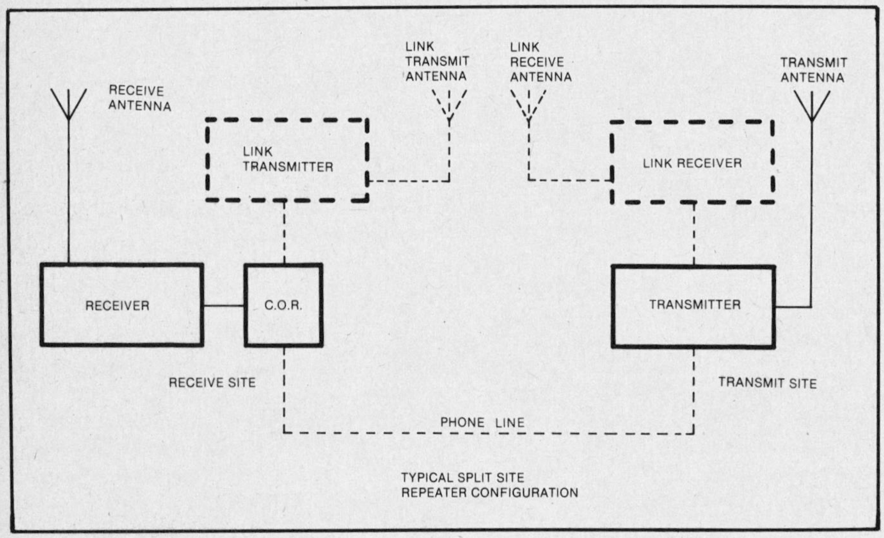

Fig. 3 — A split-site repeater system, which may be linked by means of a telephone line or a link radio system.

output frequency). In nearly all fm transceivers, the transmitter and receiver frequency tuning is accomplished simultaneously; that is, one control tunes both your transmitter and receiver to the repeater's "pair" of frequencies.

When you depress the press-to-talk switch (PTT) on your microphone, the transmit/receive switching circuit transfers your antenna from your receiver to your transmitter and a signal is transmitted on the input frequency. This signal is heard by the repeater's receiver and detected by the COR, which turns on the repeater's transmitter and repeats the signal on the output frequency.

When you release the PTT switch, the transmit/receive switching circuit transfers your antenna back to your receiver and monitors signals on the repeater's output frequency. Usually, after you release the PTT switch after transmitting, the repeater's transmitter does not immediately cease transmission. Rather, the repeater continues to transmit an unmodulated carrier for a very short time after you ceased transmitting. This short carrier and noise burst (the "squelch tail") indicates to repeater users that they are successfully accessing the repeater. Whenever you stop making a transmission and hear the carrier and squelch tail, you can be sure you are accessing the repeater.

Now that you know how a repeater basically works, a little more knowledge about repeater systems is handy. Whenever a transmitter and receiver operate simultaneously on nearby frequencies, as is the case in all repeater systems, there is a problem that must be overcome. Some of the transmitted signal leaks back to the receiver and desensitizes it. This "desense" decreases the receiver's sensitivity, with the result that weaker stations that normally could be copied by the receiver are not heard. There are several ways to alleviate this problem. Split sites are one solution (see Fig. 3). The repeater's receiver and transmitter occupy separate locations. Signals heard by the repeater's receiver are transferred to the repeater's transmitter by means of a telephone line or a transmitter/receiver link on a different vhf or uhf band. (Being on different amateur bands, the frequency separation between the repeater and link frequencies is such that the link itself does not cause desense.)

The second solution to desense is by the use of a "duplexer" (see Fig. 1), a device that provides isolation between the transmitter and receiver and also permits the use of the same antenna for both receiving and transmitting. Duplexers are expensive and difficult to build, but are practical when compared with the expense of maintaining split sites, links and separate antennas. Most repeater builders have eliminated some equipment and the desense problem by taking the duplexer route.

With this background information about repeater basics, you are ready to get a start in the world of fm and repeaters. Have fun!

License and Equipment

The Technician class is the minimal license needed to operate on the vhf and uhf repeaters. Technicians are not permitted to use 10-meter repeaters, however, nor may they use a vhf or uhf repeater while such a repeater is linked to permit transmission on an hf band. The exception to the rule is that Technicians are permitted to operate through OSCAR; that is, Technicians can transmit on the satellite's input on 2 meters and be repeated by the satellite on 10 meters. Of course, none of these restrictions apply to hams holding a General or higher class of license.

With license in hand, it is time to obtain some equipment. (When using mobile or portable equipment away from the shack, remember that the FCC requires you to carry your original license or a photocopy.) The array of gear available for repeater operation is vast. There is more repeater-oriented gear to choose from than all the other types of amateur gear combined. If you don't know what you are looking for, you may be swamped by all of the available choices.

Transceivers

Your favorite ham radio dealer will typically have three types of transceiver gear available (separate transmitters and receivers are not available): mobile, portable and base.

Since mobile transceivers are designed to operate from motor vehicles, they usually require a dc-power source. However, this does not prevent the use of mobile gear from inside the shack; many hams use their mobile rigs in both the car and the home. Adequate power supplies may be built or purchased inexpensively.

Also because they are designed to be used in the car, mobile 2-meter transceivers have only a few operating controls. The basic mobile rig usually has controls for power, volume, squelch and frequency tuning. These controls need no explanation. There is usually a simplex/repeater switch that causes confusion, however. This switch allows you to choose between operating in the repeater or simplex mode. In the repeater position, there may also be a choice of frequency shift, either up or down. This permits you to conform to the local band plan. For example, on 2 meters, above 147 MHz, you must transmit 600 kHz higher than your receive frequency, while below 147 MHz you must transmit 600 kHz lower. The frequency-shift control permits you to shift in the correct direction (many rigs provide the correct shift automatically).

Mobile transceivers usually include a hand microphone and an automobile-mounting bracket. Average power output is 10 to 15 watts.

Frequency is determined in three ways in mobile transceivers: by means of crystals, diodes or synthesizers. Crystallized rigs require either one or two crystals for each repeater channel you may wish to use. The average crystallized rig has space for 12 channels. Why one or two crystals? Some units require separate crystals for the input and the output frequency, while other units require only one crystal (usually for the output frequency), with the unit itself providing the necessary shift for the other frequency.

The crystallized transceiver typically costs less than a synthesized unit. If you plan to purchase crystals for only two or three repeaters, the crystallized unit, economically, is the way to go. If you were to purchase crystals for each channel in a 12-channel rig, however, you would have spent enough money between the crystals and transceiver to purchase a synthesized unit that would provide you with substantially greater frequency coverage.

The diode-matrix transceiver is priced about the same as a crystallized unit. However, nothing more needs to be purchased to fill all the available channels in the rig. These transceivers contain a handful of diodes that provide the desired frequencies. If you wish to change one of the diode-programmed frequencies, all you have to do is reconfigure the diodes. Modifications of the most popular diode-matrix units have permitted them to cover all of the frequencies on the band.

The synthesized transceiver provides the most versatility. This type of rig is ready to use right out of the box; nothing has to be purchased or installed. Simply dial up the frequency of your choice and you are on the air. The price of a synthesized radio is higher than the others, but if you want to use a lot of repeaters on different frequencies, then the synthesized transceiver is what you need. Most synthesized units have some kind of digital frequency readout which facilitates operation. Usually, dialing in the repeater's output frequency is all that is necessary to get on that repeater.

Portable transceivers are very popular items. Today, it is impossible to go to a hamfest and not see a variety of these units carried by the hams in attendance. Portables range in size from over-the-shoulder units to tiny hand-helds that are no larger than two packs of cigarettes.

Most portables are powered by nickel-cadmium batteries (NiCads). They may be recharged hundreds of times and provide years of use. To achieve extended life, however, these batteries must be recharged properly. Manufacturer recommendations should be followed to the letter. One general recommendation: Don't overdo it. If you constantly recharge NiCads after short periods of use, the batteries develop a memory and their recharge-to-recharge life decreases noticeably. The more you charge NiCads, the more you will have to charge NiCads!

Portables are available in crystallized and synthesized versions. Generally, the crystallized units have space for five channels. Average output power is in the 1.5- to 2-watt range. The operating controls on the portables are similar, if not the same, as those on the mobile transceivers. Portability means that these transceivers have their own antennas. Most allow you to choose between a metal quarter-wavelength telescoping antenna or the ever-popular foreshortened "rubber duck" antenna.

Besides being popular with those attending ham gatherings, portables are also used extensively by amateurs involved in emergency communications. What better way is there to relay emergency information from the site of a disaster than with a compact, self-contained transceiver? Many lives have been saved using portables in emergency situations. Their versatility is unmatched.

Few repeater-oriented base transceivers are available. The majority of base transceivers include the ssb and cw modes, as well as the fm repeater mode. They are usually frequency synthesized and have a power output of 10 to 15 watts. The controls are often as numerous as those you will find on the average low-band transceiver. Some base rigs include a discriminator and an S meter. Many may be powered by both ac and dc power sources.

Base units are the Cadillacs of repeater radios. They are the most expensive and also the most versatile. If you are interested only in fm repeater operation, a base rig would tie up a lot of your money in ssb and cw capabilities that you may never use. You would probably be better off purchasing a mobile rig and dc power supply for the shack. If you desire to own the dream machine of repeater transceivers, however, a base unit is what you are looking for.

Antennas

Antennas for repeater operation range from stubby rubber ducks to 16-element rotatable beams. One thing that all fm repeater antennas have in common is vertical polarization. The reason is that omnidirectional coverage, the goal of most repeater systems, can be more easily and economically achieved using vertically polarized antennas. Mobiles using repeaters have found that vertical antennas are much easier to install and maintain than a horizontal antenna and so, vertical it is (although a few repeaters use horizontal polarization for one reason or another).

A complete 220-MHz and repeater system (sans antenna and feedline), WA1IXU/RPT was based on a Japanese 220-MHz transceiver that was literally cut in half, rebuilt and expanded into the units illustrated here. The rack on the right holds the transmitter and duplexer; the other rack contains a receiver and the repeater-control circuitry, plus a link receiver for 450 MHz.

There are usually only two choices of antennas to use with portable transceivers: a quarter-wavelength telescoping whip or a rubber duck. The quarter-wave will give you more gain, while the rubber duck is more convenient to use. (On the uhf frequencies, a quarter-wave antenna may be a rubber duck). With a rubber duck installed, you can carry your portable rig on your belt or in your pocket and monitor all that is going on as you walk around the flea market at the local hamfest. On the vhf frequencies, you would find it difficult and hazardous to walk around with a fully extended aluminum quarter-wave telescoping whip.

Quarter-wave and five-eighth-wave whip antennas dominate the mobile repeater installations. These antennas may be permanently mounted or temporarily mounted (using a magnetic base). The advantage of a temporary mounting is that when you park and leave the mobile, you can remove the antenna. Also, you won't have to drill holes in your automobile. Whether permanent or temporary, the best place to locate the mobile antenna is in the center of the highest part of the vehicle. In automobiles, that would be the center of the roof. Omnidirectionality is best achieved at this location.

The Great Frequency Hunt — Users

If you are planning to use fm repeaters for amateur communications, you will need to know what frequencies are being used by the repeaters in your area. A good source of this information is *The ARRL Repeater Directory*, which lists the frequencies (along with other informative data) for thousands of repeaters in the U.S. and Canada. You may also wish to contact your section communications manager (see page 8 of every issue of *QST*), who can tell you the frequencies of the active repeaters in your area.

If you own a crystal 2-meter transceiver and do a lot of traveling, you would find it very expensive to "crystal-up" for all of the repeaters you are liable to encounter in your journeys. Following is a list of the 10 most popular repeater frequency pairs in the U.S. and Canada. With these channels in your transceiver, the chances are very good that you will be able to find an accessible repeater wherever your travels take you.

The top-10 frequency pairs are:

146.34/146.94
146.16/146.76
146.25/146.85
146.13/146.73
146.28/146.88
146.22/146.82
146.07/146.67
146.37/146.97
146.31/146.91
146.40/147.00

The Great Frequency Hunt — Builders

Amateurs who are comtemplating putting a repeater on the air may wonder how to obtain the frequency pairs for their repeaters. *Frequency coordination* is the recommended method. Across the U.S. and Canada, volunteer individuals and groups have taken on the task of recommending frequencies for proposed repeaters within their "jurisdiction." These frequency coordinators keep extensive records on repeater operation in their area, as well as adjacent areas. With this information, they are usually able to recommend the best pair of frequencies for the proposed repeater's location.

Because of the amount of work involved in finding the best frequency pair, and the fact that this work is done on a volunteer basis, there is a time lag between the request for a frequency pair and receipt of the coordinated channel. So if you plan to request a pair of frequencies, do it early.

Consult the latest edition of *The ARRL Repeater Directory* for the listing of frequency coordinators, or contact your section communications manager (see page 8 of *QST*) or ARRL hq. to get the name and address of your area's frequency coordinator.

There is a wide selection of antennas available for base installation. Vertical antennas come in ground plane and collinear varieties. For directionality, 4- to 16-element beams are available. When using a beam in the repeater mode, remember to mount it vertically. Theoretically, there is a 20-dB signal loss when antennas are 180° out of phase; the gain obtained from a beam is lost via cross-polarization.

Power amplifiers are available for fm repeater operation at power levels ranging from 10 to 180 watts output. Amplifiers facilitate continual repeater communications on the fringes of the repeater's coverage area. If you are in an area where your transceiver is no longer consistently accessing the repeater, but you are still receiving the repeater well, turn on the amplifier and you are communicating again.

One word of caution: Use amplifiers only when necessary. It is possible to access more than one repeater on the same frequency when transmitting with high power. You may not be able to hear the second or third repeater that you are keying-up, but they are hearing you, so use the amplifier with discretion. The FCC requires that the minimum amount of power necessary be used to maintain communications, so operate legally and prevent unintentional interference.

If you'd like to homebrew equipment for repeater operation, a number of circuits have been published in *QST* and other ARRL publications, especially *FM and Repeaters for the Radio Amateur*. Vhf and uhf homebrewing is challenging and very rewarding. Start with something small and become acquainted with the specialized construction techniques involved, and then move on to bigger and better things.

Operating

Before you turn on your transceiver and say your first words on the local repeater, you should become familiar with some general operating procedures. Once you get a handle on these procedures and begin to turn on your radio and monitor the repeater you intend to use, you will discover how these general rules fit into the scheme of things on the repeater and you will also discover that each repeater has its own special operating procedures. If you listen long enough, these special rules will become apparent. If something is unclear, however, there is always somebody willing to explain; all you have to do is ask. Now, on to the general procedures . . . .

How do you make yourself heard on the repeater? Don't call CQ. Long CQs tie up the repeater and are unnecessary. Usually, all you have to do is announce your call sign; "WB1EYI monitoring" or "W9KDR listening" should suffice. If someone else is listening and wants to talk to you, he will initiate the conversation after hearing your announcement. If no one is on frequency to talk to, a long CQ will have no greater effect than a simple "WA1ZUY monitoring." Some repeaters have specific rules on how to make yourself heard, but, in general, the procedure outlined here should work.

If you are looking for a specific person on the repeater, simply call him once: "K1XA, this is N2YL," for example. If there is a QSO in progress, there are two ways of making that contact, depending upon the state of the ongoing QSO. If the conversation in progress sounds as if it is going to end shortly, wait until it is over before making your call. If the conversation sounds like it is going to continue *ad infinitum*, however, drop in your call sign between transmissions. When you are recognized, ask if you may make a quick call. Usually, the other stations will acquiesce. If the station you are calling answers, ask him to stand by until the present conversation is over or, if possible, ask him to move to another repeater or a simplex frequency. And, return the repeater back to the original conversationalists.

If you wish to join a conversation in progress, simply drop in your call sign between the transmissions of the other parties. Don't use the word "break" unless that is the normal procedure on that repeater. "Break" has serious connotations in some regions. Usually its use is restricted to emergency situations, when someone has to use the repeater immediately.

If you are conversing on a repeater and another ham drops in his call sign, the next station to pick up the conversation should recognize the "breaker" and let him make a call or join the conversation. It is impolite to recognize the station and not let him speak. He may need to use the repeater immediately, so let him make a transmission. It may be an emergency!

Although the repeater is there to use, it is advisable to move to a simplex frequency whenever possible. The repeater is not a soapbox; you may like to listen to yourself, but others who need to use the repeater that you are tieing up may not be as appreciative. Most two-way, base-to-base local conversations can be carried on a simplex frequency, so try simplex and you will learn to like it. QSOing on a simplex frequency offers an advantage over repeater QSOing. You can usually carry on an extensive conversation on a simplex channel without interruption. Accomplishing the same on a repeater is difficult.

Keep transmissions short. There are timers on repeaters that vary in length from 30 seconds to 3 minutes. If you transmit beyond the timer's limit, the repeater will "time-out" and shut-down for some designated period of time. The purpose of repeater timers is twofold. It prevents extraneous signals from keeping the repeater on continuously, which could damage the repeater's transmitter. The other purpose of the timer is to allow everyone a chance to speak. If the length of each transmission is limited, more users will be able to get in their two-cents' worth. Long winds are for sailing; keep it short on the repeater. Learn the length of the timer on your favorite repeater and stay within its limits.

Leave spaces between transmissions. Don't hit your microphone as soon as someone else releases his. Another station may want to use the repeater and if your exchanges are too quick, you can prevent that party from breaking in. There may be a life and death situation involved, so leave that space.

This problem has become so prevalent that many repeaters have installed "courtesy beepers" to control the situation. As soon as one party releases the microphone, there is a short pause and then the beeper sounds, indicating that the next person can transmit. If the next station is too quick and begins transmitting before the beeper sounds, the repeater will indicate the violation in some way, sometimes by shutting down the repeater. The short pause before the beeper allows other stations to break into the repeater and make themselves heard.

Identify legally, but don't overdo it. You must transmit your call sign at the beginning and end of each contact, *not* each transmission. Also, you must identify every 10 minutes in between the beginning and end of each contact. At the end of a contact, you must also identify the station you were in contact with. If you are checking out of a roundtable QSO, identify only one of the stations in the group; identifying them all wastes time and is not necessary.

It is illegal to key up a repeater without identifying. If you want to check on whether or not you are able to access a particular repeater, but do not necessarily wish to engage in conversation, simply say "K1WJ testing." You've accomplished what you wanted to do, legally.

If you are operating from a base station, it is advisable to stay off the repeaters during the morning and evening rush hours. During peak hours, a lot of mobiles may wish to use the repeater. They should even have preferred use over base stations at these times. Some repeater groups have specific rules to this effect. Simply because you are a base station, however, don't abandon the repeater completely. Monitor the mobile activity. A base station's assistance may be needed for an emergency.

There is an unwritten practice in repeater operation that says, if you are a regular user of a particular repeater, you should help to support the effort to keep that repeater on the air. The easiest way of accomplishing this is by joining the club or group that sponsors the repeater and pay the membership dues. Repeaters are expensive to build, operate and maintain. Unless the repeater has a wealthy benefactor, there are electric, phone, maintenance and site-rental bills to pay, and the club dues are used to foot the bill.

Besides monetary support, physical support is also necessary. When the new antenna is being erected or the repeater shack is getting a new coat of paint, join in and lend a hand. Your assistance will be appreciated and it will give you a chance to become familiar with the repeater itself

This 180-foot tower is chock full of amateur repeater antennas. Six and 2 meters, as well as 220 and 450 MHz, aerials sprout at WA1NQP/RPT.

and the people involved in keeping it running.

Open Vs. Closed Repeaters

The majority of repeaters you will encounter are of the "open" variety; that is, anybody with a Technician (or higher) class license is permitted to use the repeater. A minority of repeaters are "closed" or "private." Their use is restricted to exclusive groups and unless you are an invited guest, you should avoid such repeaters. Some of the "closed" repeaters even limit accessibility by requiring special codes or subaudible tones in order to gain their access. And so it goes.

Autopatch

One of the most popular specialized repeater functions is the *autopatch*. This device interfaces the repeater with the telephone system and permits the repeater user to extend his communication capabilities via the telephone system. In most repeater autopatch systems, the user must be able to generate the standard telephone company tones to access and dial through the system. This is accomplished by interfacing a telephone tone pad with the user's transceiver. These tone pads run the gamut from surplus telephone company equipment to miniature state-of-the-art devices. Whatever is used, once you have such a device installed and ready to operate, certain operating procedures should be followed.

First, you must ask yourself whether or not the call is necessary. If you are making the call with regard to an emergency situation, there is no problem. Anything less significant than an emergency, and you're in a sticky area. Some repeater groups expressly forbid use of the autopatch except for emergency situations. The FCC (along with the phone company) frowns on the indiscretionary use of autopatches. The autopatch in Amateur Radio is a privilege granted by the FCC and if this privilege is abused, it may be taken away.

If a phone call is important, there are often telephones available along our highways and byways, so pull off of the road and use the pay phone to avoid jeopardizing the autopatch privilege. If the phone call is unimportant, why jeopardize our privilege by making that trivial phone call via Amateur Radio in the first place? How many times have you heard "hubby" call "wifey" via the autopatch to let her know that he is on the way home? Why didn't he make that call before he left the office? And, just as it is illegal to conduct business on the ham radio airwaves, it is illegal to conduct business over the autopatch. Don't use the autopatch to call the local radio store to find out if they have a framistan in stock, and don't call Sally's to order a pizza. Think twice before you access that autopatch!

Accessing the autopatch is usually accomplished by pressing a designated key or combination of keys on the phone tone pad. Sometimes you must hold down the designated key for a certain length of time (the # sign and the * sign are the most common access keys). When a dial tone comes back, dial the number you wish to call.

Once you have established contact, remember not to speak at the same time as the party on the other end of the line. The autopatch system is a "half-duplex" operation, and the conversation may go in only one direction at a time. Also, remember that some autopatches have timers, so if your conversation isn't completed in a specified time, your call may be cut off.

Autopatch traffic must be logged because it is third-party traffic. Many repeaters have a tape-recording system operating in conjunction with the autopatch to log the calls. Such a system may require the ham making the call to announce his call sign, the date and the time of the call at the end of the completed autopatch, so that this information can be recorded in the log. Sometimes this information must be announced before deaccessing the autopatch, sometimes after deaccessing.

After your phone conversation is completed, you must shut off the autopatch. This procedure is similar to accessing the autopatch — a key or combination of keys must be pressed and the repeater is returned to the normal repeater mode. Autopatch users should consult the repeater group sponsoring the autopatch for specific information about the access

Repeat/Speak

Definition of the words used every day on fm repeaters:

Autopatch — a device that interfaces the repeater system with the telephone system to extend ham communications over the telephone communications network.

Breaker — a ham who interjects his call sign during a QSO in an attempt to get a chance to communicate over the repeater.

Coverage — the geographical area in which the repeater may be used for communications.

Channel — the pair of frequencies (input and output) a repeater operates on.

Closed repeater — a repeater whose use is limited to certain individuals.

Control operator — an individual ham designated to "control" the repeater, as required by FCC regulations.

COR — carrier-operated relay, a device that, upon sensing a received signal, turns on the repeater's transmitter to repeat the received signal.

Courtesy beeper — a tone sounded shortly after each repeater transmission to permit other stations to gain access to the repeater before the tone sounds.

Crossband — communications to another amateur frequency band by means of a link interfaced with the repeater.

Desense — degradation of receiver sensitivity caused by emissions from a nearby transmitter.

Diode matrix — a scheme of frequency generation in transceivers that uses plug-in diodes to program the desired frequency.

Duplexer — a device that permits the use of one antenna for both transmitting and receiving without degradation to either the incoming or outgoing signal.

Frequency synthesis — a scheme of frequency generation in transceivers using digital techniques.

Full quieting — a received signal that contains absolutely no noise.

Hand-held — a portable fm transceiver that is small enough to use and carry in one hand.

Input — the frequency the repeater receiver is tuned to; the frequency that a repeater user transmits on.

Intermod — interference caused by spurious signals generated by intermodulation distortion in a receiver front end or transmitter power-amplifier stage.

Key-up — turning on a repeater by transmitting on its input frequency.

Machine — the complete repeater system.

Mag-mount — a mobile antenna with a magnetic base that permits quick installation and removal from the motor vehicle.

NiCads — nickel-cadmium batteries that may be recharged hundreds of times, used to power portable fm transceivers.

Omnidirectional — an antenna system that radiates equally in all directions.

Output — the frequency the repeater transmits on; the frequency that a repeater user receives on.

Picket-fencing — rapid flutter on a mobile signal as it travels past an obstruction.

Polarization — the plane an antenna system operates in; most repeaters are vertically polarized.

Reverse autopatch — a device that interfaces the repeater with the telephone system and permits users of the phone system to call the repeater and converse with on-the-air repeater users.

Reverse split — a split-channel repeater operating in the opposite direction of the standard 30-kHz repeaters to avoid cochannel interference.

RPT, R — abbreviations used after repeater call signs to indicate that the call sign is being used for repeater operation.

Simplex — communications on one frequency, not via a repeater.

Split frequency — 2-meter repeater channel 15 kHz above or below the standard 30 kHz-spaced channel.

Split sites — the use of two locations for repeater operation; usually, the receiver is at one site and the transmitter at another, and the two are linked by phone or radio.

Squelch tail — the noise burst that follows the short, unmodulated carrier following each repeater transmission.

Timer — a device that limits the length of a single repeater transmission.

Tone pad — a device that generates the standard telephone system tones used for controlling various repeater functions.

and deaccess codes, as well as the timer specifics and logging requirements.

A related repeater function is the "reverse autopatch." This device permits anyone to use the phone system to call the repeater and converse with on-the-air hams via the repeater autopatch system. All one needs to do is dial the repeater autopatch telephone number via any handy telephone and the repeater control operator will answer and put you through to the repeater so that you may speak over the air. The control operator is necessary to control the possible nonamateur transmissions originating from the telephone — transmissions that go over the amateur frequencies. In addition, each incoming call must be screened by the control operator before it is allowed to go on the air. Unattended reverse autopatches are illegal, as are reverse autopatches that generate a tone over the air to let the repeater-users know there is an incoming call.

Traffic

Traffic handling is a specialized aspect of Amateur Radio that has found an ideal home on local repeaters. Because of the limited coverage of most vhf and uhf repeaters, the traffic nets operating on repeaters are similarly limited and consist primarily of local nets. Despite this limitation, local nets operating on repeaters provide outlets for traffic never dreamed of on the hf nets. Repeater users are everywhere and provide potential outlets for traffic destined everywhere. The success of repeater traffic nets attests to this.

Checking into a repeater traffic net and handling its traffic is similar to traffic handling on the low bands (consult the traffic-handling chapter for full details). However, there are a couple of idiosyncracies of repeater operation that affect the normal traffic procedures. When relaying a message, keep in mind the length of your transmission. Repeaters have timers and if you transmit too long without a break, you will time out the repeater (and hold up the net operation). Because of the audio quality of fm operation and its lack of interference and noise, the use of phonetics and repetition, standard procedures on the hf nets, is not necessary on repeaters. If the receiving station missed something, repeat it then, not beforehand. And phonetics should not be necessary except for spelling out words containing similar-sounding letters (b, d, p, etc.). Very efficient traffic handling can be achieved via fm repeaters.

If you are interested in joining a traffic net, consult the latest edition of the ARRL *Net Directory*. If there are no repeater nets operating in your area, your area needs a repeater traffic net, so why not start one? It will be a very interesting experience.

Grab-Bag

There are many other specialized aspects of repeater operation. Some repeaters are devoted to special modes or Amateur Radio activities. There are repeaters for radioteleprinter (RTTY) operation (on 2 meters, the frequency pair of 146.10/146.70 is popular for RTTY activity). In the uhf spectrum, there are machines devoted to repeating amateur television transmissions (ATV — fast scan, not slow scan).

Links are very popular. A repeater on one band may be linked with one or more other bands, permitting crossband operation throughout the ham radio spectrum. Accessing a link is usually accomplished via the same tone pad used for autopatch operation, so you can see why these pads are very popular items.

Devotees of various aspects have made their own homes on fm repeaters. There are repeaters devoted to DX chasers, computer hobbyists, public service performers, contest competitors . . . the list goes on and on. If a group of hams has a special interest, before long it will have a repeater on the air where its members can talk all day long about their favorite things. Would you believe a beer can collector repeater?

Operating Conditions

The higher the repeater, the better the coverage. That seems to be a simple rule of thumb. In reality, things are different. A repeater atop a 500-foot tower in the middle of the Great Plains can have similar or even better coverage than a repeater on top of an 800-foot hill in the middle of New England. The rolling hills of 1-land degrade repeater coverage, while the flatlands of the Midwest offer perfect sites for repeaters — there are no hills in the way.

A 10 GHz experiment is coordinated via a battery-powered 2 FM transceiver. (*Bruce G. Graham, photo*)

Leaves also degrade coverage. In the spring and summer, the leaves on the trees adversely affect repeater coverage, especially in the uhf spectrum. Those who use 450-MHz repeaters will attest to improved coverage as the leaves fall from the trees in the autumn months.

Weather is another factor in repeater coverage. Moving weather fronts and inversions can greatly affect vhf and uhf propagation. In the early and late summer, you may find yourself talking through repeaters 300 to 500 miles away as the bands open up, and low power is all that is needed!

Everyone wants to know how he or she is "making the repeater." Your transceiver's signal-strength meter indicates how well you are receiving the repeater, not how well the other station is accessing the repeater. So S-meter readings have been discarded and a new language of signal reporting has grown up for repeater operation. The repeater-signal reporting system is related to the degree of "quieting" that the transmitting station has achieved; that is, how well the incoming signal quiets the internal noise of the receiver. If a station using the repeater has absolutely no noise on its signal, it is said to be "full quieting." Anything less than full quieting is "X% full quieting." For example, a station with a trace of noise may be said to be "90% full quieting"; a noisy signal may be "50% full quieting." Admittedly, any signal report less than 100% full quieting is very subjective. Although the system is not perfect, this method will give you a general idea of how well you are accessing the repeater.

If someone tells you that your mobile signal is "picket fencing," this means that your signal is dropping in and out of the repeater in a rapid manner. "Fluttering" is another word used to describe this condition, which is usually caused by some temporary obstruction that you are passing during your journey.

Interference on repeaters is rare, but it does exist in a number of forms. A station running high power can access more than one repeater and interfere with communications on a repeater that he has no intention of accessing. Most of the users of the local repeater should be strong enough to overcome ("capture") the interfering signal. The main problem with this kind of interference is that it may cause the local repeater to time out, so if you ever encounter this problem, make sure that the interfering signal "drops out" (and resets the timer) of the local repeater before continuing with your transmission.

Another form of interference is "intermod" (intermodulation distortion in a receiver front end or a transmitter power amplifier stage, which causes spurious signals). These "spurs," which may be on the local repeater's input frequency, can cause the same timing out problem as caused by unintentional interference. Usually, the source of the problem can be traced and the problem can be cured, but until it is the same precautions must be taken to prevent the repeater from timing out.

The worst form of interference is intentional interference, and the best cure for it is to ignore it. If you talk about the intentional interference on the air, the jammer is receiving the recognition he is seeking. But if you ignore him, he will go elsewhere to find an audience. If all repeater users everywhere ignored all interference, jammers would be completely frustrated and would fade away. In fact, if a jammer tries hard enough he will receive recognition somewhere and will continue to play his game. To alleviate this problem, many repeater groups have successfully used direction-finding equipment to track down jammers, and have put the problem to rest once and for all.

Join the Fun Mode

Fm repeater operation provides its users with versatility unmatched elsewhere. All of the operating enjoyment you may seek can be found on the local repeater, and the best news is that you can bring the fun along with you wherever you may go, whether traveling in your car or walking down your street. The world of fm repeaters means mobile communications for a mobile society.

Band Plans

"Band plan" is a term you may encounter during your sojourns in the world of repeaters. It refers to an agreement among concerned vhf and uhf operators as to how each vhf and uhf Amateur Radio band should be utilized. The goal of a band plan is to systematically minimize interference between the various modes sharing each band. Besides fm repeater and simplex activity, you can find cw, ssb, a-m, OSCAR and R/C operations (among others) on these bands.

Most of you who use fm or plan on using fm will probably be using 2 meters, so a description of the 2-meter band plan will be useful. First, the band is divided for various modes according to the following plan:

Frequency (MHz)	*Mode*	*Emission*
144.0-144.1	weak-signal	cw
144.1-144.5	weak-signal	ssb, a-m
144.5-145.49	linear translators	all
	repeaters and simplex	fm
145.8-146.0	satellite	cw, ssb
146.0-148.0	repeaters and simplex	fm

The fm portion of 2 meters is organized in the following manner:

Frequency (MHz)	*Operation*
144.61-144.89	repeater inputs
144.90-145.10	simplex
145.21-145.49	repeater outputs
146.01-146.37	repeater inputs
146.40-146.58	simplex
146.61-146.97	repeater outputs
147.00-147.39	repeater outputs
147.42-147.57	simplex
147.60-147.99	repeater inputs

In the 144.61-145.49 MHz range, the spacing between the repeater channels is 20 kHz, so the first input frequency is 144.61 and the next is 144.63 (144.61 MHz + 20 kHz = 144.63 MHz).

In the 146.01- to 147.99-MHz range, the spacing between the repeater channels is 15 kHz, so the first input frequency is 146.01, the next is 146.025 and the next is 146.04.

And according to this plan, the spacing between each repeater input and output frequency is 600 kHz. For example, the input frequency of 146.34 has an output frequency of 146.94 (146.34 MHz + 600 kHz = 146.94 MHz).

The simplex frequencies between 144.90 and 145.10 are nonchannelized, while those between 146.40 and 146.58, and between 147.42 and 147.57, are channelized. The spacing between these simplex channels is 30 kHz.

The most common of the number of variations to this plan in the United States and Canada is the reversal of the input and output frequencies of those repeaters operating on 5 kHz, such as 146.625 and 147.315 (not 146.85 and 147.15). The purpose of this reversal is to eliminate co-channel interference. Under the implementation of this plan, the repeater pair of 146.01/146.61 would have a normal low input/high output split; that is, the lower frequency (146.01) would be the input and the higher frequency (146.61) would be the output. However, the pair of 146.025/146.625 would have a reverse split; that is, high input/low output (146.625 would be the input; 146.025, the output).

Chapter 10

VHF/UHF Operating

Mention vhf and uhf and most hams think only of fm repeaters. Almost every ham has tried fm, but few have even listened to vhf/uhf ssb and cw. If you haven't tried the narrow-band modes above 50 MHz, you don't know what you're missing! You might think you can't work more than 50 miles on vhf, but even low-power stations find 1000-mile (1600-km) DX isn't uncommon. Here's the best part: You don't have to take out a second mortgage to buy the antenna hardware. Vhf antennas are smaller by far than their hf (below 30-MHz) counterparts. You can stack them up for more gain and still turn them with a small rotor.

Table 1
Amateur VHF/UHF Bands

Band	*Frequency Range*	*Weak-Signal Work Centers At*
6 meters	50-54 MHz	low end
2 meters	144-148 MHz	low end and just above 145.5 MHz
1-1/4 meters	220-225 MHz	low end (222.0 MHz on West Coast)
70 cm	420-450 MHz	432.1 MHz
23 cm	1240-1300 MHz	1296 MHz
13 cm	2300-2450 MHz	2304 MHz
10 cm	3300-3500 MHz	3456 MHz
5 cm	5650-5925 MHz	5768 MHz
3 cm	10,000-10,500 MHz	10,240, 10,368 and 10,270 MHz
15 mm	24,000-24,500 MHz	

Vhf operators aren't much different than their hf cousins; they like to ragchew, chase DX, and maybe build some of their own gear. Once you had to be a builder; commercial gear wasn't as plentiful as now. Today you can put together a top-of-the-line station on 50, 144, 220 and 432 MHz, using only commercial equipment. But because most vhf gear is designed to work on one band, it's easier to build a linear for 432 MHz than for 20 meters.

If it seems it's "all been done" on hf, you're ready to try vhf. New discoveries in equipment design, antenna theory and propagation modes are constantly being made on the higher bands. Vhfers like their ham radio a little different, a little more challenging. New Englanders never tire of being in contact with a New Jersey station, about 150 miles (240 km) away — this is a typical 2-meter ssb contact. The band often opens up to Kentucky, Tennessee, or even Texas and Oklahoma. The band opening usually lasts long enough for everyone to work the DX, even the 10-watters the DX has to dig a little to hear. And vhfers are always willing to take time to listen for a weaker station, just as the locals are glad to stand by while you work him. This is a far cry from the madness of an hf DX pile-up!

Vhf DXing

Hams will probably always be interested in seeing how far they can be heard. The situation is only slightly different at vhf. The successful vhf DXer is often conversant in astronomy and meteorology, not to mention the mechanics of ionospheric propagation. On vhf, the DX struggle is with the forces of nature, not against other hams. You probably won't find a more cooperative segment of Amateur Radio. Ideas are freely exchanged through a network of small newsletters, on-the-air ragchews and nets and at the growing number of vhf conferences around the country.

Vhf DX is usually counted in miles (or kilometers), rather than by states or countries, but you'll no doubt keep a tally of states, call areas and even countries. States-worked lists are published frequently in *QST*, in the "World Above 50 MHz" column, to spur activity and provide recognition to successful DXers. On the higher vhf bands you might keep a record of the number of stations worked, if for no other reason than to change the

Not every vhf station is a room full of equipment. This is club station OK1KFW in Prague, Czechoslovakia, with OK1JH operating. The neat package to the left is a homebrew solid-state 2-meter transceiver with 1 watt of cw, ssb and a-m output. In general, vhf and uhf activity is higher in Europe than in North America.

subject when some non-vhfer claims "there's no one up there."

What Happens When the Band is Dead?

Of course there's more activity when the band is open, but normal conditions (called "bad" by DXers) bring out a fair sampling of stations, too. Remember, if no one's operating, there's no way to know when the band *is* open. Some stations keep regular schedules with others slightly more distant than their normal working range and keep track of signal strengths and propagation conditions. The information gathered tells them a great deal about how propagation works and how well their stations perform. Skeds are also useful for testing equipment and antennas and they help stations far removed from population centers make contacts. Once in a while, swing your antenna around in a direction where you usually don't hear activity and call a brief CQ. You'll be amazed at how often you'll turn up a new contact. Because vhf antennas are so directive, you can miss plenty of activity if you point your antenna only where you expect to hear someone!

Under normal conditions, ragchewing is as popular on vhf as on hf and the subjects are just as varied. Since the advent of multimode (fm/ssb/cw) rigs, many hams are trying ssb and cw for longer-distance, hassle-free contacts not afforded on repeaters. The directivity of vhf antennas makes long-distance roundtables a bit tricky, though. Some juggling of antenna pointing directions will usually keep things in order.

Contests

The League and several other groups sponsor contests designed to encourage vhf and uhf activity. Some contests are scheduled to take advantage of good weather conditions and many stations take to the hills. Operating portable from mountaintops is a tradition as old as radio itself. (Actually, the term "portable" is slightly inaccurate when truckloads of equipment are carried up twisting mountain roads, but that's another story.) It happens that the best propagation conditions also take place during the summer and fall when the weather is good. The ARRL June and September VHF QSO Parties have been popular with vhfers for years, and their scoring is designed to encourage operating on as many bands as possible. In June it's rare for 6 meters *not* to be open for distances up to 1000 miles (1600 km), at least part of the time. Home stations are just as effective during these openings, but on the higher bands, mountaintop stations have the edge.

Many stations go mountaintopping for the ARRL UHF Contest, held in August. Just started in 1978, this contest is attracting a great deal of attention. The ultrahigh frequencies are generally con-

Pickup-truck-mounted campers are favorite vehicles for transporting vhf-contest gear to high-altitude locations. During the contest the camper serves as a rustic shack. In this photo, WD4MBK is shown assembling the antennas used by W4ATC/4 on Whitetop Mountain, VA, during the 1979 ARRL UHF Contest. The antennas shown are for 220 and 432 MHz. *(N4BYK photo)*

If you can't find a mountain, climb a tower! That's what W9IP, K9AKS and W3EP did during the 1979 ARRL UHF Contest. On top of this 120-foot (38-m) tower near Champlain, IL, are antennas for all the bands from 50 through 432 MHz. *(W3EP photo)*

sidered to start at 300 MHz, but to encourage 220-MHz operation, ARRL chose to count 220 as a uhf band. An interesting aspect of the UHF Contest is the multiplier. Rather than states or ARRL Sections as multipliers, a four or five-digit number is used, giving the location in longitude and latitude, rounded *down* to the nearest degree. For example, if you live in Cabot, AR, which is at exactly 92° West longitude, 35° North latitude, you send a signal report followed by "9235." Not everyone you meet during this contest is likely to know his or her latitude and longitude, but it's all right to get the name of the town and look up the coordinates yourself.

A perennial favorite contest is the ARRL VHF Sweepstakes, held in January. Since fewer people want to go mountaintopping in the winter, the Sweepstakes rules don't favor operation on the higher bands as much as the June and September QSO Party rules do.

Another recent addition to the ARRL contest calendar, the EME Contest, for moonbounce contacts only, is described in the "Mounbounce" section of this chapter. It's held on two weekends, one in April and one in May.

Vhf contest operating isn't very different from that on hf, although (especially on 432 MHz and above) some cw stations use crystal control, and you should tune around your frequency after you call CQ. Many a-m stations (yes, a-m is alive and well on vhf!) are crystal controlled. They'll often mention the part of the band on which they're listening for answers: "This is W1XZ, tuning 145.5 to 145.55."

Even if contests aren't your major interest, they're good times to be on the lookout for new states and call areas. Especially during the summer/fall season, many high-power stations operate from otherwise-rare locations. Except for 6 meters, vhf contesting is much more relaxed than on hf. During a band opening, 6 can become almost as hectic as 75 meters on a Saturday night.

What Are the Bands Like?

We can't possibly discuss in detail all the types of vhf band conditions in this small space, but let's take a quick turn

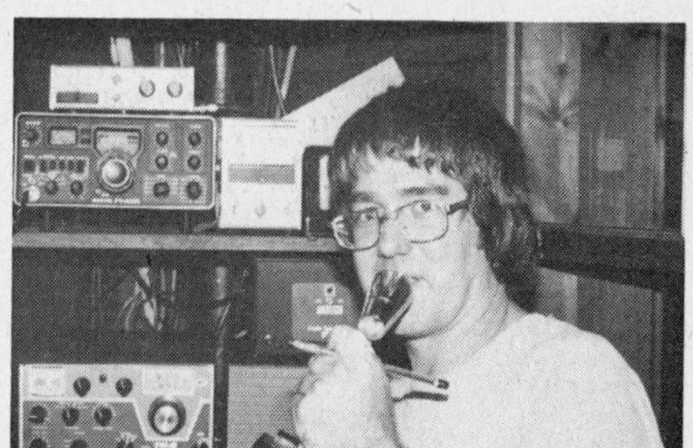

WB1FUB used this neat collection of commercial vhf gear to take second place in his ARRL section in the 1979 ARRL September VHF QSO Party. The transceiver on the top shelf is a multimode 2-meter rig, while the bottom shelf contains a 6-meter ssb/cw transceiver.

through the vhf/uhf bands and see what a variety of propagation modes they offer.

6 Meters (50-54 MHz)

This band has to be the most varied of all amateur bands. At different times of the year and of the sunspot cycle, 6 meters often provides spectacular contacts all over the world. During the 1979 sunspot maximum, the muf (maximum usable frequency) often exceeded 50 MHz, and communications via the F layer of the ionosphere were an everyday occurrence. Solar astronomers tell us that it may be 40 years before we again see such activity on the sun, but activity on 6 meters will continue nonetheless.

Sporadic E

During the early summer months especially, clouds of ionized atoms are generated in the lower latitudes and drift in a northwesterly direction. Because the clouds are formed at the altitude of the ionospheric E layer (about 50 miles or 80 km above the earth), they give rise to a propagation condition called "E skip." No one totally understands how or why the clouds are created. Because of their erratic comings and goings, the scientific name for E skip is "intense sporadic E," or just "sporadic E," abbreviated Es. The muf of sporadic-E clouds often exceed 50 MHz, frequently extending into the 2-meter band.

An opening may last from a few seconds to several hours. Signals reflected from cloud to cloud in a longer opening have permitted contacts from the East Coast to Hawaii, and from the West Coast to Japan. When a good opening comes along even QRP (low power) stations can work DX. Running on a few milliwatts of double sideband to a 2-element beam, W7ZOI in Oregon has worked several New England stations.

Auroral Propagation

If you live in Canada or the northern half of the U.S., you've probably seen the *aurora borealis* or Northern Lights. Some of the invisible radiation from the sun is responsible for excitation of the ionosphere allowing long-distance contacts on the hf bands. Following a solar flare, large quantities of high-energy particles are thrust out from the solar surface. When these particles, which travel at less than the speed of light, approach earth, the geomagnetic field deflects them toward the solar regions. The intense ionization which results often creates the shimmering fluorescent curtains called aurorae. Frequently, the ionization is so great that the aurora is capable of reflecting radio signals. Contacts via auroral reflection are common on 6 meters, but the other propagation modes on 6 will provide the same range. The qualities of an aurora-reflected signal are described in more detail in the 2-meter section of this chapter.

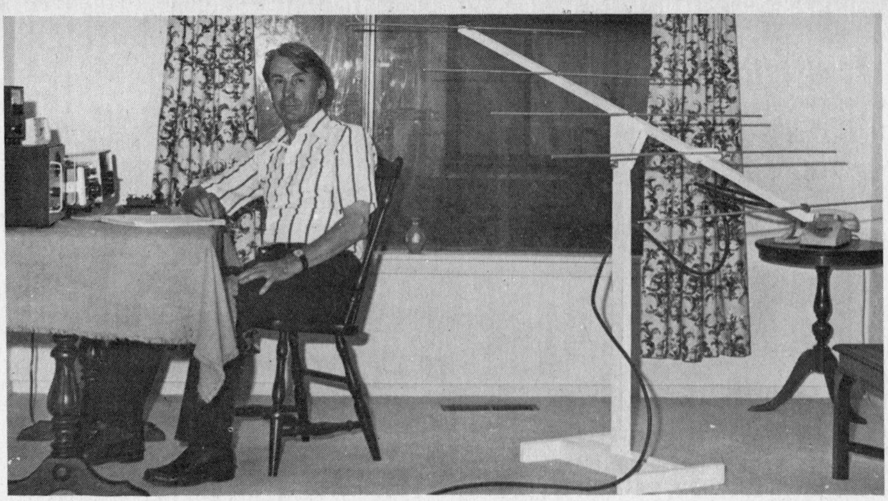

You don't need a big, high antenna to "get out" on vhf. Using a multimode transceiver, homemade 30-watt amplifier and the homemade 5-element beam shown, G8VR/W1 has worked many states on 2 meters, including some on aurora. The beam tilts up for working through amateur satellites. *(G8VR photo)*

Ionospheric Scatter

For well-equipped stations, ionospheric scatter provides a means of working stations from 600 to 1200 miles (960 to 1900 km) away. Signals are extremely weak and cw is the preferred mode. You and the station you're trying to work will call each other at alternate one-minute periods for one hour, or until signal reports and acknowledgements are exchanged. Accurate clocks and frequency calibration are a must.

Transequatorial Propagation

During the spring and fall months, when the F-layer is just below 50 MHz, it's often possible to work stations in the Southern Hemisphere during periods of *ionospheric tilt*. The tilt occurs when the part of the ionosphere above the equatorial region settles slightly. A signal which is slightly bent by the northern part of the tilt and which might normally continue off into space, is reflected back to earth by the tilted southern part. Because the angles of incidence and reflection are nearly the same, the stations in contact are usually almost the same distance on either side of the equator. There's plenty of 6-meter activity in South America, so if you hear a fluttery signal signing a Latin American call, don't be too surprised.

Other Modes on 6

Tropospheric propagation and meteor-scatter are two modes commonly used on vhf bands, but they work on 6 as well. They're discussed further in the section on 2 meters.

6-Meter Equipment

Using a-m on 6 meters was frequently the cause of many TVI complaints, mostly in areas where a channel 2 TV station was on the air. TV channel 2 is directly adjacent to the 6-meter band. Ssb rigs and their associated linear amplifiers have greatly reduced this problem. While a-m is still in use, the dominant mode on 6 is ssb. A number of 6-meter-only rigs have been marketed, most of them providing power inputs in the 150-watt range, but the newer solid-state transceivers run only about 20-watts input. An ssb transceiver for 6 is perhaps the most convenient method of getting on the band. If you already have an ssb rig, you might want to consider a *transverter*. To transmit, a transverter mixes the low-level output of an hf transmitter with an internal crystal oscillator. The resulting vhf signal is amplified and sent on, either to the antenna or to another linear amplifier. Received signals are similarly heterodyned down in frequency, to be detected on the hf receiver.

Most of the activity on 6 is very near 50.1 MHz. Cw, the only mode allowed below 50.1, is rarely heard on 6, an unfortunate circumstance considering its usefulness in weak-signal reception. Most operators monitor 50.110 MHz, with 50.105 another favorite. More cw is heard on the band during band openings (which can happen at any time) and during contests, usually around 50.090 MHz.

ARRL 600 Club

To stimulate activity on 6 meters, ARRL sponsors the 600 Club Award. To qualify, applicants must submit QSL cards confirming contacts with enough stations to earn 600 points. Points are determined as follows: Each QSO, 2 points; each different ARRL Section, 6 points; each new country, 25 points. VE8 and each of the Maritime provinces count as separate ARRL Sections for this award. Applicants living in the U.S. and Canada

The well-laid-out combination of homemade, surplus and commercial gear used at W1XM, Massachusetts Institute of Technology. Shown in the photo is equipment for all bands from 50 through 1300 MHz. *(NØND photo)*

must be League members. Repeater contacts don't count, and you can work a station only once.

6-Meter TVI Problems

Because 6 meters is so close to TV channel 2, the band has gained an undeserved reputation as an interference problem. If channel 2 is received weakly in your area, you may cause interference requiring careful filtering at the TV receiver. If you have no channel 2 in your area, or if the signal is strong, it's unlikely that you'll cause TVI. At one time or another, every ham receives an interference complaint, and most of us overcome them. If you want to get on 6, by all means do so. Don't be discouraged by a problem you may not have.

How To Use A Calling Frequency

The concept of calling frequencies is almost as old as Amateur Radio itself. Basically, the idea is this: Very few amateurs are equipped to monitor more than one frequency at a time in a given band. Therefore, unless everyone tunes up and down the band constantly, transmissions such as CQ calls made on randomly selected frequencies are going to be heard only by a few of the people listening. The solution is to have everyone listen on a *calling frequency*, and to make CQ or other calls on that frequency when a contact is desired. Once contact is made, the stations can agree on another frequency, where they will continue the contact.

In theory, this sounds like a great idea. In practice it often falls short of its potential, because operators don't always move off the calling frequency after they make contact with another station. This has three unhappy consequences. First, operators get tired of listening to idle chatter on the calling frequency and turn off their receivers. Second, unnecessary interference results from having everyone sit on the same frequency. Finally, the calling frequency is tied up for long periods of time by a couple of stations, making it impossible for others to use it for its intended purpose.

2 Meters (144-148 MHz)

Two meters is frequently the first vhf band many hams encounter, because of the popularity of fm and repeaters, which totally dominate the upper half of the band. Unfortunately for them, many fm operators think the only DX possibilities on 2 are through repeaters. This is definitely not the case, as we'll soon discover.

The diminutive size of a 2-meter beam makes operation on this band attractive to mountaintoppers and those with limited roof space. While there's always been activity on the bottom half of 2 meters, two products of modern technology — Amateur Radio satellites and multi-mode transceivers — have brought many newcomers to the band. Amateur satellites use the 2-meter band as both an input and output frequency for their transponders. Many satellite users have been persuaded to try terrestrial communications on 2 meters when no satellites were overhead, and have become ardent vhf enthusiasts. An entire chapter in this book is dedicated to satellite communciations so we won't discuss the subject in detail here.

Most sideband activity on two is below 144.5 MHz, or between 145.5 and 145.9 MHz (satellites operate between 145.9 and 146 MHz, and the new repeater subband covers 144.5-145.5 MHz). A-m stations operate in the higher segment as well. We suggest 144.2 MHz as a general ssb calling frequency, although much activity centers around 144.110 MHz. By keeping 144.2 MHz clear of ragchews it's possible to hear DX stations calling when the band opens. Cw only is permitted below 144.1 MHz. There's much more activity on 2 than on 6, usually above 144.080 MHz. Moonbounce and other very-weak-signal stations operate down around 144.0 MHz.

Although sporadic E is not as common on 2 meters as on 6, contacts in excess of 1000 miles (1600 km) may be made when the Es muf gets up to 144 MHz. As we found on 6, high power isn't needed during an Es opening: W1FB in Connecticut has worked Minnesota while running 10 watts input. The antenna at W1FB was a 10-meter beam!

Inversions and Ducts

When referring to band openings on 2 meters, one word often bandied about is "tropo." Band openings are generally the result of a *thermal inversion*, a warm air mass sitting above an area of colder air. The boundary area between the two bodies of air may refract, or bend, radio signals far beyond the radio horizon. The best indication of an inversion is a high-pressure system that stays in one place for more than a day. Smaller inversions occur just after dawn and just after sunset, mostly during the summer months. Actually, the term *tropo* is a bit misleading here, as *all* normal vhf signal propagation is through the troposphere (which extends up to about eight miles, or 13 km, above the earth's surface).

The term is more correctly applied when describing another similar type of propagation: *tropospheric ducting*. Ducts in the troposphere are probably the result of wind-shears in the lower stratosphere and upper troposphere, which separate two masses of air. One end of the duct may be at a high elevation while the other end is close to the ground. A mobile operator driving along the California coast may suddenly hear a Hawaiian mountaintop repeater coming through like it's in the next town. So he can work fixed stations in Hawaii, the mobiler might decide to drive up a nearby mountain, only to find the band has gone "dead." In fact, the duct probably didn't extend very far inland or very high in elevation on the California end.

Ducts appear to be frequency selective, unlike inversions which will propagate signals on many bands. A duct which works at 2 meters may not pass signals at higher frequencies. While thermal inversions may be predicted by keeping an eye on the weather maps, ducts are less obvious. The best way to find one is to maintain schedules with stations at the opposite end of a suspected duct path. After years of trying by many stations, WB6NMT worked KH6HME on 432 MHz in 1979, setting a new record for that band. Two-meter openings between Hawaii and California have been noted for several years.

Solar-Related Propagation

We mentioned auroral propagation in the section on 6 meters, but it's more commonly used on 2 meters. Auroral reflec-

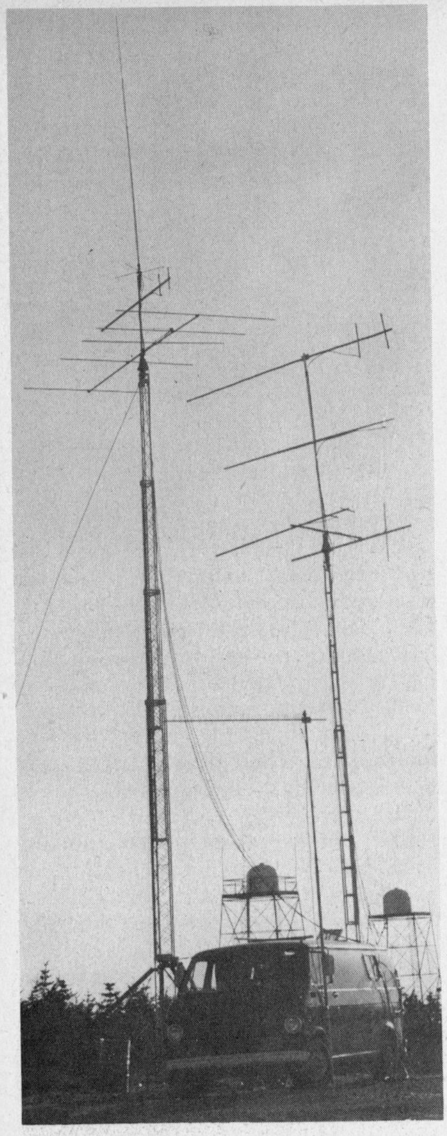

N6NB's East Coast portable station (he uses another one on the West Coast). Shown in position on Mount Equinox, Vermont, the van carries complete stations for 50 through 1300 MHz. Much of Wayne's gear is homebrew, including the trailer which is used to haul the antennas. The van is stored at an undisclosed location between contests. *(N6NB photo)*

tion is a popular mode for Canadian and Northern U.S. stations. Instead of aiming your antenna at the station you want to work, point it north. Even when there is no visible aurora, a radio aurora may exist. Radio aurorae also form during daylight hours when a fluorescent aurora is invisible because of daylight. An auroral curtain is in constant, rapid motion, causing a Doppler shift in the signal frequency and giving auroral signals an unmistakable quality. A cw signal is usually just a hiss, and ssb signals sound like unearthly whispers. No matter how pure the note, a cw signal will sound rough, so the letter is substituted for the number denoting tone in the signal report; a signal of readability 5 and signal strength 8 would be reported as RST 58A.

Table 2

Major Meteor Showers

Date	*Name*
January 3-5	Quadrantids
April 19-23	Lyrids
July 26-31	Aquarids
July 27-August 14	Perseids
October 18-23	Orionids
October 26-November 16	Taurids
November 14-18	Leonids
December 10-14	Geminids
December 22	Ursids
*May 19-21	Cetids
*June 4-6	Perseids
*June 8	Arietids
*June 30-July 2	Taurids

All showers occur in evening except those marked (*), which are daytime showers. Evening showers begin at approximately 2300 local standard time, daylight showers at approximately 0500 local standard time. The average shower lasts about 15 hours. A more complete meteor-shower list appears in the ARRL *Electronics Data Book*.

Depending on the width of the curtain and the level of ionization, contacts of up to about 2000 miles (3200 km) are possible on 2 meters. High-gain, highly directive antennas are a disadvantage, because the reflecting area between you and the other station is narrow and constantly in motion. It's easy to work stations with only 50 watts and a single Yagi antenna. The effects of Doppler shift make it impossible to copy a-m or fm signals reflected from an aurora.

Contacts between Puerto Rico and Argentina, Japan and Australia, and Cyprus and Rhodesia, distances of over 8000 miles (13,000 km) would seem to be impossible on 2 meters, yet several have been made in the last few years! While some theories have been proposed to account for this phenomenal range, no one is really sure how these signals, unassisted by moonbounce or satellite, were propagated. Most of the contacts were made an hour or two after sunset, which occurred at about the same time at both ends of the path. Signals have been heard at 432 MHz, although no contacts have been made on that band — yet. Stations in the southern U.S. stand an excellent chance of taking advantage of this exciting and little-understood transequatorial mode. Some of the contacts were made by stations running about 100 watts and a single beam antenna, so you don't have to be a "big gun" to try it. There's still plenty to learn at vhf!

Meteor Scatter

In addition to solar radiation, our atmosphere is constantly being penetrated by meteors. Some of the larger objects appear as "fireballs" or "shooting stars," as friction from the earth's atmosphere causes them to burn up. Smaller meteors leave only a trail of ionized air as they burn up, but this trail is still capable of reflecting radio signals. Meteors are constantly entering the atmosphere and meteor-scatter contacts are possible the year round.

At certain times during the year, large clusters of meteroids (they're not properly called meteors until they enter the earth's atmosphere) in orbit around the sun enter the atmosphere in large numbers. When these meteor "showers" occur, schedules are arranged between stations wishing to work each other. Meteor "bursts" are usually of short duration, so high-speed cw or ssb is used. An occasional long burst may allow a quick contact, but these are rare. To prevent the misfortune of having both stations transmitting or listening at once, the schedules are arranged so that one station transmits during the first half of a one-minute period, the other during the second half.

The big event for "ping jockeys" (short bursts of signal sound like pings) is the Perseids meteor shower in late July and early August. (A table lists the approximate dates of the largest meteor showers.) As the earth rotates, more meteors enter the atmosphere in the morning, and that's the best time for most meteor-scatter contacts. High-power and high-gain antennas are useful for meteor-scatter work, but as meteor trails are spread over a large area of the sky, too much antenna directivity can be a handicap. If you want to work the big guns, you can do so with about 100 watts output and a good single Yagi of 12- to 15-dB gain. Contacts are almost always prearranged, and may take several hours. Here's how a typical schedule might go:

1) Just before schedule time, both operators set their clocks by WWV or CHU and carefully calibrate their transmitters and receivers. A frequency error of 500 Hz may spell the difference between success and failure.

2) They have agreed to follow the common practice of using 15-second sequencing of calls, so that if a good burst occurs they will have a chance to finish the contact quickly. From past experience they know there is nothing more frustrating than to listen to a long, loud burst from the other station which disappears before you have a chance to transmit! The eastern station will listen during the first and third 15-second periods of each minute, and will transmit during the second and fourth periods. When his turn comes, K1FO begins transmitting on ssb: "KØMQS, K1FO; KØMQS, K1FO; KØMQS, K1FO" and so on for exactly 15 seconds, which is time enough for about eight sets of calls. Extraneous words are eliminated; it is clear from the operator's voice inflections which station is being called and which is transmitting, so there is no need for "this is" or "from" to be inserted. Also, there is no need for phonetics.

3) The pattern continues, with only an occasional "ping" to reassure the

operators that there is anyone on the other end. About 20 minutes into the hour-long schedule, K1FO hears a half-second burst: ". . . 1FO . . ." While it is there the signal is quite strong; but when it disappears, a tantalizing bit of signal just a little too weak to copy remains for a second or two.

4) Another five minutes go by, and through additional short bursts K1FO has received both sets of call signs, complete. Now he can begin transmitting a signal report: "KØMQS K1FO S2." Meanwhile, out in Iowa, KØMQS has also heard both calls and is transmitting the signal report in a similar pattern. As soon as he hears K1FO send the "S2" report he switches to this pattern: "Roger S2; roger S2; break." He can transmit a half-dozen of these in his 15-second period, listening very quickly in between to see if he can catch the final "Roger" from K1FO which will complete the contact.

5) K1FO hears the acknowledgement and the report, and shifts to a new pattern: "Rogerrogerrogerrogerroger, break." It takes only a couple of minutes for MQS to hear enough, on a short burst, to know the contact is complete. They've been lucky; most schedules aren't that successful.

This is a tough way to make contacts! Even more challenging, but rapidly gaining popularity is the practice of working meteor scatter without schedules. It's only realistic to try this during a major shower, such as the Perseids. While a big antenna and high power aren't required to make meteor contacts, you need patience and skill. KØMQS used meteor scatter primarily to become the first person to earn Worked All States on 144 MHz. What about the states too far away to work on meteors? He worked *them* on moonbounce!

Before we leave 2 meters, one note regarding antenna polarization: For cw and ssb, horizontal polarization is universal. To work both ends of the band (fm *and* ssb/cw), you'll need two antennas, or one with two driven elements and two sets of parasitic elements, one in each plane. Although horizontal polarization is inconvenient for mobile operation, it provides superior results over long distances.

1-1/4 Meters (220-225 MHz)

In the last few years, 220 has probably been the subject of more heated discussions than any other single amateur band. Lacking a nearby commercial two-way radio band to provide surplus rigs suitable for conversion, 220 has never shared the popularity and respect accorded the 50-, 144-, and 420-MHz bands. Despite this handicap, hams who have spent any amount of time on 220 are enthusiastic about its potentials.

Signals on 220 seem more favorably suited to inversion openings than 2-meter signals, and the ambient noise level on 220 is less. With the exception of the hundreds of active repeaters, activity on 220 is lower than on the adjacent bands. Amateurs have been unwilling to commit themselves to a band they feared might be taken by FCC and given to another service. The best way to keep a band is to *use* it! 220 seems safe for the time being, but we shouldn't become complacent. In the last couple of years, several new transverters and receiving converters for 220 have entered the market, so if you live within 150 miles (240 km) of a metropolitan area you'll probably find some activity, usually centered around 220.1 MHz.

Moonbounce and meteor-scatter contacts on 220 aren't much more difficult than on 2 meters, but antennas of equivalent gain are 25 percent smaller on 220.

70 cm (420-450 MHz)

DX and other weak-signal work on this band is conducted near 432 MHz. This choice of frequency was not arbitrary: It is the third harmonic of 144 MHz, and many transmitters employ triplers from the 2-meter band. The area below 432.1 MHz is generally avoided by casual operators, to give the moonbounce stations plenty of interference-free room. In some countries, only a narrow portion of the band is available for amateurs use, often near 432.0 MHz.

Commercial frequency triplers, good for use on fm or cw but *not* ssb, are available, as are transverters and receiving converters. Several manufacturers build high-quality beams for 432 (and 50, 144, 220 and 1296 MHz), but why not build your own? A 15-dB gain Yagi for 432 MHz has a boom length of about 10 feet (3 m), and is easily made.

Most amateur moonbounce activity takes place on 432, because of the relative ease of building high-gain antennas of reasonable size. Power amplifiers for 432 MHz are, as we said earlier, easier to build than their hf equivalents. Why? Because a sheet of copper or aluminum can be fabricated to act as both parts — inductor and capacitor — of a tuned circuit. Smaller pieces of metal, called "flappers," are used for fine tuning. Consequently, you don't need expensive coils and tuning capacitors, as you do at hf. Some hams are intimidated by vhf/uhf equipment because it doesn't *look* like hf gear, but once you peer inside, you see it's easier and cheaper to build. (To you vhfers reading this: Sorry we're destroying some of your mystique. We're sure you understand!)

While thermal-inversion conditions on the 70-cm band are similar to those on lower frequencies, the "path loss" or attenuation is greater. Nevertheless, contacts of several hundred miles are not uncommon. Meteor-scatter contacts on 432 MHz are more difficult than on the lower frequencies, because the duration of that part of a meteor trail capable of reflecting signals this high in frequency is short. Similarly, the ionization of the auroral curtain is not often sufficient to reflect 432-MHz signals.

Here's a good example of radio amateur ingenuity: The WA6EJO homebrew 1296 station mounts directly behind the antenna system, a quad array of helicals. Placing the equipment at the antenna eliminates feedline loss, which is quite high on the 23-cm band.

Perhaps because of the ready availability of commercially manufactured transverters, amplifiers and antennas, local 432-MHz activity nights are nearly as well attended as those for 2 meters. The portability of competitive 432-MHz stations is best exemplified by N6NB's pickup-truck moonbounce station with the antenna mounted on a lightweight trailer.

23 cm (1240-1300 MHz)

Although we're starting to get into true pioneer territory, commercial triplers, transverters and antennas are available for use on "1296." The DX record here is over 1200 miles (2000 km), and the stations who set the record were running only 20 watts to small antennas. Thermal inversions provide the most consistent DX openings on 23 cm, but many stations are equipped for moonbounce. The best way to work new states on 23 cm is to operate during the VHF QSO Parties and the UHF Contest, when well-equipped stations are taken to good locations. Activity on this band is from 1296.0 to 1296.3, concentrating around 1296.1 MHz. The biggest problem is knowing the exact frequency you're on. The simplest way is to listen for the harmonic from a calibrated 144- or 432-MHz rig.

13 cm (2300-2450 MHz)

Imagine a single amateur band 150

Microwave equipment is light and simple to operate. This photo shows a pair of 10,250-MHz transceivers fitted with 22-dB gain horn antennas. They were used by WA6EJO during the 1979 ARRL UHF Contest. *(WA6EJO photo)*

Even moonbounce antennas are small enough to fit on a city-dweller's lot. K1FO is making adjustments to his 2-meter EME array. Gain of this array is about 20 dB. *(W1XZ photo)*

MHz wide! Activity on the 13-cm band centers around 2304 MHz, the 16th harmonic of 144 MHz. Solid-state receiving converters and preamplifiers are economically feasible, and component prices are dropping as the technology needed to make them improves. Some amateurs are even experimenting with microwave-oven magnetrons, attempting to phase lock them to generate high power at 13 cm. Difficulties in constructing efficient transmitting amplifiers and low-noise converters at 13 cm are partially offset by the small size of a high-gain antenna. Future amateur satellites will carry transponders operating in the 13-cm band, which should provide an impetus for further development of the band. Commercial transverters and converters are available from sources advertising in *QST*, but receiving equipment is not hard to build. *Striplines*, which act as tuned sections of transmission line, may be etched directly on copper-foil printed-circuit board. You then mount a couple of trimmer capacitors and the transistors and tune it up. Plans for stripline converters have appeared in Amateur Radio magazines. Building your own is a good way to learn something about how circuits operate at microwaves while saving money.

10 cm (3300-3500 MHz), and 5 cm (5650-5925 MHz)

The current activity level on these bands is quite low. When surplus klystron tubes became available after World War II, activity on 5650 surged, then declined. The 10-cm band has never been heavily used. As the prices of microwave semiconductors decline, equipment for the 3300-MHz band will probably resemble that on 1296 and 2304 MHz. Surplus waveguide mixers suitable for use at 3300 MHz are sometimes available at flea markets. They were formerly used for microwave telephone links in the 4200-MHz range, but some have been converted by hams for use on the 10-cm band.

3 cm (10,000-10,500 MHz) and 15 mm (24,000-24,250 MHz)

If you've ever fussed with a drifty hf vfo, it may seem impossible for *anything* to work this high in frequency! The truth of the matter is, operation on these bands is much simpler than on the bands below. War-surplus klystrons, such as the 723A/B, can be tweaked into operation on 3 cm, commonly referred to as "X band," its military designation during World War II. Gunn-diode oscillators, originally designed for radar applications at 10,525 MHz, will work well on the 3-cm band. Similar "K-band" devices will work fine on 15 mm. At least one manufacturer provides Gunn oscillators with varactor diodes for tuning and Schottky-barrier diodes for receiver mixers, set up for operation on either band (see *QST* ads). Hams all over the world are using these units with separate receiver i-f strips and low-level modulators, and are pumping new life into our two highest bands. Distances of over 500 miles (800 km) are possible on 3 cm, slightly less on 15 mm. The power output ranges from 15 to 50 mW! Here's the ideal way to try out the microwave bands. A complete transceiver and horn antenna will fit in a briefcase. The Gunn oscillator requires about 10 Vdc, making operation from a car battery a snap.

An interesting result of 10-GHz experiments was the discovery, by British amateurs, of long-distance ducts close to the surface of large bodies of water. First observed along the English Channel, ducts have aided the making and breaking of 10-GHz records across the Channel and the Mediterranean Sea. It isn't necessary to go mountaintopping to work microwave DX. Some European stations have worked over 10 countries with their QRP 10-GHz gear!

Moonbounce Communciations

Because this is such a "far out" mode of communication, we decided to give it a special section. Moonbounce is the most challenging area of Amateur Radio. Unless you have access to a giant commercial or military parabolic "dish" antenna, your power amplifier must be the most efficient you can build and your receiving system must be as sensitive as you can make it. Despite these apparent obstacles, dozens of hams all over the world have joined the EME (Earth-Moon-Earth) or moonbounce ranks.

Most EME activity takes place on 144 and 432 MHz, although a number of stations are operating on 220 and 1296 MHz as well. In the past, all contacts were by prearranged schedule, but now it's not uncommon to hear someone calling CQ — and being answered by more than one station! The signal attenuation over the round trip from earth to moon and back would be bad enough, but the signal is also subject to absorption by the lunar surface, amplitude modulation by the moon's shaky movement (called libration fading), frequency shift by the relative motion between earth and moon (Doppler shift), and polarization rotation as the signal passes through the earth's ionosphere (Faraday rotation). Add the difficulty of mounting and accurately pointing a large, high-gain antenna array, and you can see that moonbounce is more than just a weekend project!

But vhfers aren't easily daunted. N6NB — a lawyer, not a graduate engineer — has built a couple of moonbounce stations

Prominent Japanese moonbouncer JA6CZD, shown at his handsome station console. His homemade 432-MHz amplifier is behind him on the right.

Some of the "sophisticated" test equipment used at antenna-measuring contests. The large instrument is a ratio meter purchased for $15 at an auction. WA2ZZF converted the beer cans to serve as high-Q cavities. *(W1XZ photo)*

designed to be taken to rare states and set up in a few hours by one person. He mounts the antennas on a trailer, which is pulled behind the camper containing the station. The Mount Airy VHF Society, a Philadelphia-area club, transported a complete 432-MHz EME station to the South American country of Colombia, to give stations around the world a chance to work a (then) new continent. Since then, other South Americans have gotten on EME. Two-meter and 432-MHz Worked All Continents and Worked All States awards are almost commonplace these days.

Because a number of stations have large antenna systems, it's just possible to make contacts with one Yagi on 144 MHz and four on 432 MHz. If your receive converter has a noise figure of 1 dB or less, and you have a 15-dB gain antenna, you should be able to hear EME stations with one antenna. On nights when the moon is at perigee (nearest earth), and above your horizon, listen between 144.0 and 144.05, or 432.0 and 432.060 MHz, while manually pointing your antenna. If you can hear signals (they'll be *very* weak), you've got half the battle won. To compensate for fading, slow-speed cw is often used, but high-speed cw and even ssb are more common at better-equipped stations — usually those with the best receiver preamps and bigger antennas.

EME schedules are very similar to those of meteor-scatter operators, although different signal-reporting methods are used. Once contact is made, the schedule may be discarded, and necessary information quickly exchanged while conditions are favorable. Better-equipped stations have even been heard rag chewing.

The best time to listen is during the two weekends of the ARRL EME Contest, held in April and May of each year, although there will probably be someone on any night when the moon is at or near perigee. To be a successful moonbouncer you have to be sort of a Renaissance person, knowing something of celestial physics, mechanics, electronics — and Murphy's Law. But then, that's what vhf — and Amateur Radio — is all about.

Exchanging Information

QST features two columns of interest to those interested in narrow-band vhf communication: The World Above 50 MHz and The New Frontier. The latter is dedicated to microwave work, above 1240 MHz. By reading the columns you can find out who else in your area is interested in a particular band or mode, or who might have just what you need to get on a particular band.

Antenna-gain-measuring contests are a popular vhf event. W3HMU is shown testing the performance of four 432-MHz Yagis at the contest held at Trenton State College in 1977. *(W1XZ photo)*

There are numerous newsletters published by individuals and groups around the world pertaining to special areas of vhf interest. You can find out about publications of interest to you by writing to the editors of the two columns mentioned. Be sure to enclose a large s.a.s.e. Don't feel shy about approaching these vhf groups — everyone begins as a beginner — they'll be glad to learn of your interest.

Major vhf conferences are usually publicized in *QST*. The West and Northeast meetings are usually held in May, the Central in August, and the Middle Atlantic group meets in October. The best way to meet vhfers in person is at one of these conferences — most of the time they're at home working on station improvements! The technical sessions offer knowledge direct from the experts, and there are often antenna-measuring sessions and flea markets. If you can possibly attend one of these conferences, do so by all means.

The Future

Ten years ago, few vhfers would have predicted the interest in, and activity on, our highest bands. EME was just coming into its own, and the Gunn diode was just coming out of the lab. Where will we be in another 10 years? All our microwave bands will be in active use, the lower vhf bands will be swimming with stations and the trailblazers will be experimenting with Venus bounce. Along the way we still have a great deal to learn about circuit and antenna design, and propagation. Hams will be using computers to take care of such mundane things as antenna pointing, weak-signal enhancement and timing during EME and meteor-scatter schedules. By no means has it "all been done." There's still plenty of time to join the pioneers. But don't wait too long — you don't know what you're missing!

Chapter 11

Satellites

Just as Amateur Radio has expanded beyond being an experimental curiosity into a sophisticated communication service, the amateur satellite program has progressed from a 2-meter beacon barely able to send its own name into a modern communications system. Satellite communication reliability is well recognized by commercial communications systems and is the staple of their long-distance circuits. Our own satellite system reflects this thinking. While experimentation and scientific inquiry certainly underlie part of what passes through the satellite transponders, amateur satellites are just beginning their reign as a major communications service within Amateur Radio.

As the satellites have become increasingly complex and the service more sophisticated, ground stations and their operation have had to keep pace. Technically the stations have improved immensely, to the point where their sophistication rivals that of some commercial installations. But you don't *need* super-sophisticated gear to communicate through them.

As the satellites and ground stations improve, increasing numbers of users utilize the orbiting spacecraft daily. The potential user-population has expanded to the point where careful coordination of satellite operations is now necessary for the full potential of these spacecraft to be met. The details of satellite operating will be covered in this chapter; the technical aspects of setting up a station are covered in the *ARRL Handbook* as well as in *QST*.

W3PK (left), K1JX and W3GEY (right) prepare the Phase III spacecraft and test cables for environmental tests. (*W4PUJ/AMSAT photo*)

OSCAR as a Communications Tool

Radio waves travel in a straight line unless some intervening medium — like the ionosphere — refracts (bends) or reflects them into a different direction. The ionosphere has little effect on signals at 144 MHz or above. There is very little to refract these vhf radio signals, unlike on the hf bands. As our earth is round, vhf radio waves continue off into space, where at least no earthling can intercept them for reception. This drastically limits reliable communications distances above 144 MHz.

Fm repeater groups get around this limitation by placing their transmitting and receiving site as high above the terrain as possible, so as to be within "straight-line" view of as many users as possible. While increasing reliable communications distance to encompass large communities, repeater locations are confined by the practical bounds of geography.

Orbiting satellites take the repeater-on-the-mountain-top concept one step further. The highest altitudes at which the satellites orbit place them within view of a large segment of the earth at one time. Communication with the satellite is thus possible over great distances. The increased path loss brought by the satellite being much farther away than the usual fm repeater is more than compensated for by the lack of obstructions between a ground station and the satellite.

Things Even Your Best Friends May Be Afraid to Tell You

"Boy, I called that K1XA for almost the entire Mode A OSCAR 8 pass last night. He kept calling CQ. He sure was loud. Doesn't that guy have a receiver or is it me?"

Sound typical? Perhaps the biggest complaint, and biggest problem in the history of amateur satellite work, is typified by the last comment. For some unexplained reason, hams have invested a lot of time and effort in their transmitting set-ups, but are willing to use almost anything for reception. The frustration dealt out to other hams trying to contact a station who receives poorly is only a small portion of the problem. In their frustration of receiving their own downlink signals poorly or not at all, a lot of hams turn up the EIRP, figuring that will make their downlink signal stronger. It does up to a point. Beyond that point where a downlink signal is louder that the transponder beacon signal, the internal automatic gain control in the satellite transponder reduces the transponder receive sensitivity. Not only does this prevent the offender from getting much louder, it also prevents *all* the other stations in the passband from being heard. The effect is very similar to when your next-door neighbor ham fires up at 7124 kHz while you're listening to 7210 kHz. The whole receiver desenses, and what little you can copy isn't very readable. The situation on the satellite really deteriorates when everyone turns up their power in an attempt to be heard when the satellite desenses.

Is there a solution? There is a very effective remedy, which also happens to be very simple. Unfortunately most hams consider the solution as a moralistic edict, heaped upon them by an unfeeling, ignorant *Big Brother*. The cure: effective receiving set-ups. Everyone who is able to hear very effectively will find it unnecessary to run excessive power to attain a satisfactory signal-to-noise ratio. The transponder won't desense and everyone will be happy.

"Alright then wise guy," you're probably saying, "how do I get an *effective* receiving set-up?" Obviously, there is no one simple answer; a lot depends on the individual circumstances. There are a lot of little things that will improve almost every station that doesn't already have first class receiving capabilities. If the satellite beacon is not several dB out of the noise (as much as 15 or 20 dB on an overhead pass), then these suggestions will be of use to you.

1) *Use a different 10-meter receiving antenna.* That tribander at 40 feet may be just the ticket for working DX via the ionosphere, but really is an inferior satellite antenna. Why? The tribander probably has a terrific lobe at 17° above the horizon, but not much elsewhere. The vertical beam width of the tribander and ground reflections combine to produce nulls all through the vertical pattern of the antenna system. This isn't very pronounced when you're working a station via the ionosphere; usually you're aiming at the horizon. Higher angles of radiation are just wasted for long distance ionospheric contacts. But, the satellite only spends a small portion of its orbit actually at your horizon. When the satellite passes through these nulls, it may fade into oblivion.

There are two solutions. The first is to raise your tribander to a height of say 200 feet or so, where the ground reflection interference is minimal and mount the antenna on a fixture such that you can steer it both in azimuth and elevation. For most people that solution is somewhat less than practical. A second solution is to take advantage of the ground reflections rather than avoid them. An antenna with a very broad pattern, such as a horizontal quad loop or a dipole placed 0.375 wavelength (about 12.5 or 3.85 meters) above the ground has a pattern suited for Mode A reception.

2) *Add a 10-meter preamp.* Many receivers destined for the ham-bands are optimized for the most used sectors of the bands. Since 10 meters is so wide, it's very difficult to get good performance at 28.5 MHz and still perform well at 28.0 or worse yet 29.5 MHz. One solution is receiver modification but a better avenue is the use of a low-noise preamplifier. Most receivers are least sensitive on 10 meters, at least compared to the lower bands, so a preamp will be of benefit to almost any receiver. There are commercial units available from a couple of manufacturers and very simple but effective ones are published from time to time in the ham magazines as well as in the *Radio Amateur's Handbook*. For the ultimate, install the preamp at the antenna. A preamplifier combined with the aforementioned antenna may well bring the owner into an entirely new world of satellite communications. One note of caution: If you're using a transceiver for 10-meter reception, be absolutely sure that it's impossible to transmit into the preamp. Otherwise, it's bye-bye preamp.

3) *Use circular polarization on 2 meters and 70 centimeters.* Most of the satellites' antennas are circularly polarized, so you'll gain 3 dB over a linear antenna. More importantly, fading is drastically reduced. On the Phase III satellites, using a linear polarized antenna will most surely be a disappointing experience. Spin modulation, a form of very fast fading, will make the satellite signal sound like it's coming from the whirling blades of a helicopter. The ability to switch circular sense, from left hand to right and vice versa from inside the shack can also be a valuable asset. Not only do different satellite modes require different senses, but ionospheric effects occasionally favor one sense over the other. Switching polarizations might yield an improvement of as much as 20 dB.

4) *Use a 2-meter preamplifier.* One of the big factors in popularizing 2-meter sideband and cw in the past few years has been the advent and proliferation of commercially made multimode transceivers. Drawing from their experience in designing hf sideband equipment as well as their vhf fm work, the manufacturers have combined the two yielding transceivers with few compromises. One of the compromises does affect satellite users, though. The receiver "front ends" are designed with the best attributes of fm transceivers destined for use in a highly populated area, such as Japan. The receivers are very resistant to overload on the fm channels from both nearby stations and from out-of-band commercial signals. Unfortunately, the noise figure of these transceivers is less than optimum. They have been measured as having noise figures between 5.5 and 10 dB. Add a couple dB of feed line loss to this and your 2-meter downlink capabilities are severely limited. Just as with 10 meters, the solution is to use a preamplifier. Again, the magazines and the *Handbook* offer simple designs and inexpensive units are commercially available. The caveat regarding transmitting into the preamp applies here. The preamp should be connected literally to the transceiver where it is only in the receive line. Some manufacturers have provided an alternative; preamps with built-in switching to automatically bypass themselves on transmit. Mounting the preamp at the antenna will circumvent the feed line's contribution to receive noise figure.

5) *Use a 70-cm preamplifier.* Most converters, commercial and otherwise, are deficient in noise figure. The same principle applies here as on 10 and 2 meters.

6) *Install a band-pass filter at the 435 MHz front end.* A severe problem with operating through the 146 to 435 MHz satellite systems occurs at the ground station. The third harmonic from the 2-meter transmitter can desense the 70-cm receiver. This is partly a technology problem, owing to the difference in semiconductors suitable for use at 435 MHz versus those for use at 146 MHz. Assuming that the transmitter itself doesn't emit excessive third harmonic energy (which would be a violation of the law, incidentally) the solution is to place a filter in the line between the receiver front end and the antenna. Not only will this clear up the desense problem from the transmitter, but it may reduce some of the interference caused by radar and some commercial services. The filter should be low loss so as to not degrade receiver sensitivity. Commercial units are available and a suitable design is shown in the *Handbook*.

7) *Minimize system losses.* As was mentioned earlier, any losses between the receiver front-end and the antenna degrade system performance. Transmission line losses, connector losses, even connection losses all reduce effectiveness. Perhaps the soundest investment a ground station builder can make is in quality, low loss, coaxial line and high quality connectors. Properly installed and weatherproofed, an installation like this should provide years of trouble free and satisfactory service. Make good, clean connections to the antennas (use steel wool), and weatherproof them with RTV silicone rubber or spar varnish. Tape the feed lines to solid tie points to prevent placing undue strain on the cables; also tape up the coaxial connectors. Uhf connectors have no place at 435 MHz; use N type or other constant-impedance connectors. They're also worth using at 146 MHz. Don't use cheap relays; they're false economy.

8) *Use az-el mounts for directive antennas.* Also make sure that you can accurately point the antennas.

9) *Have accurate orbit information.* Whether you are using a plotter, a calculator, a computer print-out or even a computerized system to point your antennas, use the most accurate information you can get. Since the orbits are somewhat dependent on solar influences, don't rely on year-old data. Don't even rely on three-month-old data, especially on Phase II satellites. Being two minutes off in antenna pointing out of an eight-minute pass negates all the work you've done to set up your station.

10) *Use a reliable means for adjusting transmitter output.* As was mentioned, nobody's your friend if you overload the satellite. On most modern sideband rigs, turning the microphone gain up or down is a suitable means to adjust output power. Other means might be necessary when a converted fm rig or other class C type transmitter is used.

Being able to communicate with a satellite is terrific, but most satellites have a very limited vocabulary — and they only talk about themselves. Their language is called telemetry and is transmitted over a satellite's various beacons. More on that later. Most satellite users wish to communicate with other humans based on the ground.

OSCAR (Orbiting Satellite Carrying Amateur Radio) 3 brought a new concept to the amateur world, the orbiting repeater. Signals sent from ground stations were received at the satellite and retransmitted back toward the ground.

Glossary of Satellite Terminology

AMSAT — The Radio Amateur Satellite Corporation, a nonprofit organization located in Washington, DC; has overseen the OSCAR program since the launch of OSCAR 5. (AMSAT, P. O. Box 27, Washington, DC 20044.)

AOS — Acquisition of signal — The time you can first hear satellites, usually just after it rises above the horizon.

Apogee — That point in a satellite orbit where it is farthest above the earth.

Area Coordinators — Volunteers in the AMSAT organization who coordinate satellite activity in their regions. Most states have at least one; many countries are also represented.

Ascending Node — The point where the satellite crosses the equator travelling from the south to the north.

ASSC — Amateur Satellite Service Council — A coordinating body comprising equal representation from Project OSCAR, AMSAT and ARRL.

Az-el Mount — Antenna mount that allows antenna positioning in both the azimuth and elevation planes.

Azimuth — Direction (side-to-side in the horizontal plane) from a given point on earth; usually specified in degrees (N = 0°, E = 90°, S = 180°, W = 270°).

Circular Polarization — A special case in which the electric field component of a transmitted radio wave is equal in both the vertical and horizontal planes and effectively rotates. The sense of polarization, whether right-hand circular or left-hand circular, is determined from behind the antenna, looking out along its axis of propagation.

Codestore — A special system that allows digital information (Morse code, e.g.) to be placed in onboard electronic memory storage by ground stations for later retransmission.

COSMAC 1802 — A CMOS, 8 bit microprocessor made by RCA. Its low power consumption, high noise immunity, wide temperature tolerance and flexible (to the programmer) architecture, make it well suited to use in orbital hardware. The 1802 is the heart of the onboard Phase III-A computer.

Descending Node — The point where the satellite crosses the equator travelling from north to the south.

Desense — A problem characteristic of Mode J operation in which the strong 2-m uplink signal overloads a low-noise 70-cm preamplifier or converter.

Doppler Effect — An apparent shift in frequency caused by satellite movement toward or away from your location.

Downlink — The frequency of signals transmitted from the satellite to earth.

Earth Sensor — A device to be used on the AMSAT Phase III-A satellite that will enable the computer to determine the spacecraft orientation to the earth.

Eccentricity — That orbital parameter used to describe how much an elliptical orbit deviates from a circle; eccentricity values vary between 0 and 1: e = 0 for a circle.

Elliptical Orbit — Those orbits in which the satellite path traces an ellipse with the earth at one focus.

Elevation — Direction (up-and-down in the vertical plane) from a given point on earth usually specified in degrees (0° = plane of the earth's surface at your location; 90° = straight up, perpendicular to the plane of the earth, overhead).

EQX — Equator crossing, usually specified in time (UTC) of crossing, and in degrees west longitude (0-360°).

Erp — Effective radiated power — system power output after transmission-line losses and antenna gain are considered.

ESA — European Space Agency — Will launch AMSAT Phase III-A.

Experiment Day — Routinely scheduled days during which the satellite is closed to casual use and reserved for scientific and educational experimentation; a secondary purpose is to allow for battery recharge.

Falsing — Unscheduled, unintentional and undesirable mode switching initiated by conditions in the spacecraft.

Geostationary Orbit — An orbit at such an altitude (22,300 miles) and in such a direction (W to E) over the equator that the satellite appears to be fixed above a given point.

Groundtrack — The imaginary line traced on the surface of the earth by a satellite subsatellite point.

IHU — Integrated housekeeping unit — Phase III's on-board computer that will manage many of the routine in-flight tasks automatically.

Inclination — The angle at which the satellite crosses the equator at its ascending node; also the highest latitude reached in an orbit. An orbit crossing directly over the North Pole would have an inclination of 90°, east of the pole less than 90° and west of the pole greater than 90°.

Increment — The number of degrees longitude that the satellite appears to move westward at the equator with each orbit, caused by the earth's rotation under the satellite during each orbit. (The earth rotates 360° in a 24-hour period.)

Kick Motor — A one-shot motor on-board the Phase III-A satellite that will be fired at perigee a few weeks after launch to adjust the orbit to the desired final perigee and inclination.

LHCP — Left-hand circular polarization — counterclockwise.

LOS — Loss of signal — The time when the satellite passes out of range.

Mode A — Transponders with 2-meter uplink and 10-meter downlink.

Mode A/J — Simultaneous operation of the Mode A and Mode J transponders on AMSAT-OSCAR 8.

Mode B — Transponders with 70-cm uplink and 2-meter downlink.

Mode C — Equivalent to Mode B with less power output; of no discernable difference than Mode B to the user.

The idea is similar to the usual fm repeater with two important differences. Rather than using single channel fm receivers, the satellite *transponders* have a receiver designed to process any type of input. Whatever mode of transmission is sent to the satellite is detected for retransmission. Additionally, the transponder does not restrict itself to reception at one frequency; it covers a *passband* (or range of frequencies) which may be as wide as 150 kHz. Any and every signal within that passband will be detected (if strong enough). The entire transponder input passband is then amplified and changed in frequency to another band for retransmission. A typical example is the Phase III transponder. It receives signals over a 150 kHz passband near 435.2 MHz, converting them to 2 meters. The identical 150 kHz wide set of signals is retransmitted near 145.9 MHz.

The transponders are called linear transponders because the retransmitted output is a faithful reproduction of its input. Whatever you transmit will be retransmitted exactly as it is received. This is not a case of putting a 2-meter transmitter's microphone next to the 70 centimeter receiver's speaker; the entire transponder input passband is converted to the output passband frequency. Moving your transmit (uplink) frequency will move your retransmitted (downlink) frequency.

As is often the case, there are exceptions. On certain satellites the downlink passband is inverted from the uplink input. With an inverting transponder, like on Phase III-A, moving higher in the uplink passband places you lower in the downlink passband. Similarly, single sideband signals will "flip" over; lower sideband becomes upper sideband and vice versa. Incidentally, the convention is to use ***upper*** sideband on ***uplink***. If the transponder is noninverting, the downlink will be upper sideband; an inverting transponder will respond with lower sideband. Initially, you'll find this a little confusing, but with experience, doubts will be eliminated.

With all the fm transceivers available and the resultant activity, why is all the activity restricted primarily to ssb and cw? The reason is spacecraft power allocation. Fm is a constant carrier mode of transmission, draining battery power at a much higher rate than low-duty-cycle modes such as ssb or cw. Use of fm by a number of users could discharge the spacecraft batteries during even a single orbit. A-m and RTTY, also continuous carrier modes, should be avoided. On-off keying, however, provides an alternative to standard RTTY. Normally both the mark and space tones are transmitted. If only the mark tone is transmitted, the duty cycle drops to half the constant carrier value. Simple modification of ground station equipment makes the previous proposal a feasible venture. Though the bit error rate is substantially increased, at high signal-to-noise rations this should not be a problem. High speed ASCII data transmission, on the other hand, requires constant carrier for acceptable bit error rates, but normally these transmissions are short in duration. Similarly, slow-scan television is constant carrier, but each frame only lasts but several seconds, so the total carrier time is small.

Phase II vs. Phase III Satellites

The most significant differences from a user's point of view between a Phase II satellite and a Phase III satellite is orbit. Phase II satellites reside in nearly circular,

Mode D — Battery-recharge mode; transponders off.
Mode J — Transponders with 2-meter uplink and 70-cm downlink.
NASA — National Aeronautics and Space Administration — Government agency that has provided "piggyback" launch opportunities for AMSAT OSCARs 5, 6, 7 and 8 in recognition of the OSCAR program's contributions.
OSCAR — Orbiting Satellite Carrying Amateur Radio; there have been eight Amateur Radio satellites named OSCAR at the end of 1979 and two Soviet Amateur Radio satellites designated RS1 and RS2.
OSCAR Education Program — A special program that brings live demonstrations of the OSCAR satellites to classrooms, helping teach students physics, space science, astronomy and related subjects. Teachers use ARRL curriculum materials to structure their courses around the OSCAR satellites.
OSCARLOCATOR — A satellite tracking device consisting of a ranging oval and groundtracks superimposed on a polar projection map.
Pass — An orbit of the satellite.
Passband — The range of frequencies handled by a satellite transponder.
Perigee — That point in a satellite orbit where it passes closest to earth.
Period — The time it takes for a complete orbit, usually measured from one EQX to the next. The higher the altitude, the longer the period.
Phase I — The term given to the earliest, short-lived OSCAR satellites that were not equipped with solar cells. When their batteries were depleted, they ceased operating.
Phase II — The term given to low altitude, long-lived satellites. Equipped with solar panels that powered the spacecraft systems and recharged its batteries, these satellites have been shown to be capable of lasting up to five years. (OSCARs 6, 7 and 8.)
Phase III — Extended-range, high-orbit satellites, typically in either elliptical orbit, as AMSAT Phase III-A, or in geostationary orbit.
Power Budget — A determination of how much power is actually available to operate the on-board satellite systems, taking into account such things as solar cell surface area, solar cell efficiency and angle toward the sun. A positive power budget means that ample power will be available to power the desired systems; a negative power budget means that periods of shutdown and recharge must be periodically scheduled.
Precession — An effect that will be characteristic of the AMSAT Phase III-A orbit; the satellite apogee after the firing of the perigee kick motor will occur at about 26° N latitude, but will gradually rise higher to 57° N latitude within a few years. Then, gradually, it will move lower in latitude until after five years or so, the apogee will occur near the equator.
Project OSCAR — California-based group, among the first to recognize the potential of space for Amateur Radio; responsible for OSCARs 1 through 4.
QRP Test — Special orbits set aside for operating through the satellites while using a maximum of 10 watts erp; output powers of less than 1 watt have proven effective in some cases.
Reference Orbit — The orbit beginning with the first ascending node during a given day UTC.
RHCP — Right hand circular polarization — clockwise.
RS1 and RS2 — The first two Soviet Amateur Radio satellites.
Secondary Payload — Usually smaller packages that in essence share the launch with the primary payload which is the main purpose for the launch. OSCAR 8, for example, was a secondary payload "hitchhiker" to the LANDSAT C Earth Resources Technology Satellite primary payload; secondary payloads, in effect, substitute for ballast weight.
Spin Modulation — Periodic amplitude-fade and -peak resulting from Phase III-A 60-rpm spin; the effect is a 3-Hz "modulation" of the passband.
SSC — Special service channels — Frequencies in the downlink passband of AMSAT Phase III-A that are designated for authorized, scheduled use in such areas as education, data exchange, scientific experimentation, bulletins and official traffic.
SSP — Subsatellite point — That point directly beneath a satellite on the surface of the earth at a given instant; usually defined in terms of latitude and longitude.
Sun Sensor — A device to be used on AMSAT Phase III-A to determine the spacecraft orientation to the sun.
Sun-synchronous — A type of orbit that approximates the sun's apparent movement; satellites in sun-synchronous orbit, such as OSCARs 6, 7 and 8, can be heard at one location at about the same times each day.
SYNCART — Synchronous Satellite Carrying an Amateur Radio Transponder.
TCA — Time of closest approach — a datum used in determining orbital characteristics by plotting Doppler shift against time.
Telecommand Station — Designated ground stations assigned the responsibility for commanding the satellites into scheduled or protection modes.
Telemetry Beacon — The transmitters aboard each satellite that enable ground stations to monitor vital functions.
Transponder — The "repeater" aboard the satellites (OSCARs 7 and 8 have two) that retransmits signals it receives on a different frequency band.
UOSAT — Present designation of the University of Surrey's educational and scientific Amateur Radio satellite.
Uplink — The frequency at which radio signals are transmitted to the satellites.
UTC — Universal Coordinated Time, the time referenced to the Greenwich Meridian.

low-altitude orbits. The altitude, which is about 925 kilometers for OSCAR 8, has several consequences. A station's range circle is relatively small, which means that the maximum time any one station can access the satellite is about 20 minutes for an overhead pass. Access during more distant passes is correspondingly shorter. The maximum distance any one station can communicate is typically about 3000 miles. While adequate for short QSOs within relatively limited areas, these orbits place limits on organized communications networks.

Phase III satellites are characterized by high orbits. With an apogee of approximately 36,000 kilometers and a perigee of about 1500 kilometers, each orbit will be about 11 hours long. For the first few years of satellite life, the apogee will occur over the earth's northern hemisphere. At apogee, where its coverage will be greatest, the satellite will travel slowest. It will spend the greatest portion of its orbit where communications range is greatest. At apogee and for a few hours before and after, the satellite will be accessible to nearly half the earth. Any station sharing access will be able to communicate with any other. Though this is ideal for scheduled, long term communications schemes, it may lack the spontaneity and "mystery" some amateurs find appealing. But the reliability of AMSAT-OSCAR Phase III-A will prove much more important.

Using a Phase II Satellite

For the remainder of this chapter, it is assumed that you have a working ground station or are in the process of assembling one. We'll concentrate on what you have to do with that station to contact other hams.

The first step in working the satellite is to find out where the satellite is. As the spacecraft is constantly moving, some type of tracking calculator is needed. The simplest type of tracking device is an OSCARLOCATOR, available from ARRL. Briefly, you use one reference point each day, usually the first time in a given (UTC) day that the satellite crosses the equator in a northerly equation. The equator crossing (EQX) time and longitude, when and where the satellite passes over the equator, is available from numerous sources such as W1AW Bulletins and the various amateur publications. With this information, you can easily determine the satellite's approximate location and where to point your antenna at any given time from any point in the Northern Hemisphere.

More-sophisticated calculating techniques are also possible. Actual antenna bearings can be mathematically calculated and recorded for a large number of points during a satellite pass. Programmable calculators and even computers facilitate this approach.

As home computers become more popular, mechanized tracking systems will doubtless become standard, further simplifying the tracking routine. Several hams already use computerized systems, and the heavy involvement of computers in the Phase III project will bring more computers into use in the ham shack.

Regardless of the approach taken, it is a good idea to document your calculations well in advance of the pass. Jot down all the information in a logical fashion on a work sheet that is easy to read. You will have enough to do during a pass, with turning your receiver and transmitter, aiming antennas, logging and actually

Satellite Ranging Experiment

Have you ever wondered how far OSCAR is from your QTH as it speeds from horizon to horizon? If you can measure the time it takes a signal to make the round trip, it's a simple calculation from there.

An oscilloscope is one way to make the measurement. By triggering its sweep with the transmitted pulse and displaying the received pulse returning from OSCAR, you can read the elapsed time. This figure will consist of the time it takes the signal to travel from your station to OSCAR and back, plus any phase delays in the equipment. Knowing that the delays in your transmitter and in the transponder onboard OSCAR are negligible (measured in microseconds), you need only measure the phase delays in your receiver to make the calculations. Here's how.

Connect your station as shown in the diagram. If you do not have a dual-trace oscilloscope, a single-trace scope with external trigger, or events-per-unit-time counter will do well.

Feed the keyed dc pulses from the key *input* to your transmitter to channel A on a dual-trace scope (such that the pulse triggers both sweeps A and B simultaneously) to the external trigger on a single-trace scope, or to the start connection on an events-counter. Feed the received pulse from your receiver audio output to channel B, or to the signal input on a single-trace scope, or to the events-counter's stop connection. Set the horizontal scale of the oscilloscope to 5 or 10 milliseconds per division. Then, transmit a series of pulses to OSCAR and observe the display.

The time between the initiation of the transmitted pulse (at the start of the sweep) and the leading edge of the received pulse (that point at which it first rises) can be read directly off the oscilloscope display. This is your total time figure.

To measure the phase delay inherent in your receiver, you'll need a signal source at its frequency. An rf generator or a receive converter that converts your transmitter's output to your receiver's frequency will work equally well. In much the same way as before, trigger the sweep with the dc pulse that turns your signal source on and off, and feed the received audio pulse to the signal input on the oscilloscope. The time between the beginning of the trace and the leading edge of the received pulse is the total phase delay in your receiver. It should be about two milliseconds.

The calculations are simple. Subtract the phase delay from the total time and you have the time it took your signal to make the journey up and back. To calculate the range, you only need to know that radio waves travel at the speed of light: 300,000 km/sec. Then using the formula: 300,000 km/sec. × elapsed seconds = distance, and distance traveled/2 = range, you have the answer.

OSCAR 8 orbits at an altitude of approximately 910 km. Thus, the minimum expected range, when it is directly overhead, will show approximately a 6 ms difference between transmitted and received signals. The maximum range at the horizon is about 3200 km which will show a maximum difference between transmitted and received signals of about 21 ms. Therefore, during a nearly overhead pass, you should expect to see the time between transmit and receive vary from around 20 ms at access of signal to about 6 ms halfway through its journey over your QTH, and back up to about 20 ms at loss of signal (LOS).

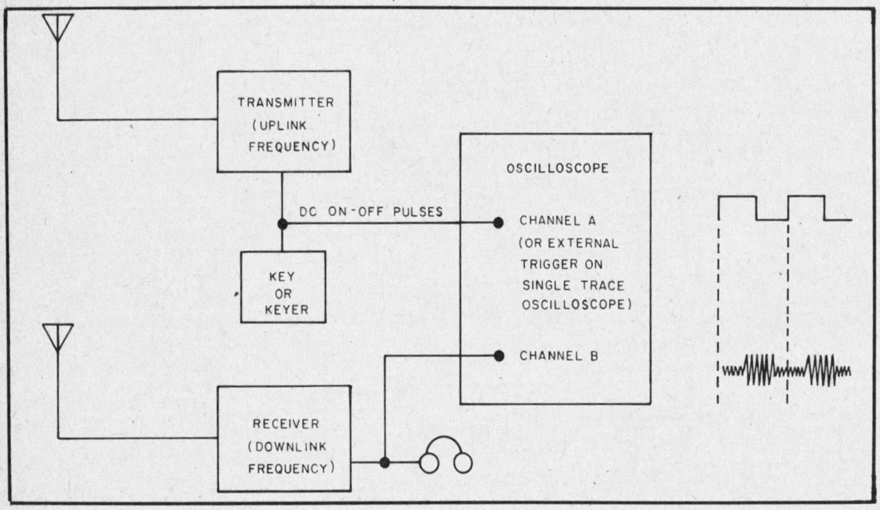

Block diagram of the ranging experiment. It can be done either with a single-trace oscilloscope or a dual-trace oscilloscope.

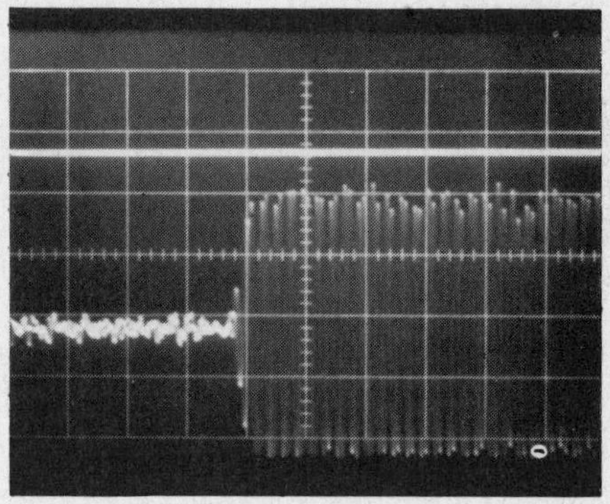

Ranging experiment through OSCAR 7. The horizontal scale is set at 5 ms per division. The upper trace shows a portion of the dc pulse which keyed the transmitter and triggered both sweeps. The lower trace shows the received audio. There is an elapsed time of about 19 ms from the rise of the transmit pulse to the leading edge of the received pulse.

making contacts, without adding the burden of calculating the orbital data at the same time!

Using your system, whatever it may be, figure out when AOS is for your location and the azimuth (compass heading) to where the satellite will rise. Since the satellite will come into "view" on the horizon, you automatically know that the required antenna elevation at AOS is Ø degrees; in other words the antenna points parallel to the ground. Record on the chart your calculated AOS time, and the associated antenna azimuth and elevation. Now, recalculate the antenna readings for the time one minute after AOS. Record these values on the chart. Continue for every one-minute interval until LOS. If you later find that your antennas are very sharp in beamwidth and require greater pointing accuracy, reduce the interval to 30 seconds.

Using all this information is much simpler than it sounds. At the times indicated on your chart, simply point your antennas in the azimuth and elevation shown. It's that simple. Stick with the chart. Often times satellite users won't have calculated too well, or they figure that they don't need advance planning. Signals start fading or are weak to begin with, so they wildly start rotating their antennas in an attempt to find the satellite. Quite obviously, this is not the most effective approach. Find a tracking system you are comfortable with and plan the pass in advance.

Now that the preliminaries are completed, the actual fun of working through the satellite can begin once it's in range. The best procedure is to have your antennas pointed toward the AOS point and begin listening at the transponder beacon frequency several minutes before you expect acquisition. You'll probably have to tune a little off the published frequency as Doppler shift makes the signal appear higher or lower than expected. If you don't hear the beacon within a minute or so of what your chart predicts, something is wrong. Likely possibilities are miscalculations on your part, inaccurate orbital data, listening to the wrong transponder for that day, equipment failure, or even a miss-set clock. The keynote of satellites is reliability; failure is rare compared with various equipment failures and mistakes encountered on the ground.

Once the satellite is accessed, you can set about the task of making a contact. The toughest part is setting your transmitter frequency so that it is retransmitted on the proper downlink frequency. Calculating the expected downlink frequency for a given uplink frequency will put you fairly close. An alternative is to use a reference chart, which plots output frequency versus input frequency. With either of these methods, Doppler shifts and equipment frequency inaccuracies will only place you close to the correct frequencies, but close enough so that sending a few words of sideband or a few dits will enable you to pinpoint yourself. The key word is few. Unfortunately, you'll often hear a series of dits swishing up and down the passband. Not only is this inconsiderate, it also wastes time, which is in very short supply during a pass. Initially this may seem like an impossible task or one best suited to an octopus. With a little practice, however, the longing for an extra set of hands gradually subsides.

Since you're listening on an entirely different band from that on which you're transmitting, full duplex operation is possible: you hear what you're sending through the satellite as you send it! (Traffic handlers often go to great effort to achieve this.) The ability of the station you're working to interrupt you while you're transmitting greatly improves the natural flow of the contact. It is even common for a CQ to be answered while it's being sent. Once again, anything to save time is to your advantage.

Typical contacts through the satellite are short contest-type exchanges. Exchanging calls, signal reports, locations and names is the extent of most contacts. Time is of the essence: Long-windedness doesn't work. Like an fm repeater, the satellite eventually times out, in this case, by going out of range.

A typical contact might go like this:

CQ CQ CQ DE W1AW W1AW W1 . . .

W1AW DE K1XA 579 CONN BOB

K1XA DE W1AW 569 CONN CHUCK 73

W1AW DE K1JX . . .

There is one other very important aspect for the satellite user to consider. The transponder's internal automatic gain control system reduces the receiver's sensitivity as the signal input level increases. This is done for two reasons. The first is to prevent receiver overload which might produce spurious signals within the transmitter passband. Second and more important, is that theoretically a very loud signal would be retransmitted with as much power as the transponder's transmitter can muster. If this were to occur, serious damage could be sustained by the satellite. Even with the agc system, transmitter current drain rises with very strong input signals, placing unnecessary and undesirable burden on the spacecraft batteries. It is every satellite user's responsibility to ensure that he or she does not transmit too great a signal to the satellite.

The proper transmitter power level is easy to determine. Since the downlink signal you receive indicates the effectiveness of your uplink, adjustments of your uplink power are reflected in the strength of the downlink signal strength.

Azimuth-elevation mount used to track Amateur Radio satellites. The antenna on the left is a commercial 2-meter antenna with switchable circular polarization. On the right is a homemade nine-turn helix antenna. This helix is featured in May 1980 *QST*. (*W9KDR photo*)

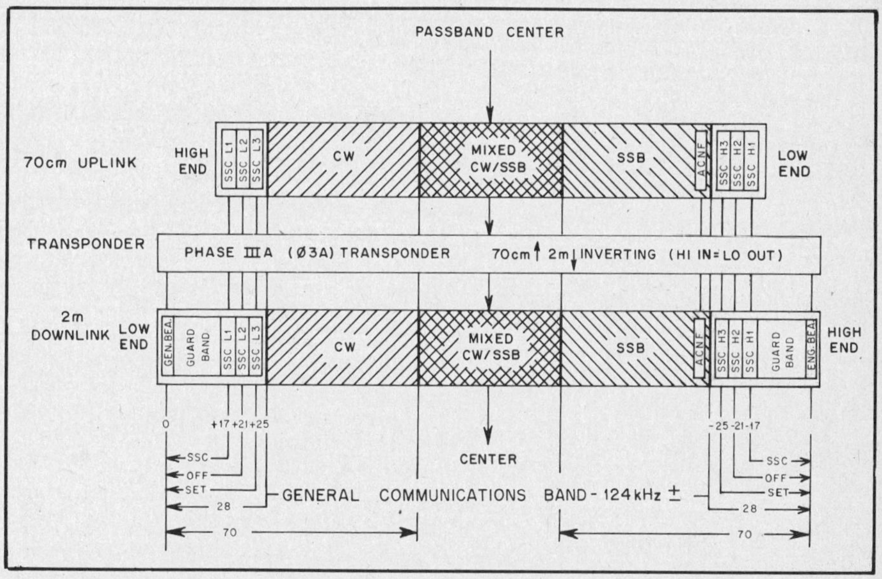

AMSAT-OSCAR Phase III-A Bandplan: ACNF = AMSAT Coordination and Network Frequency; SSC = Special Service Channel; Beacons: General (GB) = 145.810 MHz; Engineering (EB) = 145.990 MHz; Mixed cw/ssb subband tentatively 40 kHz wide. Lower edge = f_{GB} + 70 kHz, upper edge = f_{EB} − 70 kHz. All frequencies subject to final calibration. (Thanks to WA2LQQ for his assistance.)

Your downlink strength should be no louder than the transponder beacon. Following this simple rule prevents receiver desense and guarantees easy transponder access for others as well as preventing satellite damage.

Using the Phase III Satellites

Many of the techniques just described for Phase II satellites apply as well to the Phase III satellites. The big difference is orbit. Unfortunately, the very virtue of the Phase III orbit is its failing when it comes to tracking simplicity. Simple plotters and calculations will not work for elliptical orbits. Both the size of the range circle and the satellite's velocity constantly change. These combine to make simple OSCARLOCATORs inadequate for Phase III.

Through the efforts of amateurs such as K2ZRO, K2UBC and W2GFF, there are now plotters available for Phase III use. These are the least expensive approach to

Doppler Shift

Doppler shift is caused by the relative motion between you and the satellite. The relativistic velocities between the wavefront of the signal traveling to or from the satellite and the satellite's movement toward or away from the observer either add or subtract, making the observed frequency change. As the satellite moves toward you, the frequency of the downlink signals will increase by a small amount (the velocity of the satellite relative to the observer *adds* to the velocity of the wave front).

As the satellite passes overhead and starts to move away from you, there is a sudden drop in frequency of a few kilohertz, much the same way as the tone of a car horn or train whistle drops as the vehicle moves past. This Doppler effect is different for stations located at different distances from the satellite, because the velocity of the satellite relative to the observer is dependent on the observer's distance from the satellite. The result of all this is that signals passing through the satellite transponder move slowly around the calculated downlink frequency. Locating your downlink signal is more than just a simple computation, as tuning is needed to compensate for Doppler shift.

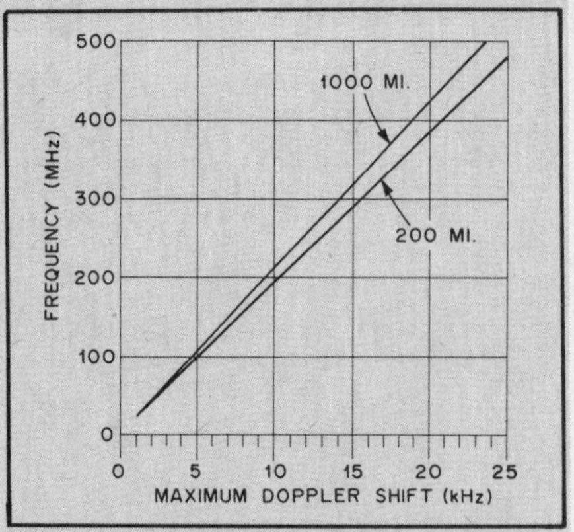

Satellite transmitter frequency versus Doppler shift for satellite in 200- or 1000-statute-mile orbits. For a translator, use the difference between uplink and downlink frequencies as the "frequency."

Doppler shift through a transponder becomes the sum of the Doppler shifts of both the uplink and downlink signals. In the case of an inverting type transponder (as in OSCAR 7 Mode B or OSCAR 8 Mode J), an increase in the uplink frequency causes a decrease in downlink frequency, so the resultant Doppler shift is the difference of the Doppler shifts, rather than the sum. The shifts tend to cancel.

solving the tracking problem. Alternatively, the use of a computer to aim the antennas automatically is ideal. Both techniques reduce tracking difficulty to the level of that encountered with using a Phase II satellite.

The Phase III satellite will open up a new vista for amateurs. Large areas of the earth (up to an entire hemisphere) will be workable up to 10 hours per orbit. The communications will be absolutely reliable. No longer will amateurs be subject to the vagaries of the ionosphere when trying to maintain scheduled communications. The potential user population at any time is staggering.

To take full advantage of all these blessings with the limited transponder bandwidth will require coordination techniques not yet employed before within Amateur Radio. Additionally, certain hardware features on board the satellite will present new challenges and benefits to hams.

The universal bandplan mentioned earlier is the basis for the Phase III bandplan. The bandplan is based on the downlink passband. This means that the downlink passband is the frame of reference for passband partitioning. The user is thus obliged to control his uplink frequency to assure that the resultant downlink frequency falls within the recommended allocation for that mode.

On Phase III-A, the 180 kHz downlink passband is partitioned into three major usage bands. These are further divided into subbands according to emission type. The three major usage bands are described below.

The general communications band is tentatively 124 kHz wide with edges at: 145.838 and 145.962 MHz. This band is further divided into three subbands by emission type as follows:

Cw: 145.838-145.880 MHz
Mixed cw/ssb: 145.880-145.920 MHz
Ssb: 145.920-145.962 MHz

This partitioning is essentially identical to Phase II satellite bandplans. The "special" emissions currently used (i.e., RTTY, SSTV) are allocated as shown in the bandplan.

All Special Service Channels are 4 kHz wide, and their placement is at the passband extremities just above and below the beacons and their associated guard-bands. The various SSC functions are explained later.

Beacons and Guard Bands

The final portion of the 2-meter downlink passband is assigned to the engineering beacon and the general beacon. Protecting these beacons from interference are 17 kHz wide guard bands. The engineering beacon will be sent in a computer compatible format while the general beacon will transmit information in both RTTY and Morse Code.

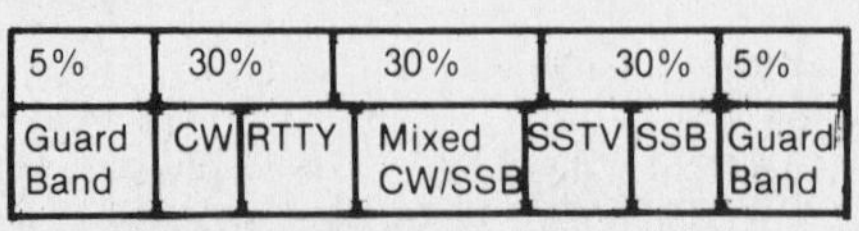

The percentage numbers along the top line represent portions of the passband allocated to the mode indicated below. RTTY and SSTV are usually confined to the frequencies directly adjacent to the boundaries between the 30% wide band segments. The so-called guard bands protect the beacons from being interfered with by signals in the passband. These guard bands should be avoided entirely.

The Channelized Special Services Band

The second major band within the 2-meter downlink is allocated to channelized special services. This is a new direction for amateur satellites, as it is in marked contrast to the non-channelized communications predominating the amateur bands (with the notable exception of the vhf/uhf repeater bands). The six Special Service Channels (SSCs) are as follows: (L = Low end of 2-meter passband; H = High end).

L1 — Scientific channel.

L2 — AMICON channel (AMSAT International Computer Network).

L3 — NTS (National Traffic System) channel. Formal cw record traffic handling and/or its international equivalent.

H1 — Code practice and cw/RTTY bulletins.

H2 — Education channel.

H3 — Voice bulletin channel.

The scientific channel (L1) will be available to any amateur or noncommercial group (with a ham sponsor) on a scheduled basis, to perform serious scientific experimentation. Typical experiments might include satellite-assisted ELT location, mobile long distance EKG transmission, propagation study and long based interferometry.

The AMICON channel (L2) is dedicated to providing a means for licensed amateurs who are also computer hobbyists to network their computers through the satellite. Typical applications might be the establishment of an electronic post office, linking several computers to expand memory capabilities, multiple batch processing, real time data analysis from multiple locations and software exchange. The real potential for this channel has yet to be conceived.

L3 (NTS) is dedicated to daily formal message handling for the National Traffic System, taking advantage of the satellite's reliability. This reliability will hopefully lead to the realization of two goals long sought by amateurs.

The first is round-the-clock message traffic capability, now limited by ionospheric propagation considerations. This would greatly facilitate the handling of emergency communications, common during disasters. On-the-scene communications from the disaster site will be greatly enhanced both because of the ultimate reliability of the system and the ease with which an effective ground station can be set up. No longer will propagation play a role in limiting the communication effectiveness from a remote disaster site. NTS will serve to provide long-range communications to the ARES.

The second goal which might be

reached is a progression from the National Traffic System to an *Inter*National Traffic System. The ITS would provide for an organized link between North America and Central America, South America and whatever other areas wish to become involved. SSC L3 is the culmination of the activity that first tied amateurs together: relaying of messages.

H1 will provide amateurs and would-be amateurs all over the world with code practice, as well as information bulletins on cw (also a form of code practice) and radioteletype.

The Education channel (H2): scheduled demonstrations to educational groups, such as school groups, will be conducted on this channel. Not only will scientific phenomena such as Doppler shift be demonstrated, but student exchange on a variety of subjects will be featured. The use of RTTY and SSTV will enhance the exchange capabilities between different student groups.

H3, the voice bulletin channel, will be employed (similar to H1) to disseminate information of interest to the amateur population as a whole. Information pertinent to the operation of the satellite, such as orbital data, will be broadcast. Bulletins of general interest, comparable to those presently sent by W1AW and GB2RS, will be transmitted on both a real time and prerecorded basis.

Telemetry and What it Means

As mentioned earlier, the satellite transmits a continuous report of its internal conditions in the form of telemetry. Even OSCAR 1, though primitive by today's standards, sent telemetry by varying the speed of its Morse code identifier. Phase II satellites send their information either via RTTY or cw. From labeled code groups, a formula can be applied to calculate actual values of the parameters measured.

The Phase III satellites will send telemetry in three forms. The first and least complicated form is Morse code sent through the General Beacon. These groups will also be repeated on RTTY on the General Beacon. More detailed data will be transmitted in computer-compatible ASCII on the phase-shift keying modulated Engineering Beacon. Special hardware will be required to receive and decode information sent on the Engineering Beacon. One major change over the Phase II satellites is that the computer on-board the Phase III spacecraft will perform the calculations necessary to scale the telemetry parameters to their actual values.

The information sent by the satellite is not only for the benefit of the people responsible for successful spacecraft operation. The average ham can learn astonishing amounts of information from the telemetry. Spin rate of the satellite, battery condition and dozens of other

Station Requirements

Most long-distance amateur communications depend on the application of natural phenomena such as ionospheric refraction or tropospheric scattering. Station design to take advantage of these phenomena is largely a matter of personal preference; bigger stations usually mean bigger signals. Operation through a satellite is somewhat different; fixed conditions such as transponder receiver sensitivity, transponder power output and path loss are the dominating factors in establishing minimum ground station requirements. Unlike other areas of amateur interest, having an inadequate station won't merely limit satellite communication effectiveness, but might prohibit success entirely. Technical considerations in station design for satellite operating are more than just personal preference.

Satisfactory satellite communications might be possible with your existing station, but planning your satellite ground station around certain parameters will enhance your effectiveness and enjoyment. The orbit of the Phase III satellites imposes greater demands on a ground station, so if you build your station to the rigors of working through Phase III, only slight adjustments in transmitter output power will be necessary for work through a Phase II satellite.

Here are the parameters you should build your station around, for present and planned satellites.

1) 146-MHz antenna gain of at least 13 dBi (decibels over isotropic), circularly polarized. The polarization should be switchable between left- and right-hand circular. The antenna should be accurately pointable in both azimuth and elevation. More antenna gain is always desirable; when the downlink is on 2 meters, signal strengths are increased with greater gain. When the uplink is on 2 meters, lower transmitter power is necessary to achieve a given effective radiated power. A disadvantage of higher gain antennas is that they have narrower beamwidths and must be aimed more accurately.

2) 435 MHz antenna gain of at least 13 dBi, circularly polarized, again switchable. The very same considerations apply to 435 MHz as just described for 146 MHz.

3) 10-meter antenna gain of 2 dBi. The key factor here is not high gain, but rather effective coverage. Antennas placed above ground (not being in free space) are subject to ground-reflection interference, causing nulls in the vertical pattern of the antenna. Graphical presentations of this effect are shown in the *ARRL Antenna Book*. The best bet here is to use a virtually non-directional antenna placed at a height above ground to take advantage of favorable ground reflections.

4) Feed line loss of less than 2 dB.

5) 146 MHz EIRP (*E*ffective *R*adiated *P*ower over an *I*sotropic source) of 100 watts. Your EIRP can be calculated by multiplying your transmitter power output in watts, by your antenna gain minus feed line loss. Example:

10 watts transmitter output
12 dBi antenna gain (absolute gain = $10\,\frac{\text{gain dB}}{10}$; $G = 10^{12/10} = 15.85$)
2 dB feed line loss (same as above, but loss is "negative gain"; $G = 10^{-2/10} = 0.63$)
so
$10\ W \times 15.85 \times .63 + 100\ W$ EIRP

6) 432 MHz EIRP of 800 Watts.

7) 146 MHz receive noise figure of 4 dB. Any loss between the antenna and the receiver input is directly added to the receiver noise figure (i.e., 2 dB line loss plus 3 dB receiver noise figure equals 5 dB receiver noise figure). Higher noise figures reduce signal to noise ratios.

8) 432 MHz receive noise figure of less than 4 dB.

9) 10-meter receive noise figure of 4 dB or less.

10) Frequency readout resolution at least 5 kHz, on both transmit and receive. 1 kHz readout is better. Accurate frequency determination facilitates finding your own signal through the satellite without causing needless QRM.

11) An accurate means of determining antenna pointing directions at all times during an orbit. This could be in the form of calculations, a plotter, a computer printout or even computer controlled antenna pointing.

Like any other aspect of hamming, trade-offs can be made which will minimize both expenditure of time and money. Up to a point, antenna gain is easier to attain than transmitter power, and a lot cheaper. It's easier to put up a bigger Yagi to increase your EIRP than it is to build or buy an amplifier. The antenna also works on receive, doubling the benefit of your antenna efforts. All things considered it's much more cost effective and desirable from a performance standpoint to optimize the receiving set-up before investing in bigger transmit ware. All too many stations suffer from "alligatorism," (big mouth, little ears): don't be the cause of frustration to others, as well as yourself.

AMSAT Nets and Bulletins Schedules

The following AMSAT Nets meet regularly to disseminate information to newcomers and to keep regular satellite users in communication with one another.

USA-East Coast Net	Wednesdays	0100Z	3850 kHz	Net Control WA3NAN
USA-Mid States Net	Wednesdays	0200Z	3850 kHz	Net Control WØCY
USA-West Coast Net	Wednesdays	0300Z	3850 kHz	Net Control W6CG/W6DOW
International Net	Sundays	1800Z	14,280 kHz	Net Control WA3NAN
International Net	Sundays	1900Z	21,280 kHz	Net Control WA3NAN
AMSAT Calling Frequency		24 hours	28,880 kHz	

Bulletins of general interest to those interested in amateur satellites are transmitted regularly on OSCAR-7 and OSCAR-8 reference orbits, at approximately 10 minutes after ascending node. These bulletins, transmitted on a downlink frequency of approximately 29,490 kHz, 145.960 MHz and 435.160 MHz can be received over most of eastern North America.

The Amateur Satellites: A History

OSCAR 1, the first of the "Phase I" satellites, was launched on December 12, 1961. It sent information concerning its internal temperature back to earth on the 2-meter band; the new traditional cw identification HI varied in code speed as the temperature changed. The 0.10-watt transmitter onboard discharged the batteries after only three weeks, but the mission was so successful that construction of a second satellite was begun.

OSCAR 2 was launched on June 2, 1962. Virtually identical to OSCAR 1, Amateur Radio's second venture into space lasted 18 days.

OSCAR 3, launched March 9, 1965, was the world's first free-access communications satellite. During its two-week life, over 100 pioneering amateurs in 16 countries communicated through the 1-watt linear transponder on 2 meters.

OSCAR 4 flew from Cape Kennedy on December 21, 1965. More advanced in design than OSCAR 3, this satellite had a 2-meter to 70-centimeter linear transponder with an output of 3 watts. Unfortunately, a launch vehicle defect placed OSCAR 4 into a high elliptical orbit, which prevented widespread amateur use. A handful of hams did communicate through it, however, including the first U.S.-to-U.S.S.R. satellite contact of any kind.

OSCAR 5 was built by students at Melbourne University, Australia. A new amateur organization, *AMSAT,* prepared the satellite for launch and coordinated ground activities for this satellite. OSCAR 5 transmitted telemetry about its operating parameters on both 2 meters and 10 meters. Its batteries lasted for over a month.

OSCAR 6 was the first in the second series of satellites, referred to as Phase II. Launched on October 15, 1972, it carried a 1-watt, 2- to 10-meter linear transponder. Not only was the satellite magnetically stabilized, but the internal battery package was continuously recharged by solar cells. Tens of thousands of contacts were made during the nearly five-year lifespan of this satellite.

OSCAR 7 was built by hams from West Germany, Canada, Australia and the United States. Two elaborate transponders each ran 2 watts output; one was a 2-meter to 10-meter unit like the one on OSCAR 6, the other a 70-cm to 2-meter linear transponder. Internal circuitry, as well as ground commands, controlled this sophisticated satellite's functions. OSCAR 7 was launched on November 15, 1974.

OSCAR 8, a cooperative effort of United States, Japanese, German and Canadian amateurs, was launched on March 5, 1978. The two transponders on board are a 2- to 10-meter unit as used before, and a 70-cm to 2-meter unit. Both have an output of 2 watts.

Radio Sputniks 1 and 2 were launched from the Soviet Union on October 26, 1978. Each satellite carried a 2- to 10-meter transponder with extremely high sensitivity. Like their OSCAR counterparts, the RS satellites transmitted telemetry data relating to their well-being and were commandable from the ground.

General Bulletin Time Formatting

Time	*Interval*	*Duration (Minutes)*	*Content*
TØ	Ø-1	1	Cw i-d preamble
T1	1-3	2	Basic orbital data
T2	3-6	3	Cw telemetry data
T3	6-21	15	Cw bulletin board
T4	21-26.5	5.5	RTTY rescan of above
T5	26.5-30	3.5	Cw telemetry
T6	30-31	1	Cw i-d and preamble
T7	31-33	2	Basic orbital data
T8	33-36	3	Cw telemetry data
T9	36-51	15	Cw bulletin board
T10	51-56	5	RTTY week's orbits
T11	56-60	4	Fill to the hour with cw telemetry

parameters are sent every minute. Many amateurs have made studying spacecraft telemetry a hobby unto itself. Much can be learned about space science by correlating the natural events surrounding the satellite with its telemetry.

DXing through the Satellite

Working DX through the satellites requires planning. On Phase II satellites, communicating over the longest distances, requires operating when the spacecraft is just above the horizon. Since acquisition time is very short at this point, very alert operating and careful planning is necessary. Even with all the handicaps, several dedicated satellite DXers have achieved satellite DXCC.

Phase III will be entirely different. Communication over most of the Northern Hemisphere will be routine when the satellite is near apogee. However, communication with points outside this area will be possible by using the satellite when it is far from apogee. Virtually every DXCC country will be workable from the United States over the life of the satellite. Admittedly, the targets of opportunity for working certain areas will present themselves for only a few minutes every few weeks, but that's the challenge, isn't it?

Successful DXing requires an effective station, but more importantly, it necessitates a keen knowledge of the satellite's orbits. Knowing when the "band is open" is far more important than owning an expensive station. The use of an OSCARLOCATOR for Phase II satellites or any of the more sophisticated plotters for Phase III will be adequate for planning DX work.

The Space Age

The space age of Amateur Radio is here — and you're invited to take part. Amateur satellite engineers from four continents have had their creations put into earth orbit. More are sure to follow.

Chapter 12

Visual Communications

Among the most notable features of Amateur Radio is its operating diversity. With increasingly sophisticated solid-state technology continuing to be developed, the greatest diversity for Amateur Radio has proven to be in the field of visual communications.

In the early days, "spark gap" radio amateurs communicated strictly by Morse code. Later, telephony was developed. Shortly after WW II, radioteletype was introduced, allowing operators to really see what was being said. Today, RTTY is among the most popular "exotic" modes of Amateur Radio. A spinoff from the early days of space exploration has been facsimile. And during the past decade the video forms of Amateur Radio communication — slow-, medium- and fast-scan television — have come about.

Radioteletype

Appearing forth with automated precision, radioteletype offers many attractions to Amateur Radio operators. A person can delve into the mechanical workings of a surplus commercial machine and have it "on-line" again, handle message traffic 24 hours a day without being near the terminal, compose artistic creations with letters and numbers, or just watch characters appear on a video screen as the text is sent at 60, 67, 75 or even 100 words per minute.

Radioteletype (RTTY) uses a typewriter-like keyboard to encode the characters into a standard code for transmission by conventional amateur equipment. It is received, decoded and appears on a similar typewriter-like machine. Watching a 'TTY machine print all by itself can be a bit unnerving — it is like watching a player piano perform, except that the keys do not move.

The technical standard is known as the International Telegraphic Alphabet No. 2. Popularly known as the Baudot code, each character is represented by a combination of five *message units*. The transmitting machine sends the units in a matter of milliseconds, so to synchronize the receiving machine, a start pulse of one unit precedes each character and a stop pulse of 1.4 units concludes it, making the total sequence 7.4 units long. To have some uniformity in teleprinter speeds, yet allow flexibility for various systems, the FCC allows either 60, 67, 75 or 100 wpm, which means the baud rates (bits/second) are 45, 50, 56.25 and 75, respectively.

Much like cw or binary code, each of the five message units is either in an "on" or "off" state known as *mark* and *space*, respectively. Since the permutation of $2^5 = 32$, you can see that enough combinations are available for the 26 letters, but what about numerals and punctuation? On a conventional typewriter you would hold down the shift key while hitting the desired capital letter or punctuation. You can make a 'TTY machine double up by

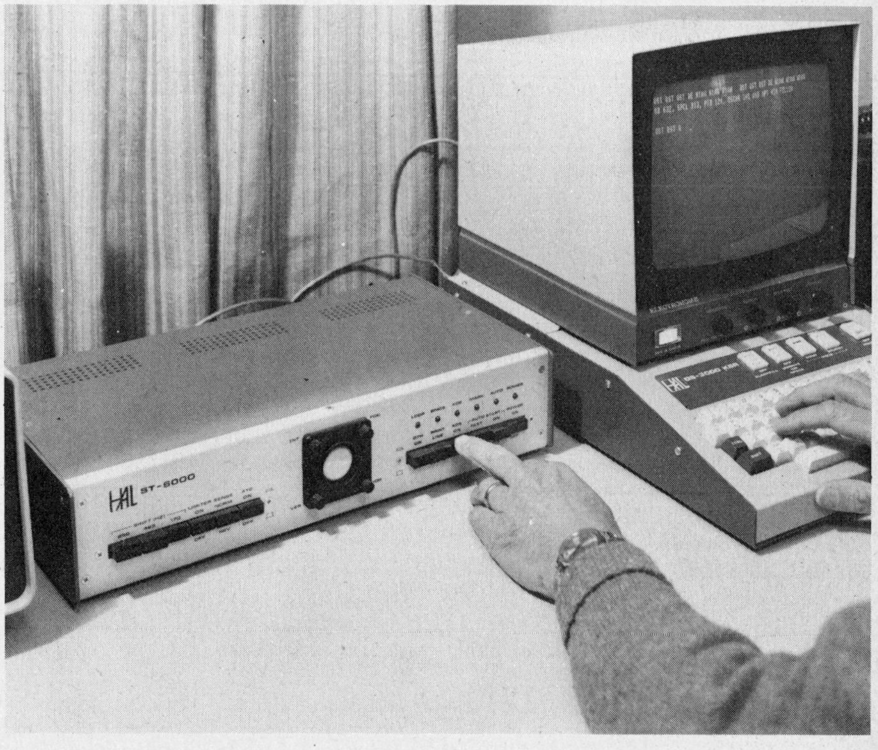

Fig. 1 — The present state of the art in RTTY encompasses all-solid-state TU, keyboard and display.

Fig. 2 — Most amateurs on RTTY still use surplus gear such as this Siemens teleprinter from Great Britain.

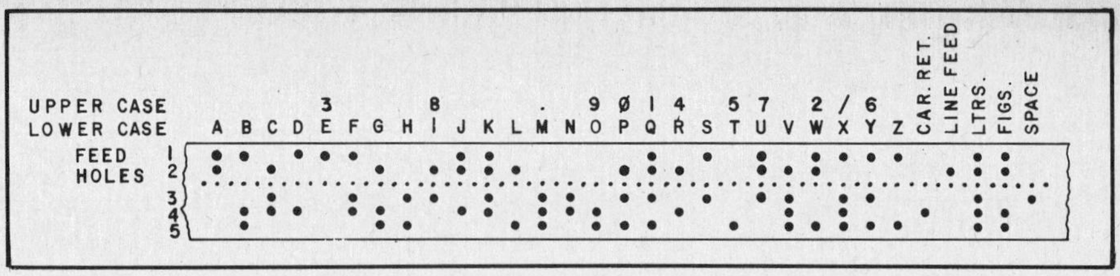

Fig. 3 — Teleprinter letter code as it appears on perforated tape. The start and stop pulses are not on the tape, but are automatically generated in the machine. The dots (perforations) indicate mark pulses. Numerals, punctuation marks and other arbitrary symbols are secured by carriage shift (FIGS key). Where undesignated positions appear in the "upper case" line of the above chart, characters may differ on various machines.

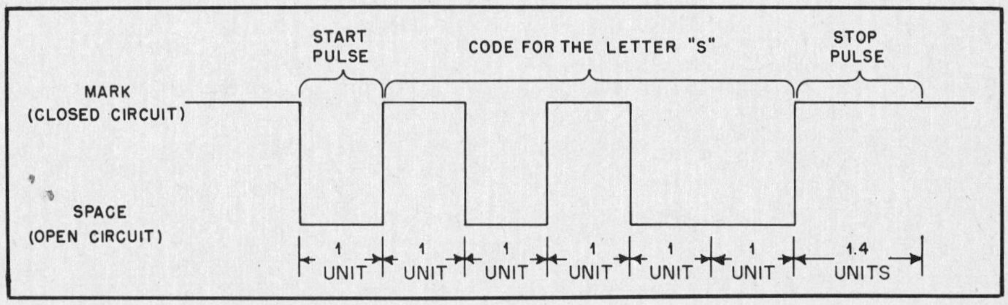

Fig. 4 — The letter S as sent on a teleprinter. At rest, a steady mark signal is sent. When a character is typed, the sending machine first precedes it with a start pulse, telling the receiving machine that a new character is starting. After that, each letter code is made up of five message units, each of which takes 0.022 second to send on a 60-wpm machine. At this speed the transmission rate is 45.5 baud (1 s/.022 s).

first hitting the FIGS (figures) key, letting go and typing the desired characters. Hit the LTRS key to return to "lower case." In practice, only capital letters are used and the "upper case" characters are arranged differently from a regular typewriter because of the 32-combination limit.

RTTY could be sent by simple on-off keying like Morse code. There is a better method, called frequency-shift keying (fsk), however. The transmitter puts out a continuous signal that is shifted back and forth between two frequencies in step with the character codes. Standard amateur practice is to send the lower frequency for space and the higher for mark. Most amateurs use a 170-Hz shift since sharp receiver filters can be used.

Setting Up

The first thing to do is to get a teleprinter. You can often obtain one of the older models such as the Model 15 or 19 for less than $50, depending on condition. If you can find an old Model 26 or perhaps one of the really ancient Model 12s, you might be able to pick it up for next to nothing. More modern machines such as the Model 28 or Model 32 usually cost several hundred dollars each. You can find further information on 'TTY machines and on technical aspects of all the modes covered in this chapter or in the *Radio Amateur's Handbook* available from the ARRL. Among the newer innovations in RTTY operation is the advent of all-solid-state "machines." They use a solid-state keyboard and "print" on a video monitor. Also, some hams who are also computer hobbyists have brought their two avocations together and use their computers to receive and transmit RTTY. Furthermore, ASCII (American Standard Code for Information Interchange), an encoding system for digital transmissions that is compatible with most personal computers, is now permitted by the FCC for use by amateurs. Chapter 13 provides insights into the union of home computers and hamming.

What about a rig? Any ssb or cw receiver can be used to receive RTTY. Assume you want to receive a 170-Hz fsk signal with its mark tone on 3600 kHz. If the receiver is adjusted for lower sideband reception and tuned to 3601 kHz, the output of the receiver will be a 1-kHz tone when the mark signal is coming in and 1.170 kHz (1170 Hz) when the space signal is being received (see Fig. 4). Standard practice is to tune the receiver so that the mark signal comes out at an audio frequency of 2125 Hz. The space tone is thus 2125 + 170 = 2295 Hz.

Most ssb receiving filters have about 2-1/2-kHz bandwidth. A cw filter with a bandwidth on the order of 400 to 600 Hz is a better choice for 170-Hz narrow-shift RTTY reception, especially when conditions are poor. Unfortunately a typical BFO is set about 1000 Hz from the center of the filter passband to give a pleasing 1000-Hz beat note when receiving cw. For RTTY, the BFO frequency must be changed to about 2200 Hz from the center frequency so that when the mark and space tones are centered in the passband, the audio output frequencies will be 2125 and 2295 Hz. If you are planning to buy a receiver, this should be taken into account. If you don't want to modify the rig, look for a receiver with a variable BFO knob or a passband tuning control.

The device that converts the audio tones from the receiver into a form decipherable by the teleprinter is called a *demodulator*. You may also see it referred to as a converter or terminal unit (TU). The standard setup is for the demodulator to supply a 60-mA current to the printer when receiving a 2125-Hz (mark) tone and no current during the space tone. A few printers may be set up to run on 20-mA current.

With any method of RTTY transmission, you should be careful not to overload the final amplifying tube or transistor and the transmitter power supply. Average power in a cw or ssb transmission is less than 40 percent of peak power, but on RTTY the figure is 100 percent. Generally speaking, you'll have to reduce power input considerably. VFO stability is more important than on cw or ssb because a frequency drift of only 50 Hz or so can cause garbled copy. These two factors might be important considerations in the choice of a rig if you plan extensive RTTY operation.

Receiver-tuning accuracy is important. Thus, a tuning aid can be a great help in proper receiver adjustment. Most tuning aids are built into the demodulator and show when the mark and space tones are coming in at exactly the right pitch. Many tube-type demodulators use a pair of front-panel-mounted neon lamps, one for the mark tone and the other for space. You just tune the receiver for maximum brightness on both lamps.

Another method uses a meter connected to the detector output. The detector produces opposite-polarity voltages for mark and space so that when a continuous signal is being sent, the average meter readings should be near zero. You tune the receiver for zero on the meter.

The popular ST-4 demodulator and others of the same series have a more accurate tuning meter system. In this circuit the meter deflects in the same direction for both mark and space signals. If the receiver is tuned correctly, the meter will read a constant value. Off-frequency RTTY will cause the meter to flicker whenever the sending station is typing.

The most accurate popular tuning aid uses a small oscilloscope tube. The mark signal is displayed as a horizontal line, and the space signal as a vertical line (see Fig. 5A). Theoretically this should appear as a "plus sign" on the 'scope face, but in most practical demodulators, it looks more like a pair of crossed bananas. When the receiver is tuned off frequency, the display looks like Fig. 5B, with the two lines shorter and tilted. If the other station's frequency shift is incorrect (or your demodulator is misadjusted), then you won't be able to get a good plus sign — if one tone is tuned in correctly, the other will be wrong (Fig. 5C).

One station accessory in the "very nice to have" category is provision for sending prerecorded messages with perforated paper tape. With this you can type taped messages ahead of time for retransmission later — at a constant 60 wpm (or whatever). Some machines (such as the Models 19 and 28ASR) have provision for punching tape by typing on the keyboard while receiving an entirely separate signal on the printer! This is very handy because you can be typing your next transmission of the QSO while simultaneously receiving the other fellow on the same machine. When he signs over to you, you just turn on your transmitter and start the tape. If you're not done typing, you can finish punching the late part of the tape while the first part is running through the "TD" (transmitter-distributor, or tape reader).

Hello CQ . . .

Once you have a station set up and have done enough listening around on RTTY to be confident you know how to operate the equipment, your next move will probably be to pick a frequency and call CQ. High-frequency fsk is okay anywhere in the cw-exclusive bands (not legal on 160 meters), but you will find most QSOs take place within 10 or 20 kHz of 3620, 7090 (sometimes 7040), 14,090, 21,090, and 28,090 kHz. Both 80 and 20 seem to be the most popular bands.

Calling CQ is quite similar to cw procedure, except that most RTTYers use a CQ tape to lessen wear and tear on the fingers. After listening first to make sure the frequency is not in use, you might send your CQ something like this:

(cw identification)
CQ CQ CQ DE KAØXYZ KAØXYZ KAØXYZ
CQ CQ CQ DE KAØXYZ KAØXYZ KAØXYZ
CQ CQ CQ DE KAØXYZ KAØXYZ KAØXYZ
THIS IS KAØXYZ JOE IN SIOUX FALLS SD K
(cw identification)

Note the cw identification. FCC regulations require i-d by voice or code at the beginning and end of each QSO and at least once very 10 minutes in between, no matter what mode you are using. This means that on hf RTTY you must i-d by Morse code. The cw may be sent either by on-off keying or by fsk; 170 Hz is the

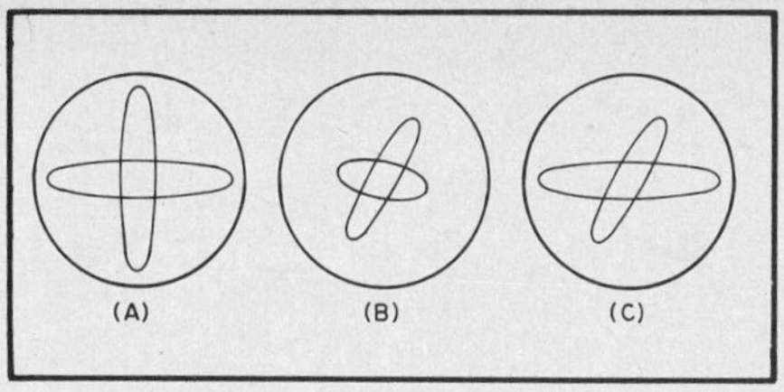

Fig. 5 — Tuning indicator oscilloscope patterns. Part A shows a properly tuned-in RTTY signal. At B the receiver is slightly off frequency. The received signal displayed at C has a different shift than what the demodulator is set for. The mark (horizontal) line is normal, but the space line is tilted and smaller in amplitude, indicating that it is coming in at the wrong frequency.

most common shift used and there is general agreement not to use less than 100 Hz. If sent automatically, it must have a speed not exceeding 20 wpm. On vhf, the identification is usually by voice. You don't have to identify on the printer, but it is common practice to send both calls on the keyboard so that anyone copying will know who is on the frequency.

Answering a CQ

If you want to answer another station's CQ, first carefully zero beat his mark frequency. It is especially important that everyone be on exactly the same frequency. If more stations join the QSO later, it becomes very annoying to retune the receiver for every station in the group. After the CQer finishes his cw i-d, give him a short call:

N1AL N1AL N1AL DE
KAØXYZ KAØXYZ KAØXYZ K
(cw i-d)

Sending his call three times gives him time to get his receiver tuned in, and sending your call three times gives him three chances to copy it through any noise or interference that might be on frequency.

If you want to break into an ongoing QSO, do it with care so as not to "step on anybody's toes." Wait until one station finishes transmitting and sends his cw i-d. Then *quickly* send BK DE KAØXYZ followed by your cw identification. If acknowledged, make a short transmission and pass it along to the next station in the group.

Now that you've made contact, the next question is "what do I say?" The answer is, "basically the same sort of thing you'd say on any other mode." Signal reporting is similar to cw — the RST system is valid for fsk, and chirps and clicks are possible on RTTY too! There is no need to over-abbreviate as is done on cw to save time:

VY FB DR OM TNX CL GLD 2 WK U AGN.

It looks a lot more cryptic typed out than it sounds on cw, and at 60 wpm this sort of thing doesn't take much less time to send than the same thing written in English. Used in moderation, abbreviations do save time, however, and it does help to economize on words — especially if you are one of those who type by the "hunt and peck" method.

Mastering the Green Keys

The touch system is the best way to type. It is not too difficult to learn. It is, in fact, quite a bit easier than learning to copy code at 13 wpm. Don't feel self-conscious, though, if you are slow at first — most beginners (and more than a few old-timers!) type with two fingers.

Typing by touch (without looking at the keys) is mainly a matter of finger placement. Look at Fig. 6. The "home" position for the fingers is with the left forefinger on the F key, and the other fingers of the left hand on D, S and A. The four right fingers are on J, K, L and CAR RET. The thumbs are used for the space bar only. You hit the Q, A, Z and FIGS keys with the little finger of the left hand. The left forefinger's territory covers R, T, F, G, V and B. Similarly the right forefinger is used for Y, U, H, J, N and M and the right little finger for P, CAR RET and blanks. All other fingers cover their home key and the keys to the upper left and lower right. Work on learning the position of two or three keys at a time. The hardest part is willing yourself not to peek! Many high schools offer typing classes for students and often have night classes for adults. But all that is really needed to teach yourself is practice.

One fact you'll discover pretty quickly is that it is impossible to type any faster than 60 words per minute on a 60 wpm 'TTY. If you try to type even two letters too quickly, the second key will lock up until the first letter is finished. In order to send with any kind of speed you have to develop a steady rhythmic typing technique — or use tape!

You have to hit two separate keys, line feed (LF) and carriage return (CR) to start a new line on a 'TTY machine. It's a good idea to get into the habit of typing two carriage returns and one line feed followed by one or two LTRS at the end of each line. The two CRs guard against the other fellow's printer losing a single CR in a burst of static which would cause the entire next line to be typed all in the same place on the right-hand edge of the paper. The LTRS ensure that his machine won't get stuck in upper case, and they also give the machine time to settle down after the carriage return. (On some of the older teleprinters the carriage bounces when it hits the left-hand stop.) Don't forget to do the CR CR LF LTRS LTRS bit at the beginning and end of each transmission. It's especially impolite to leave the other fellow's printer in the middle of a line while you send your cw i-d — sometimes his machine will decode your cw as a carriage return and then type gibberish over the last line of his copy.

Many 'TTY machines have a feature called "unshift on space." This causes the machine to go to lower case automatically whenever the space bar character is sent or received. It eliminates the possibility of getting an entire line of upper-case gibberish if a spurious FIGS character is received. You have to take this into account when typing. If you have a non-unshift-on-space machine and type YOUR RST (FIGS) 579 579 579 it may come out on the other fellow's printer as YOUR RST 579 TUO TUO. To make sure all machines will copy properly, send a FIGS before every group of figures and a LTRS after. YOUR RST (FIGS) 579 (FIGS) 579 (FIGS) 579 (LTRS) HERE IN SIOUX FALLS (FIGS), (LTRS) SOUTH DAKOTA

Some teleprinters also have a "nonoverline" feature. An automatic LF is sent every time a carriage return is received. If you have one of these machines send the LF anyway so the other fellow won't print all the lines on top of one another. Keep in mind that the maximum number of letters in a line is 72 on most machines. If yours has more, you have to get used to typing less than a full line.

After a mistake, the "oops" signal is XXXXX or EEEEE. No need to overdo this, though, as generally the other op will figure out what you meant to say anyway, and a page filled with XXXXX can be confusing. Another thing that drives some RTTYers crazy is the curious practice of typing CW ID TO FOLLOW at the end of a transmission. Most operators already know when to send the cw i-d!

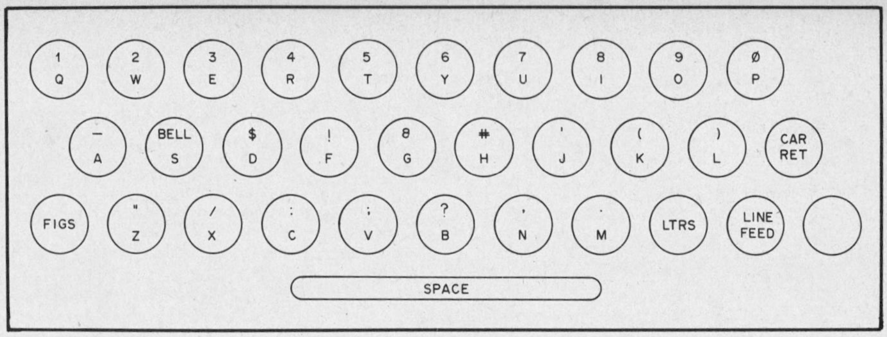

Fig. 6 — Communications-type keyboard. The arrangement resembles that of a regular typewriter, but it has only three rows of keys instead of four. All letters are capitalized, there being no lower-case letters in RTTY. The characters appearing above the alphabet letters are produced when preceded by the FIGS key, shifting the carriage.

Tape

Unless you are an exceptional typist, you can send much faster using tape than you can using manual typing. Even more important, tape-sent RTTY gets through with less garble when conditions are poor because of improved synchronization of the 'TTY machine's decoding mechanism.

Most people have several standard prepared tapes for information that is sent often. The "brag tape" generally gives a description of the station equipment and antennas and a little about the interests and achievements of the operator. Some of these can be quite humorous — let your sense of creativity be your guide!

Most CQs are sent by tape. As with any CQ, keep it brief — no need to send line after line of CQ CQ CQ . . . . If you make a mistake while punching tape, backspace the perforator over the error and type LTRS. Since all five holes are punched out for this character, it effectively erases the error.

An endless loop of RYRYRYRY is very useful for making adjustments to the teleprinter. The letter Y uses holes 1, 3 and 5 of the 'TTY tape while the letter R uses 2 and 4. Thus an RY tape tests all five message units. To make a loop you just type two LTRS followed by a line of RYs. (Don't forget the CR CR LF LTRS LTRS at the end.) Trim the two ends of the tape and attach them together overlapping the two LTRS. With "chadless" tape (the type that leaves the little flaps, or chads, over each hole), you can just press the chads from one end of the tape through the holes of the other. "Chad-type" tape can be stuck together with common household white glue. The loop will be less likely to get snagged if you attach the end that goes through the tape head *first* on the *top*.

Unlike an electric typewriter, the keyboard and printer of a 'TTY machine are not connected mechanically. The keyboard sends an electrical signal to the printer where it is decoded before being typed on paper. This means you can wire up a switching system that allows you to type your next transmission on the keyboard without affecting what is being received on the printer. "Typing ahead" in this way can save a lot of time in a QSO. Some machines, such as the Model 19 and Model 28ASR, have this capability built in. "ASR" means "automatic send-receive." That is, you can send with tape. With the 28 KSR (keyboard send-receive) you can send only with the keyboard. You may run across one of the keyboard-less "RO" (receive only) type machines. Many Model 15ROs were used as news-printers at radio and TV stations and occasionally turn up in the used market. Some hams use them with keyboard-type perforators to obtain the transmit capability.

RTTY on VHF

With the growing popularity of fm, more and more hams have been using RTTY on vhf, especially on 2 meters. An unmodulated fsk signal can be sent at vhf, but more common practice is to use afsk with FCC-required identification on voice. You just feed 2125- and 2295-Hz tones from an afsk generator into the microphone of an a-m or fm transmitter — 170-Hz shift is generally used. Upon reception, the tones are fed into a conventional demodulator set up for these two frequencies. Much simpler demodulators are usable on these bands since noise and interference are much less troublesome. RTTY is permitted anywhere in the vhf bands except the 6- and 2-meter cw segments. A few 2-meter repeaters are set up mainly or exclusively for RTTY operation, but if there are none in your area, make sure it is okay to use a repeater for this purpose before transmitting. Use of simplex channels is also common.

Radiograms by RTTY

If you have a cartload of messages to send to another station, the easiest and fastest way to do it is by RTTY. With a tape perforator and reader, you easily can send an average of more than one message per minute. Not only is this twice as fast as a good cw or ssb op can send, it's twice as easy!

Net operation is a bit more cumbersome on RTTY than on cw or ssb because few stations have fast break-in, and unless all stations are on exactly the same frequency, you have to take time to retune your receiver every time a different station starts to transmit. RTTY traffic nets do exist, however, and much traffic can be handled in a short time if most stations have at least several messages to send. A list of all types of traffic and public service nets can be found in the *Net Directory*, available from the ARRL.

When stations have only one or two messages each, the traffic is usually handled on net frequency. Sending the stations off frequency generally would take longer than the minute or so a typical tape-send message would run on a 60-wpm teleprinter.

Message format is very similar to cw procedure covered in a separate section of this book, except that RTTY traffic is nearly always sent by tape. A confirmation line is usually sent after each message to repeat any numbers or difficult words. Remember to send a CR and several LFs between messages so that the other operator can tear them apart for filing. A typical message might look like Fig. 7.

The LTRS and spaces sent before and after each message facilitate tape splicing if the other fellow is copying with a tape perforator. To make checking the word count easier, messages are usually typed 10 words to the line with a large space

```
(10 LTRS 5 SPACES CR CR LF LF LTRS)

NR 345 R WA0XYZ 15 SIOUX FALLS SD 1030Z JAN 3

MRS SAM SWARZ
1234 MAIN ST
ANYTOWN ILL 98765

HAVING A WUNNERFUL TIME X     WISH YOU WERE HERE X
SEE YOU SOON X LOVE

MARY

CFM SWARZ X 1234 X 98765 X WUNNERFUL  AR

(CR CR 4 LFS 10 LTRS)

1800Z JAN 3 DE WA0XYZ
```

Fig. 7 — A radiogram as sent on an amateur RTTY circuit. Each line starts at the left-hand margin since indenting would waste time. The address and signature are set off by double line feeds for clarity. Numbers and difficult words are repeated for clarity in the confirmation line (CFM).

separating the first five words from the last five. Log keeping is facilitated if you type in the date, time and your call at the end of the list of messages. You will often see stations send a half line or so of RYs at the beginning of a transmission to allow the receiving operator to get his receiver adjusted properly.

If you are receiving the message for retransmission it's quite feasible to reperforate a tape of it off the air. You can thus receive and relay a message without ever touching the keyboard. This is what is meant when we say that RTTY traffic handling is easy!

Autostart

It's possible to copy RTTY when you aren't even home! The RTTY demodulator can be equipped with autostart, a device that turns the machine on only when an RTTY signal is received on frequency. To send you a message, the other operator just sends a steady carrier for a certain length of time (usually a minimum of three or four seconds) to allow your machine to turn on and then types his comments. When you come home, you have a written record of everything that was sent on the frequency in your absence. For this system to work, you need very good frequency accuracy since an error of only a few dozen Hz can be too much. For this reason, most enthusiasts use crystal-controlled transmitters and receivers tuned to one of the autostart net frequencies. The same system can be used to copy W1AW RTTY bulletins. Autostart is especially popular on vhf fm nets because the lack of noise and interference makes it easy to get set up with a simple autostart device.

RTTY Art

If you aren't afraid of getting hooked on something, you might look into RTTY pictures. An example of what can be done by talented keyboard artists appears in Fig. 8.

Actually, it takes more persistence than raw artistic talent to produce RTTY art. You start out with a snapshot or a newspaper or magazine photo of roughly the proper size and tape it to the 'TTY paper. If the copy is not the right size, it can be photo-reduced or enlarged or run through the type of photocopying machine that has provision for enlarging. It can also be redrawn either freehand or with a pantograph.

Now turn on the tape perforator, and type over the picture with the teleprinter. For dark areas use dark letters such as W or M, and use H, X, I and punctuation marks for lighter areas. For extra dark areas, you can use strikeovers (M over W,

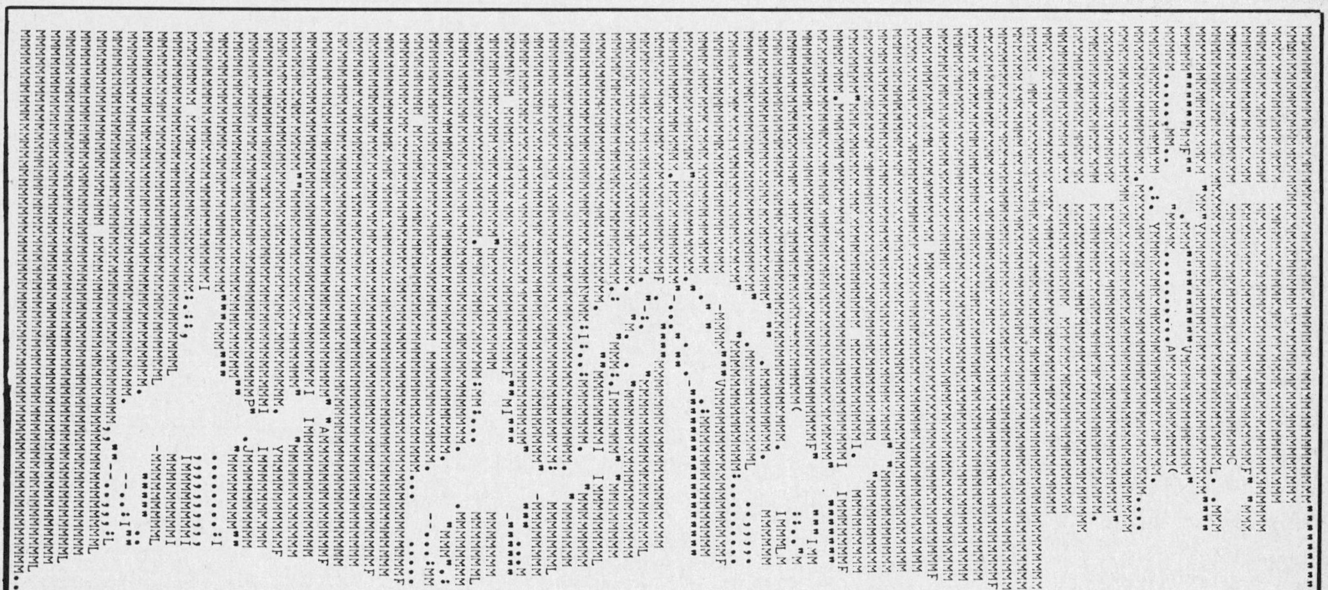

Fig. 8 — With patience and determination, talented keyboard artists can create RTTY art.

for example). Machines with non-overline provisions won't print such pictures properly, however, and strikeovers do take longer to run. Use plenty of contrast — there's a tendency to make everything too near the same shade. Areas of blank white paper are perfectly okay.

The persistence comes in doing the correcting. When you are done with the first draft, you will probably have to go back through the tape several times correcting errors and touching up. This is easiest if you are using a reperforator. A "reperf" is a device that allows you to make a tape of anything that is being printed. For example you can make a copy of a tape by running it through the tape reader. When you come to a mistake, you stop the tape, type in the correction, then restart the tape at a point beyond the error. If you are using a machine such as the Model 19 which will punch tape from the keyboard only and you don't have a separate reperf, then corrections must be made by splicing the tape (using the chad or white glue method mentioned before).

When you get the final version done, be sure to sign it. This can usually be done unobtrusively by working your call into the picture at the bottom. As a courtesy to the originator, when you relay a picture you have received on the air be sure not to omit the author's call.

Another output for ham creativity is the 'TTY QSL card. (Only an original confirmation is acceptable for DXCC or WAS, though!) People often send them as part of the brag tape, and they can be quite elaborate. To send the information that changes with each QSO (his call, RST, band, date, time), include a "bell" signal followed by a string of blanks or LTRS. When you hear the bell, stop the tape reader, type in the info and then start the reader again.

You can even send music by RTTY! By interspersing bells with figures characters to produce the proper rhythm, it is possible to give a recognizable rendition of "Jingle Bells" on a teleprinter. This tape is especially popular around Christmas time. One fellow even managed to have the machine play "Jingle Bells" while simultaneously typing out MERRY CHRISTMAS AND A HAPPY NEW YEAR!

Slow-Scan Television

Every day, all around the world, hams exchange pictures of their shacks, schematics and even themselves. They are using the most popular form of visual communication — slow-scan television (SSTV). Many DX stations are now equipped for this type of picture transmission, and more than one amateur has worked over 100 DXCC countries on SSTV! Amateurs' success with SSTV during the past decade has led to its application by industrial and commercial users as a reliable, long-range, voice-bandwidth tool for picture transmission.

Just as its name implies, SSTV is the transmission of a picture by having a TV camera very slowly scan the subject while a monitor at the receiving end reproduces it in step. Instead of viewing apparent motion from the 30 frames per second as seen in regular TV, a single SSTV frame takes eight seconds to fill the screen. In addition, only 120/128 lines of resolution are used as compared with 525 lines for commercial standards. But these disadvantages are more than offset by the fact that SSTV requires less than 1/2000th of a fast-scan TV's bandwidth. Thus the FCC permits it in any amateur phone band except 160 meters.

An SSTV signal is a variable-frequency audio tone — from 1500 Hz for black to 2300 Hz for white, with 1200 Hz used for synchronization pulses. To send SSTV over the air, just couple this audio directly into the microphone input of any single-sideband transmitter. (SSTV on double-sideband a-m or fm is illegal on the hf bands in the U.S.) To receive, you tune in the signals on an ssb receiver, feeding the tones into an SSTV monitor.

All you need to get started is an ssb station, a monitor and a video source. Like RTTY, SSTV is a 100-percent duty-cycle transmission — most sideband rigs will have to run considerably below their voice power ratings to avoid ruining the final amplifier or power supply or both. Early SSTV monitors used long-persistence CRTs much like classical radar displays. In a darkened environment the image remained visible for a few seconds while the frame was completed. Many of these are available today for reasonable prices. More recently, advances in digital techniques have led to the development of fast-scan converters. On receive, such a device converts the incoming audio to a video signal that is usable by a conventional fast-scan TV monitor. Similarly, on transmit the converter changes the fast-scan camera output to a standard slow-scan signal. Dominating the field is the Robot Research Model 400. The company, at 7591 Convoy Ct., San Diego, CA 92111, also has an information packet about the technical aspects of SSTV.

Finally for equipment, the video source can be a camera or tape recorder. Since an SSTV signal is audio, it can be recorded on a cassette machine. The frequency response is not critical, but some consideration should be given for the wow and flutter specifications to minimize picture skew. You can have a friend prerecord some pictures from his camera onto your recorder and use them for on-the-air transmissions. A simple way to avoid the expense of a camera is to build a flying-spot scanner. It sends pictures by scanning a slide transparency (with a "flying spot" of light). Even many of the SSTVers who do own cameras have an "FSS" because you can build a large library of slides, any one of which can be selected for on-the-air showing almost instantly. In addition to this convenience, having an FSS frees the camera for other uses — no need constantly to be readjusting the lighting and focus when you move the camera from subject to subject.

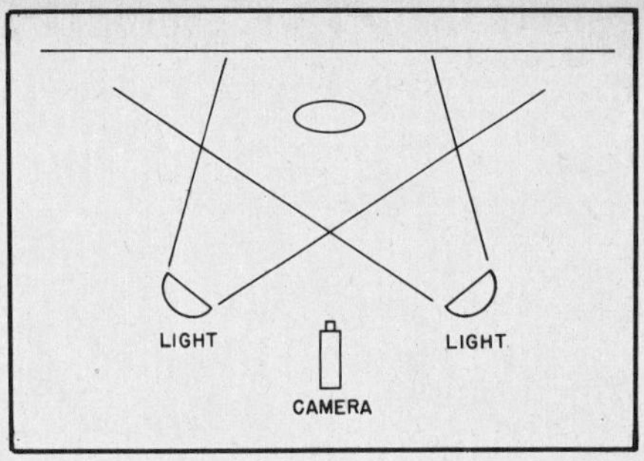

Fig. 9 — A popular lighting arrangement for ATV or SSTV pictures. Using two lights helps eliminate harsh shadows that might result from a single light source. Frequently one light is set brighter or closer than the other to provide a little shading.

Setting Up Your Studio

Use plenty of light. Standard photographic lighting arrangements work well for TV. The setup of Fig. 9 is good because it reduces background shadows and reduces harsh shadows on the subject. A pair of 100-watt bulbs with reflectors would be a minimum requirement for the typical camera. Outdoors, of course, the light level is pretty much predetermined. Care must be taken, however, to prevent direct sunlight from shining into the lens since most TV camera pickup tubes can be damaged by direct sunlight. You should note that the reproducible range of brightness is less than regular TV. Adjust the brightness and contrast controls on the camera so that the darkest screen area comes out black while the brightest areas reach maximum brightness without "blooming" on the monitor screen. This assumes that the monitor brightness and contrast controls are set correctly. The easiest way to do this is to use a test tape either obtained directly or recorded from another station off the air. Typically, these have vertical bars of various gray shades. The same brightness criterion applies to the monitor.

Depth of field is the ability of the lens to sharply focus on both near and far objects at the same time. Since the focusing range is narrowest for wide lens aperture, the most accurate focusing can be done at such a setting, the lowest f-number value. While a narrow aperture (high value) requires more light on the subject, it permits the viewing of objects at different distances without refocusing. (i.e., it has greater depth of field). Many cameras have fixed-aperture lenses with electronically controlled light sensitivity. Besides aperture variations, lenses may be purchased for various focal lengths to give wide-angle, telephoto, or zoom capa-

Fig. 10 — A way to make SSTV pictures without a camera is with a microcomputer. As described in an October 1978 *QST* article, the arrangement is useful for generating special characters, then updating the frame as changes are made. Here, a checkers game is ready to begin.

bilities. Nearly all TV cameras use the standard screw-in "C-mount" thread.

Focusing of an SSTV camera can take a few minutes to complete because you can't immediately see the effects of readjustment. A fast-scan converter, however, would allow you to make real-time changes.

Picture Subjects

What do you make pictures of? The selection of subjects is endless, but you will find that high-contrast black and white pictures that aren't cluttered up with too much detail work best. Possibilities include a close-in shot of you in front of the operating position and cartoons (commercial or homebrew). Don't forget a couple of frames with your call and perhaps name and QTH — a closeup of your QSL card is a possibility here. There's no reason why you can't also use SSTV to show off some of your choicest DX QSLs. A schematic is sometimes worth 1000 words. Illustrating a technical conversation with a simple diagram can often make clear a point that would be very difficult to get across on voice alone. There is a wealth of good SSTV material in newspapers and magazines. Color pictures usually are a poor choice since contrasting colors often come out nearly the same shade on the SSTV screen. Learn to develop an eye for potential slow-scan material whenever you read a newspaper.

Your subjects don't all have to be either "live" or on tape. Some SSTVers find it convenient to mount the camera in front of an easel into which various prepared cards can be inserted. These cards can be photographs, drawings or lettered signs. A kitchen-type noteboard with press-on removable letters is handy for making headings. You can also buy or build a keyboard-operated device that prints characters on the SSTV screen. It sends the same message over and over until you erase it or type something else.

It's fun to make up taped "programs" of related pictures. Some operators have dozens of programs each on a different cassette tape that can be selected and put on the air in seconds. Homebrew comic strips — with a couple of shots of each frame — are quite popular. People might enjoy seeing a pictorial tour of your shack and QTH. Do you have another hobby? Maybe that fellow you are working would like to learn about your model train collection. A closeup of your 1862 Cannonball Express would be a lot more interesting than a few bland comments about the weather. Use your imagination!

It's customary to send two or three frames of each subject to ensure that the other fellow gets at least one good quality picture. Many ops like to send a couple frames followed by voice comments concerning the picture just sent. Some of the more elaborate setups are equipped to send video and audio simultaneously with the voice on one sideband and the picture on the other! These "ISB" (independent sideband) signals can be copied with two receivers or a single receiver with two i-f filters, detectors and audio amplifiers.

SSTV has good potential for rendering service to the public by means of third-party traffic. For example, the scientists working on the Antarctic ice pack often don't see their families for months — except by amateur slow-scan TV. Anyone can write or call almost anywhere in the world these days, but Grandma and Grandpa would certainly enjoy seeing pictures of the kids from 3000 miles away — instantly!

Operating Procedures

SSTV is legal anywhere in the Advanced and Extra Class voice segments of the 75, 40, 20 and 15-meter bands and all voice bands above 28 MHz. The standard calling frequencies are 3845, 7171, 14,230, 21,340 and 28,680 kHz, with 75 and 20 meters being the most popular bands.

An SSTV signal must be tuned in properly so that the picture will come out with the proper brightness and so that the 1200-Hz synchronization pulses will be detected. If the signal is not "in sync" the picture will appear wildly skewed. The easiest way to tune SSTV is to wait for the transmitting operator to say something on voice and then tune him in while he is talking. With experience you may find you are able to zero in on an SSTV signal by listening to the sync pulses and by watching for proper synchronization on the screen. Many SSTV monitors are equipped with tuning aids of various kinds.

If you want to record slow-scan pictures off the air, there are two ways of doing it. One is to tape record the audio signal for playback later. The other is to take a picture of the image right from the SSTV screen. Instant-print cameras equipped with a closeup lens enable you to see the results shortly after the picture is taken. If you want to do this without darkening the room lights, you'll have to fabricate a light-tight hood to fit between the camera and monitor screen.

On any amateur transmission, the legal identification must be made by voice or cw. Sending "This is KAØXYZ" on the screen is *not* sufficient. Most stations intersperse the pictures with comments anyway, so voice i-d is not much of a problem. Otherwise SSTV operating procedures are quite similar to those used on ssb.

Medium-Scan Television

As popular as slow-scan television is, it has the disadvantage of living up to its name — there is only one frame every eight seconds. What better way to promote Amateur Radio worldwide than to be the first to have an intercontinental exchange of *moving* television pictures without the aid of a multimillion-dollar repeater satellite? This very task has been undertaken by an energetic group of experimental SSTV hams who, with ARRL help, have received a Special Temporary Authorization (STA) from the FCC for a medium-scan television system.

Granted on June 16, 1978, to W9NTP, W3EFG, WB9LVI, W6MXV and WØLMD, this STA permits the testing of a 36-kHz-bandwidth television system on 29.0-29.3 MHz. The task of transmitting moving television pictures in such a bandwidth is not easy. It rivals similar work being done in commercial and military fields.

This series of experiments was discussed in the October 1978 issue of *QST*. The article mentioned that the STA permit from the FCC was valid until July 1980. During this past period of time several tests have been conducted and several conclusions made. This update provides the latest information for those who are interested in the tests. This section cannot include all the details of the equipment so anyone not on the current MSTV mailing list should send a large manila stamped return envelope to Dr. Don C. Miller, W9NTP, R.R. 1, Box 95, Waldron, IN 46182, to keep up on the current status.

Equipment has been constructed for demonstration purposes and displayed at many hamfests and conventions across the country. Observations of many viewers and the restrictions of bandwidth on resolution have led the developers to specify a set of standards for the rest of the test period. These standards are given in Table 1.

A beacon is currently operating from Indiana on 29.150 MHz. This beacon operates on Saturday of each week, transmitting the first 10 minutes of each hour, starting when the 10-meter band opens. Several messages and tests are being transmitted. A voice-recorded message gives the present status of MSTV each week. In addition, approximately

Table 1

Medium-Scan Television Standards

These standards are being used for 10-meter transmissions during current experimentation under Special Temporary Authorization of the FCC. (See text.)

Frame rate	2 frames per second
Line rate	256 lines per second
Horizontal line resolution	128 pixels
Vertical line resolution	128 lines
Horizontal sync. pulse width	0.5 ms
Vertical sync. pulse width	3.9 ms
Base video bandwidth	16 kHz
Type of modulation	Narrow-band frequency modulation
Rf bandwidth	36 kHz

half of the time period is devoted to the transmission of a test image at the standard specifications of Table 1. If the beacon operator, W9NTP, is home, a call-in standby period will immediately follow the transmission of the recorded voice and test image.

How is MSTV Received?

In order to receive the image it will be necessary to have a wide-band (36-kHz) 10-meter receiver on 29.150 MHz. In the United States there are old two-way fm receivers available very inexpensively, but the writer has not found any type that is recommended. The development of a new economical receiver is being attempted in a short time period by one commercial company. Science Workshop is currently developing a small receiver that will be crystal controlled and will have a discriminator and sufficient bandwidth to receive the low-frequency video. In addition, W9NTP is working with a GLB front end that may convert into many of the repeater kits that use 10.7-MHz i-f strips for 2-meter repeaters. If anyone would like to work in this area to speed the development of MSTV, it would be greatly appreciated.

There are several approaches to receiving the MSTV image. The most attractive one at the moment from economic and simplicity viewpoints is to use S-100 memory computer boards. This will require two memory boards and a video interface board. The computer memory boards are standard but the video interface board is unique; the design will be furnished to anyone wishing to build the MSTV system.

Another approach which is very attractive is to use a Robot 400 monitor with two memories. The added second memory board can be purchased at cost from W9NTP. This makes the Robot 400 useful for color SSTV and for graphic applications as well as for MSTV experiments. The problem of using the Robot 400 is the requirement of slow readout of the dynamic memories which must be refreshed at a faster rate than the MSTV pixel rate. For this reason it is required to have an external static buffer memory to do the actual data rate conversion. The S-100 memory board can be used here also. Only part of the 8-K memory is used in this application.

The third approach is to use a converted SSTV P7 monitor. The deflection circuits are triggered directly with the recovered horizontal and vertical pulses. Very little actual conversion is required from the SSTV mode. The image viewed is quite good and will permit anyone to take part in the reception experiments without much expense or development time.

Since only the individuals listed on the original STA are allowed to transmit during this phase, all other participants should apply their effort to the reception of the image. There are many unknown factors in the reception of wide-band signals on an unpredictable band like 10 meters. Selective fading could easily tear the signal apart so that it would be impossible to synchronize the image. It is therefore recommended that the reader build some kind of a receiver that can receive the video and then apply the output to a P7 monitor or a digital scan converter as mentioned above. Those of you who may be too young to know about the development of SSTV in the 60s will be able to experience some of the excitement of developing a new mode that was experienced by SSTV and ssb experimenters earlier.

After the intial phase of getting the first picture across the Atlantic Ocean has passed, the experiments will include computer processing of digital pictures to enhance and simplify the data to do bandwidth reduction. WØLMD, has experimented with the transmission of digital slow-scan data and has had very good results. He believes that eventually a moving image can be transmitted with computer processing at normal TV rates in a bandwidth as low as 36 kHz. This must wait for the future. We are trying to keep the experiments going at a fast pace and record as much data as possible so that when the STA report is submitted to the FCC, the inventiveness of radio amateurs can be shown. Thus FCC will continue the STA and include more amateurs to complete the development of medium-scan television to a successful new mode of operation for the radio amateur.

Fast-scan Television

With the most resemblance to broadcast-quality television, because it uses the same technical standards, fast-scan television can be used by any ham holding a Technician or higher-class license. Popularly known as ATV (amateur television), it is permitted in the 420- to 450-MHz band and above. Because the power density is comparatively low, typically 10 to 100 watts spread across 4 MHz, reliable amateur coverage is only on the order of 20 miles (about 32 km). Nevertheless, you might find yourself exchanging pictures with stations up to 200 miles (320 km) away when tropospheric conditions are good.

ATV Information

Amateur Television Magazine
Box 1347
Bloomington, IN 47401

New England Amateur Television Group
P. O. Box 307
Peabody, MA 01960

Baltimore Radio Amateur TV Society
P. O. Box 5915
Baltimore, MD 21208

Bay Area ATV Society
c/o R. Riordan, WB6UQU
423 Donouhe Dr.
Milpitas, CA 95035

Southern California ATV Club
c/o John Ruckert, WB6ZPN
953 S. Beacon Ave.
Los Angeles, CA 90015

Cuyahoga Amateur Radio Society
c/o James A. Mackner, W8RSX
1617 Belle Ave.
Lakewood, OH 44107

Bloomington Amateur Radio Club
535 S. Walnut St.
Bloomington, IN 47401

Equipment

Getting started is not as difficult as you might think. Some enthusiasts prefer to build their cameras from scratch and design their own rigs, but most people buy a used closed-circuit television (CCTV) camera for under $100 (or a new one for about $250) and modify a surplus commercial uhf fm transceiver by adding a one- or two-transistor video modulator. The only other equipment needed is an old TV set and a frequency converter to heterodyne the amateur TV signal down to an unused vhf channel. Henry Ruh, WB9WWM, describes a complete ATV transceiver in April 1978 *QST*. It uses readily available modules for the converter, video modulator, audio subcarrier modulator, transmitter and amplifier. More information on the technical aspects of ATV can be found in the *Radio Amateur's Handbook*.

Operating Procedures

Most wide-band A5 activity occurs in the 420- to 450-MHz band. The exact frequency used depends on local custom, but a 439.25-MHz picture carrier is common

Fig. 11 — A Western Union Telefax transceiver, commonly available as surplus. The drum accommodates a sheet of paper about 4-1/2 × 6-1/2 inches (115 × 165 mm).

Fig. 12 — A weather satellite picture of Hurricane David as produced on a Telefax machine.

in the East, while 427.0 MHz is popular in California. Some population centers have ATV repeaters, typically 439.25/427.25 MHz. ATVers try to avoid interfering with the weak-signal work (moonbounce, for example) being done around 432 MHz and repeater operation above 442 MHz.

There are at least three ways to transmit the voice information. The most popular is by talking on another band, often 2-meter fm. This has the advantage of letting other local hams listen in on what you are doing — a good way to pick up some ATV converts this way! Rather than tie up a busy repeater for this, it is best to use simplex. In southern California, for example, 146.43 MHz is commonly used.

Commercial TV has an fm voice subcarrier 4.5 MHz above the TV picture carrier. This can easily be received in the usual way on a regular TV set. Many of the surplus fm rigs that are available, however, do not have enough bandwidth to pass both the picture carrier and the voice subcarrier at the same time.

Another way to go is frequency modulation of the video subcarrier. Since the video is amplitude modulated, it should not interfere with the fm audio, or vice versa. The usual way to receive this is with the fm receiver section of the surplus uhf rig. It is also possible to use a low-band police/fire monitor coupled into the TV set 44-MHz i-f.

Signal reporting on ATV differs from the RST system used on cw and phone. For example, "closed circuit" is the ATVers term for "armchair copy"! The received signal strength is indicated by the amount of snow on the screen — the more snow, the weaker the signal. Some operators use a reporting system running from P∅ to P5, roughly corresponding to readibility reports on voice or cw.

The resolution or sharpness is the number of distinguishable lines that can be displayed across the TV screen. Although commercial TV transmissions have 525 lines of resolution, acceptable pictures can be sent with systems capable of only 200 to 300 lines.

Signal reports are very useful for "talking the picture in" on the air while the sending station makes transmitter adjustments. In fact, this is the only reliable way to adjust the video modulator since receiver overloading will give misleading results when you monitor your own signal — a picture that looks fine on your own TV receiver will likely have low contrast on a distant station's set.

Camera procedures on ATV are similar to those presented in a previous section, "Setting Up Your Studio." The major difference is greater flexibility in lighting. Camera adjustments are best made with it connected directly to the monitor receiver video input. If you don't want to modify the TV set for this, most CCTV cameras have provisions for generating a vhf TV signal on one of the lower TV channels for monitoring purposes.

Facsimile

Facsimile is transmission of pictures by radio. "FAX" is similar to SSTV except that the picture definition is much better, and the image comes out on paper rather than on the face of a picture tube. Using inexpensive surplus facsimile machines, good newspaper-quality photographs can be sent on any of the amateur vhf/uhf bands. Many amateurs and other experimenters use FAX to receive government weather satellite photographs beamed direct from the satellite. On a clear day you can see forever — or a good-sized piece of North America, anyway!

Facsimile standards depend on the particular machine in use. All FAX machines make their copies on current-sensitive or light-sensitive paper wrapped around a drum. As the drum revolves, the marking device moves slowly down the side, etching a close-spaced series of parallel lines from top to bottom of the paper. The darkness of the mark depends on the picture information contained in the received signal. For transmission, a similar mechanical arrangement is used except that instead of a marking device, a photo-sensitive pickup unit scans the copy. Thus variations in lightness and darkness of the picture are transformed into variations in voltage, which modulate an audio tone. Typical standards call for 60, 120 or 180 lines per minute and 72 or 96 lines per inch. The popular Western Union Telefax unit rotates at 180 lines per minute with 72 lines per inch and produces an amplitude-modulated 2400-Hz signal that can be fed into the microphone input of any vhf a-m transmitter. (Fm FAX is not legal on the 6- and 2-meter bands.) Drum synchronization is tied to the ac house current. Further details can be found in *The Radio Amateur's Handbook*.

Chapter 13

Microcomputers

Scarcely a half dozen years have passed since the microprocessor became generally available, yet its impact on Amateur Radio is already being felt. During the next several years, the small computer will become as common an item in the average ham shack as a transmitter and receiver are today. Built-in microprocessors have made possible the design of "smart" test equipment and "intelligent" transceivers, full-function cw keyers and RTTY terminals. Stand-alone computers programmed to provide Amateur Radio-related services are re-writing the definitions of logkeeping, traffic handling, contest operating, award hunting, homebrewing and practically all other aspects of amateur activity.

Amateurs have always been quick to capitalize on the benefits provided by new technology. Not only does computer technology offer benefits to the ham, but plenty of hams will be contributing to the development of that technology as a result of their amateur work. Recent changes to FCC/DOC rules and regulations both recognize the importance of the marriage of Amateur Radio and microcomputers and encourage further development activity. It is clear that ham radio will be undergoing profound changes as a result of that activity. The intent of this chapter is to focus on some of the ways in which the advent of the small computer will shape the complexion of Amateur Radio during the next decade.

The computer terminal is right at home next to the hf transceiver. (*Photos by K1RT*)

An Unlimited Resource

A computer can be used to store information for later retrieval, reorganize or format that information and perform calculations using that information. The information may represent voltage or frequency measurements, call signs, message traffic, or just about anything else that can be expressed in terms of numeric or alphabetic characters. Before discussing some of the ways in which the manipulation of such information can be of value to the ham, let's take a brief look at the elements of the computer that make such operations possible.

Key Elements

The four key hardware elements that are present in any small computer system are the (1) central processing unit or CPU, (2) the memory subsystem, (3) input devices and (4) output devices. The CPU often takes the form of a single-integrated-circuit microprocessor. A microprocessor can perform arithmetic or logical operations on binary (base 2) numbers. Such operations may include addition and subtraction, as well as arithmetic comparisons like greater than, equal to and less than.

Programming

The order in which these operations is carried out depends on the way the processor is programmed. A *program* is a series of instructions to the microprocessor.

The program steps also are represented by binary numbers. While executing the programmed instructions, the microprocessor may change the sequence of instruction execution based on the outcome of the additions, subtractions and comparisons. It may request the input of data from some external source, or it may output the result of a calculation or a signal to control an external device.

Hams use microcomputers that range in complexity from simple one-board systems like this one built from a kit . . .

. . . to systems that include expanded memory and peripheral input/output devices.

Both the program and the information that is manipulated by the microprocessor during the execution of the program are stored in the computer's memory subsystem. The memory is usually made up of semiconductor storage elements. In these storage elements, binary numbers are represented by electrical charges. Each unique combination of 8 binary digits, or bits, can be interpreted as an alphabetic or numeric character. Typically, small computer systems require a storage capacity of at least 4000 characters to perform much useful work. Some microprocessors can accommodate over 64,000 characters of storage.

The word *byte* is often used as a synonym for character. The information stored in the memory subsystem is lost (destroyed) the instant that the power supply of the computer is shut off. Most computers include provisions for semi-permanent data and program storage external to the memory subsystem. Often, a common audio cassette recorder is used to provide inexpensive "offline" storage; the data is stored on magnetic tape in the form of audio tones. A more costly alternative to the cassette recorder is the diskette drive, which stores the data on a rapidly rotating disk-shaped piece of magnetic material. Many ham radio applications that make use of the record-keeping abilities of a computer become practical on a system equipped with a diskette drive.

Input/Output

The usefulness of a computer is somewhat limited if it has no way to communicate with the outside world. The parts of a system that provide such communication are generally called *input* and *output* devices.

One of the most common input devices is the alphanumeric keyboard. The keyboard looks quite similar to an ordinary typewriter keyboard, but when one of its keys is pressed it generates an electrical signal that can be interpreted by the computer instead of generating the mechanical movement of a type bar. The keyboard can be used to enter programs and data into the memory subsystem.

One example of an output device is a video display, which translates signals from the microcomputer into numbers and letters that appear on the screen of a TV-like video monitor. In fact, many hams modify TV sets so that they can be used as display devices. If a permanent record of the computer output is required, a printer can be used instead of a display. A printer is in essence an electric typewriter that is controlled by signals sent from a computer. The cassette recorder and the diskette drive are also considered to be input and output devices.

BASIC Language

The microprocessors made by any one manufacturer are usually designed to execute binary-coded programming statements that are unique to that particular CPU. Such statements, referred to as *machine language* instructions, direct the operation of the computer at a very detailed level. Many hams program directly in machine language (or in an *assembly language* which is closely related to machine language) to take advantage of the high speed of program execution that is possible with such statements.

Some operations, such as multiplication, division and trigonometric functions, are difficult to program at such a low level. To make it easier to perform those operations, a number of higher-level computer languages have been developed. One such language that is frequently used with small computers is called *BASIC*. Programs written in BASIC can resemble the English language sufficiently so that nonprogrammers can understand how the programs work. BASIC-language programs also have the advantage that they can be exchanged between computers that are built around different microprocessors, with little or no program changes required for compatibility. Most people learn BASIC or a similar language as a first step in working with computers.

At this time, a general-purpose microcomputer starter system will cost about as much as a good hf ssb transceiver. Such a system would feature a CPU, a keyboard, a video display, 4000 to 16,000 bytes of memory, a BASIC language interpreter and cassette storage. Expansion of the system by adding a printer, a diskette drive and more memory capacity as finances permit will broaden the scope of applications that the system can tackle efficiently. Advances in technology, market forces, and production economies of scale are driving microcomputer prices downward dramatically.

Helping Hams

Given the ready availability of inexpensive information processing for the ham shack, what kinds of Amateur Radio-related tasks are appropriate candidates for automated assistance?

Logkeeping is the one potential application that shows promise for computerization. Hams keep written records of their operating activities for a variety of purposes. Logs are kept in order to satisfy legal requirements, to chart progress toward operating awards, to record information about band condition, equipment changes and new friendships, and for the purpose of entering contests. Information entered into the computer can be stored for later retrieval, searched, sorted and summarized more efficiently than information that is recorded only on paper.

Awards Records

The tasks associated with keeping track of QSO and QSL totals for the 5-Band Worked All States award provide an example of the differences between manual and computer-assisted record-keeping. It is necessary to keep a tally by state, band and call sign showing QSOs completed, QSLs sent, QSLs received, as well as a separate list showing state and band combinations yet to be worked. On some bands, certain states may have been worked multiple times, but no cards have been received. It will be desirable to have charts showing for each of the five bands which states must still be worked. It will also be a good idea to show for each state, on which bands it has yet to be contacted, to make it possible to make schedules in advance. If the mode on which each of the contacts was made is of interest, that information can be recorded as well.

When all of the award criteria have been met, it is necessary to fill in an application form with the proper information and to submit the application with the cards for verification. This can become a bookkeeping nightmare, especially if records get out of sync because some of the information is kept in two or more places but occasionally the information gets updated in only one place.

The work involved in record-keeping can be simplified if a computer is used to keep track of the details. It is possible through suitable programming to enter a command on a computer terminal keyboard that will result in the display on the video monitor of a chart showing for each state/band combination whether it has been worked or confirmed. If more-specific information is required, entering another command on the keyboard displays a single state's data for a given band, including call signs and QSO dates for the stations worked, and QSL-sent and -received information. Whenever a new state is worked or a QSL received, it is a simple matter to enter or update the information in the computer to keep the display up to the minute. Another command brings back a summary of total cards received, total states completed and so on.

When all 250 QSLs are on hand, the computer can neatly print out the information needed for the award application. The data can be stored on a cassette tape or on a diskette when the computer is in use for another purpose. The program developed for 5BWAS tracking could easily be modified to display at a glance progress toward DXCC, working all counties and the like.

Field Day Aftermath

Eliminating paperwork is the motivation behind another Amateur Radio computer application. Every June, hams from every part of the country participate in Field Day. By the time that the smoke clears at the end of the test, it is not uncommon for a three- or four-transmitter operation to have logged several thousand contacts. For most of the participants, Field Day is over when the last antenna comes down and all of the rigs, generators and tents are packed away; but not for the one who gets stuck with checking over the logs to make sure that all of the duplicate contacts are removed. It is great fun

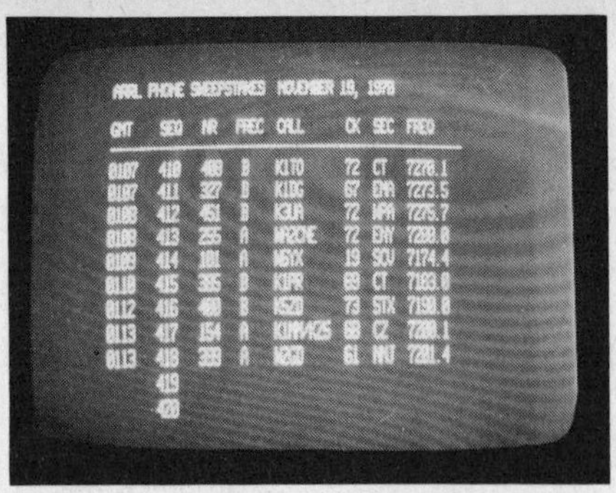

Simplify your logging with the assistance of a computer. Here is an example of how you might format the display of log information for the ARRL Sweepstakes contest on a video monitor screen.

working all of those stations, but the enthusiasm often evaporates when it is time to start the tedious process of dupe-checking.

Using the computer as an aid to dupe-checking will hardly make the task enjoyable, but it should help speed up the process and perhaps even improve the accuracy. With appropriate programming, call signs or call sign/band combinations can be entered from the keyboard. The computer maintains a list in memory of all of the calls that are entered. As a new call is keyed in, the program compares it against those that are already on the list. If the call is not found, the new call gets added to the list and a message such as "WB1AIU is a valid call sign" is displayed on the video monitor. If the call is a duplicate of one already on the list, the computer may output a message like "DUPLICATE QSO, DUMMY," or it may ring a bell or flash a light, or perform some similar attention-getting action. The program can display a running total of the number of calls entered, the number of valid calls and the number of duplicates. When all of the calls have been entered, the computer can sort the valid calls into alphabetical order and then print the list. With the wide variety of prefixes in use today, anyone who doesn't have a computer available will soon be forced to develop a three-dimensional dupe sheet just to keep up with all the call sign combinations!

Duplicate checking with a microcomputer after a contest ends can be a real timesaver, but it still entails significant effort to key in all the call signs written in the log. It may be possible to save even more time by entering the log information once directly into the computer, instead of first writing it on paper and later transferring it to electronic form. At the moment a station is worked, the call, QSO time, date, frequency, mode and QTH can be recorded. The computer can recognize a duplicate QSO right away. All of the "extraneous" information can be stored immediately on magnetic tape, leaving the list of calls worked to date in the computer's memory to use for dupe-checking purposes. If the log is stored on a diskette instead of on a tape cassette, it is possible to quickly back up and browse the information from a prior QSO right on the video monitor. After the contest, the log can be printed on paper and submitted to the sponsoring organization. There is no reason why the printer can not produce a clean copy of the log contact by contact as the contest progresses. In a few years, it will be permissible to submit a contest log in machine-readable form instead of on paper.

Most hams keep detailed records for every QSO, not just for contest QSOs. The programming for the contest-logging application described above could easily be expanded into a general-purpose automated logbook system. Key in a station call sign and presto — out pops a list of previous dates on which that station was contacted, along with any other information that was interesting enough to record. Key in a QTH and in short order a list of all previous QSOs with stations in that location appears on the video screen.

Traffic-Handling Applications

The possibilities for creative use of an electronic Amateur Radio "card file" extend far beyond simply logging QSOs. A regular net control station on a traffic net can keep a computerized listing of the net roster. During a net session, the calls of all of the stations that check in, as well as the destination of all of the traffic that is listed, are entered from the keyboard. By comparing this data with previously stored local telephone dialing area profiles for each of the net members, it is a simple matter for the computer program to generate match-ups between traffic going into a locality and stations that can handle the traffic. If a station that is not listed on the roster checks into the net, he can be included in the system just by entering his local telephone exchange along with his station call sign.

With a little bit of imagination, the program can be expanded into a complete NCS operating aid for keeping track of the net status. It should help the NCS to decide which stations to send off to which frequencies to pass their traffic in the most efficient manner, as well as providing an up-to-the-minute display of traffic yet to be cleared, stations currently passing traffic and traffic already sent toward its destination. With the aid of such a display, the NCS would avoid inadvertently overlooking one or two messages during a busy session. At the end of the net, it would display important statistics such as the number of messages passed, the duration of the net, the traffic rate and the number of messages that for some reason could not be handled.

Not only the net control can benefit from access to a small computer. Stations that handled more than two or three messages per day can really benefit, too.

What time is it? Team up a digital clock IC with your microcomputer and keep track of the exact time simultaneously in 20 cities around the world.

A microprocessor IC like this one is the heart of a great many small computers. The circuit performs the arithmetic and logical functions governed by programs stored as bit patterns in an array of IC memory elements.

Imagine how useful it would be to key in the preamble, text and signature of a message as it was being sent over the air. The program could automatically calculate whether or not the check count matched the number of words in the text. It would allow correcting spelling errors and inserting "fills." It would keep track of handling information such as the call of the station that was sending the message, the call of the station that picked it up if it was subsequently relayed, the date and time of transmission or delivery and any special operator notes. A formatted video display of the destinations of all of the messages waiting "on the hook" would come in handy when the time came to relay the traffic.

The ham acting as liaison from an NTS region net to one of the area nets might have a large stack of messages to transmit during a single session. Suppose that W2MTA is the transmit rep from the Second Region Net to the Eastern Area Net. He might have to leaf through three dozen messages, manually sorting them into piles for the first region, the third region, the Pacific Area and the like, counting the traffic in each pile to get message totals by region. This is necessary just to be able to QNI and list the traffic with the NCS. Heaven help him if he has several pieces of book traffic that must be split up among multiple regions. A computer can do a better job of selecting, sorting and counting. A "menu" screen on the video display can show, for example, that there are five messages for region one, three for region three, seven for region four and so on.

When the NCS directs W2MTA to go up frequency and send traffic to the first region rep, he enters a simple command on the keyboard, and the first message for region one gets displayed on the screen. After that one is sent successfully, the next one is shown on the display and then the next one, and so on. When all of the first region traffic has been sent, he can enter the date and time and call of the receiving station and move back to the net frequency, knowing that none of the traffic was accidentally forgotten along the way. Splitting up the book traffic by region without the problem of separating names and addresses from common message text is simplified.

Messages that can be delivered over the telephone can be printed automatically on standard radiogram blanks, even including expansion of ARRL-numbered-radiogram texts into their straight English form before printing. It's easy to see how a ham who enjoys traffic handling can increase his effectiveness and reduce his bookkeeping by doing a bit of microcomputer programming.

Other Useful Functions

Many other applications are based on the ability of a small computer to record, retrieve and display all sorts of written information. Lots of hams use their processors as electronic scratchpads. A digitized atlas is another good example — input a call sign prefix and have the computer reply by flashing the country name and beam heading on the video screen. At Amateur Radio booths set up in county fairs and shopping malls hams are using their microcomputers to display introductory information about license exams, licensing privileges and operating events like Field Day to interest the general public.

At license classes and radio club meetings, hams have used microcomputers to run radio-related quizzes, games and training exercises. All of these are fun and useful applications, but the place where amateurs have really been showing their ingenuity is in the area of the control of external devices by computer.

Telling Time

Lots of interesting Amateur Radio projects become feasible once the computer can tell time. It is not very difficult to interface a common digital clock integrated circuit to a microprocessor system so that the programs can make decisions based on the time of day when they are executing.

Right at the beginning it may be fascinating enough just to display the time on the video monitor and watch the seconds tick by, but there are other, more useful, ways to use an intelligent clock. Programming the micro to turn on the station receiver at a preset time is easy. It's a great feeling to walk into the shack to keep a schedule or work an OSCAR pass and find the rig all warmed up and ready to go. Plenty of hams prefer to wake up to the sound of cw DX on the low end of 20 meters, instead of listening to the news, weather and sports that come out of most ordinary clock radios.

A dynamic time-zone conversion aid is another project that can be programmed easily. Not only does it convert time from zone to zone, but it also keeps time to the second in each of those zones. Imagine demonstrating ham radio to a visitor by working hams from all over the world and then pointing to the video monitor which is simultaneously displaying the exact time for each of the dozen or so countries just worked. The display can show you at a glance the time of day in Tierra del Fuego or Paris. Without using a computer, it would require a whole room full of clocks to duplicate that trick.

An Electronic Scratchpad

Digital clock ICs that provide the date as well as the time are available. With one of these interfaced to a microcomputer, automatic log-keeping would be a snap. The call of each station worked would be entered from the keyboard, along with signal reports, mode, power and band

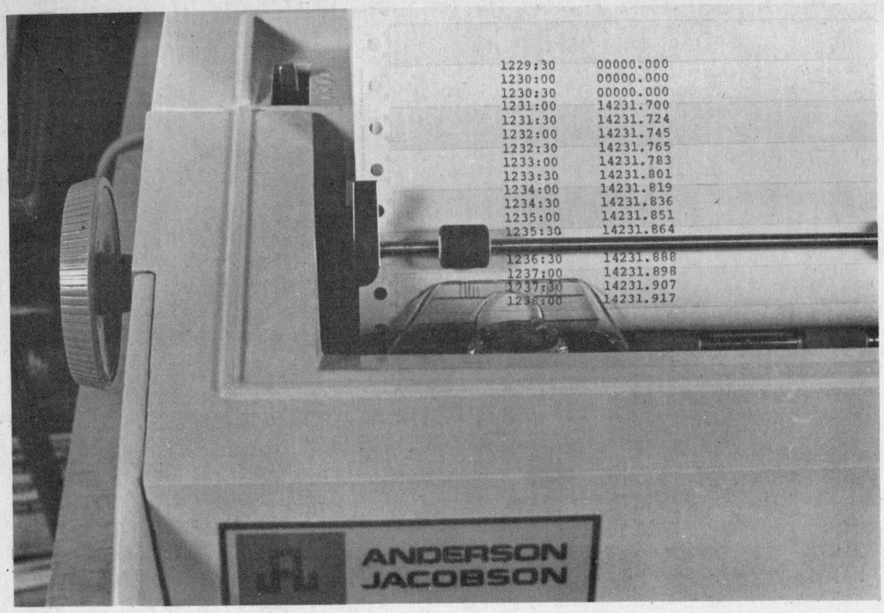

This computer printout shows the results of a drift test made on a homebrew direct-conversion receiver VFO. A frequency counter attached to the microcomputer's input port makes the frequency measurements every half minute by command from the computer, which then reads and prints out the data.

data, and the computer would fill in the date and time of the QSO. The electronic scratchpad discussed earlier takes on an entirely new dimension once it can be triggered by date. All sorts of messages can be programmed into the microcomputer — reminders of contest dates or schedules, days or weeks in advance. When the date arrives, an appropriate message can be flashed on the display screen or printed on the system printer. With the addition of about a half dozen ICs and a couple of relays, a microprocessor/clock can control anything from the push-to-talk input of a transceiver to an antenna rotator.

A Super-Display

Taking the idea of interfacing external devices to the computer for control or information-gathering purposes one step further opens a broad range of possible applications. An ordinary frequency counter or receiver "digital dial" that incorporates a multiplexed LED readout turns into an intelligent "super-display" once it is hooked up to a small computer. With such a setup, a program can be written to record and display a dial setting, and take action based on the frequency of a transmitter, receiver, homebrew VFO, or whatever, as long as it can be measured by the counter. The computer can graph the drift of a VFO versus time from a cold start until it reaches relative stability. Have it record the VFO frequency to the nearest hertz every 30 seconds, or have it record a series of elapsed times between hitting 50-Hz or 100-Hz intervals.

The ARRL Frequency Measuring Test takes on a new complexion when a system like this is available. After taking a series of measurements of the W1AW frequency, the microcomputer can calculate an average of those measurements, taking into account any built-in counter timebase errors. Writing a band-edge-monitor program that warns of operation outside of amateur band allocations or sub-bands is a trivial task. A good operator will analyze his log in detail following a contest to gain insight into operating strengths and weaknesses.

By automatically logging the transmit frequency used for each contest QSO, the computer can provide an invaluable tool for analysis. When he was having his best "runs," with rates in excess of 60 QSOs per hour, was he zipping around the band calling other stations, or was he sitting on one frequency and calling CQ? Do the best rates come while tuning up from the bottom of the band or does the opposite tactic give better results? Are there one or two frequencies in 40-meter phone band that seem to be optimal for DX stations to hear Ws and VEs? How much time is lost on average, between making the last contact on one band and the first contact on a new band? Developing information like this using ordinary means is difficult. Yet, if a microcomputer-attached frequency display is available, such information soon becomes indispensable.

Working OSCAR

The amateur satellite program has captured the interest of thousands of hams. Listening to manmade signals from space is fascinating. Communicating through an OSCAR transponder is little short of amazing. By removing the amateur's dependence on the ionosphere for long-distance propagation, amateur satellites have transformed the character of ham radio activity.

To use a satellite transponder effectively, however, it is necessary to know when the spacecraft is within communication range. Published equator-crossing data can easily be used with special-purpose graphical orbit calculators to predict the time and duration of satellite pass for the low-altitude, polar-orbiting OSCARs. But tracking satellites that have higher-altitude elliptical orbits can prove to be quite challenging with manual methods. The ham with a microcomputer has a much better solution available. A knowledge of some spherical trigonometry and the BASIC language will allow that ham to design a program that will predict acquisition- and loss-of-signal times for one or more OSCARs. With a little ingenuity, the display can be real-time; that is, minute by minute the video screen can show the subsatellite latitude and longitude, the distance (slant range) to the satellite, the altitude of the satellite, the proper beam heading, and a list of prefixes of stations that are currently within range of the transponder.

There are numerous instances where a small computer comes in handy for the OSCAR enthusiast, beyond the obvious application of satellite-orbit prediction. During a typical OSCAR pass, the optimum antenna orientation changes constantly in both azimuth and elevation. A ham really has his hands full when simultaneously aiming a pair of antenna rotators, sending cw with a keyer paddle and adjusting the downlink receiver to compensate for Doppler shift. With the aid of some programming and a bit of specialized interface circuitry between the rotator-control boxes and the computer, hands-off antenna aiming becomes possible, leaving the ham free to communicate instead of getting bogged down in operational details.

Telemetry

OSCAR satellites are equipped with on-board beacon transmitters. The function of a beacon is to identify the OSCAR transponder and to relay information about the spacecraft's environment (temperature, solar panel charging current, transmitter output power) to ground-based observers. Some of the beacon transmissions are in Morse code; others are encoded in machine-readable form. It is possible to program a microcomputer to receive and translate Morse code, although that takes half the fun away.

Once the telemetry is translated from code into decimal digits, a further translation is required to read the value in terms of engineering units (volts, amperes, degrees). A computer program can easily perform that translation, take averages over several telemetry frames, point out rapidly changing or out-of-normal-bounds parameters, and archive the information for later analysis. Future OSCARs will transmit telemetry directly in engineering units, but the computer will still be required to do the translation from machine-readable to human-readable form, as well as providing a data-analysis tool.

RTTY Renaissance

The easy availability of small computers will almost certainly revolutionize radioteletype (RTTY) operation. Lots of hams have been scared away from RTTY because of a lack of familiarity with motors and gears, a lack of room for the heavy iron equipment, and a dislike for noise, ink and grease. Yet, the idea of conducting a conversation with another ham located hundreds or thousands of miles away simply by banging away on a typewriter keyboard has much appeal.

It is not difficult to program a small computer to receive and transmit RTTY code. The incoming characters are displayed at the top of the video monitor, and the outgoing text is displayed at the bottom of the screen as it is entered from the keyboard. The keyed-in text is stored in a "buffer" in the computer's memory subsystem before it is sent over the air. This makes it possible to correct or edit the text so that no "typos" are transmitted. A renaissance of interest in RTTY using computers as terminals is underway right now. Few RTTY enthusiasts miss the noisy motors or ribbons full of ink.

Baudot Code

Since its inception, amateur RTTY has been conducted using a 5-bit code commonly referred to as the Baudot code. The choice of Baudot was based on the availability at low cost of obsolete teleprinter equipment designed for this code from the Weather Bureau, railroads and other sources. In fact, the FCC regulations previously limited all amateur RTTY transmissions to the Baudot code because their network of monitoring sta-

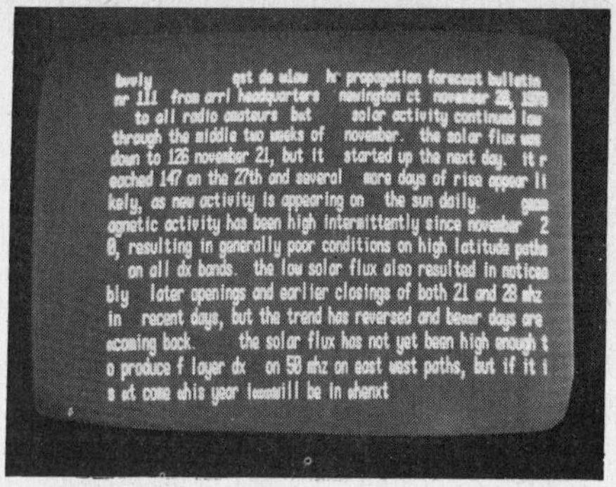

Say goodbye to your model 15 — RTTY reception on your computer terminal is noiseless, inkless and greaseless.

tions was equipped to receive only Baudot.

There are 32 possible combinations of 5 bits. If every combination is used to represent a different text character, then clearly there are not enough bit combinations to represent even the digits 0 through 9 and the English alphabet. By reserving one bit configuration to represent an "upshift" condition and another bit configuration to represent a "downshift" condition, each of the other 30 bit configurations can take on dual meanings, depending on the setting of the shift at the time they are sent. By taking advantage of this trick, the Baudot code can represent the 26 capitalized letters of the alphabet, the digits 0 through 9, common punctuation and control characters that can ring a bell, print a blank space, return the teleprinter carriage and so on.

There are still obvious limitations to this scheme. It is not possible to directly represent the lowercase alphabet, mathematical symbols, less-common punctuation, and additional control codes by using the Baudot code. The use of a 7- or 8-bit code instead of the 5-bit code allows this much wider range of characters to be transmitted. One such code is commonly referred to as ASCII.

ASCII Legalized

FCC regulations have been updated to allow hams to transmit ASCII. The ASCII code is well suited for use with small computers. The change in regulations will provide lots of room for amateurs to contribute to the development of microcomputer technology.

RTTY communication has always been plagued by transmission errors. A single burst of static may wipe out several consecutive letters in a word or a call sign. While local vhf RTTY transmissions are relatively immune to this type of interference, hf signals that are propagated through the ionosphere are prone to interference from fading, static and signals on nearby frequencies. Hams will be developing new transmission techniques that make use of error-detecting or error-correcting codes and protocols that will ensure clean copy at the receiving end. Computer-to-computer communications will become commonplace. Suppose that one amateur develops a satellite-tracking program that he wants to share with another ham located several hundred miles away. He can send a printed copy or a cassette tape of the program through the mail, or he can eliminate the delay and expense of mailing by using the unique communications mode at his fintertips to transfer the program. Not only can one microcomputer send a program directly to another, but it is also possible for the transmitting computer to cause the program to start executing in the receiving computer without requiring manual intervention at the receiving end.

Straight text and programs are not the only things that can be transmitted from one computer to another. Human speech and photographs are easily digitized and transmitted. There will be no difference to

ASCII Regulations

Part 97 of the FCC's Rules and Regulations is amended as follows:
Section 97.69 is re-entitled "Digital Transmissions" (formerly "Radio Teleprinter Transmissions") and is amended to read:
Subject to the special conditions contained in paragraphs (a) and (b) below, the use of the International Telegraphic Alphabet No. 2 (also known as the Baudot Code) and the American Standard Code for Information Interchange (ASCII) may be used for such purposes as (but not restricted to) radio teleprinter communications, control of Amateur Radio stations, models and other objects, transfer of computer programs or direct computer-to-computer communications, and communications in various types of data networks (including so-called "packet switching" systems); provided that such operation is carried out in accordance with the other regulations set forth in this Part.

(a) Use of the International Telegraphic Alphabet No. 2 (Baudot Code is subject to the following requirements:

(1) Transmission shall consist of a single channel, five-unit (start-stop) teleprinter code conforming to International Telegraphic Alphabet No. 2 with respect to all letters and numerals (including the slant sign or fraction bar); however, in "figures" positions not utilized for numerals, special signals may be employed for the remote control of receiving printers, or for other purposes indicated in this section.

(2) The transmitting speed shall be maintained within 5 words per minute of one of the following standard speeds: 60 (45 bauds), 67 (50 bauds), 75 (56.25 bauds) or 100 (75 bauds) words per minute.

(3) When frequency shift keying (type F1 emission) is utilized, the deviation in frequency from the mark signal to the space signal, or from the space signal to the mark signal, shall be less than 900 Hertz.

(4) When audio frequency shift keying (type A2 or F2 emission) is utilized, the highest fundamental modulating frequency shall not exceed 3000 Hertz, and the difference between the modulating audio frequency for the mark signal and that for the space signal shall be less than 900 Hertz.

(b) Use of the American Standard Code for Information Interchange (ASCII) is subject to the following requirements:

(1) The code shall conform to the American Standard Code for Information Interchange (ASCII) as defined in American National Standard Institute (ANSI) Standard X3.4-1968.

(2) F1 emission shall be utilized on those frequencies between 3.5 and 21.25 MHz where its use is permissible; and the sending speed shall not exceed 300 bauds.

(3) F1, F2 and A2 emissions may be utilized on those frequencies between 28 and 225 MHz where their use is permissible; and the sending speed shall not exceed 1200 bauds.

(4) F1, F2 and A2 emissions may be utilized on those frequencies above 420 MHz where their use is permissible; and the sending speed shall not exceed 19.6 kilobauds.

It is no longer necessary for amateur computer hobbyists to build their computers from scratch. Commercially manufactured microcomputers like this one (Radio Shack TRS-80) are ready to operate as soon as they are plugged into an ac socket. (*K1WJ photo*)

the casual listener between such transmissions and ordinary text. It will just require the appropriately designed equipment at the receiving end to sort out the photographs and voices from the program listings. The process of sending digitized voice over the air usually starts with a person actually speaking into a microphone. By using the computer to synthesize a voice directly, no microphone is necessary. While computer-generated speech is a little esoteric, no doubt a few hams will find applications for conversing on the air using their electronic syllable synthesizers. Computer-generated slow-scan television picture transmission is already here, and there is plenty of interest among technically minded hams in computer-assisted reception and enhancement of SSTV pictures in black and white and in color, as well as in the reception of weather photographs from orbiting satellites.

Computers and FM

Microcomputers are having a big impact in fm repeater design, with an increasing number of systems being automated under program control. Experiments with community-access computers for message switching and Amateur Radio bulletin transmission are bringing the applications of small computers closer to more and more hams every day. The Canadian amateur regulations have been amended to encourage the development of packet switching techniques. These techniques show promise of linking multiple repeaters into a coordinated network of message-relay stations with obvious implications in the areas of amateur emergency preparedness and communications reliability.

A New Frontier

In summary, the role of the microcomputer in Amateur Radio is of such importance that no ham can afford to be unfamiliar with it. The opportunities for directing the growth of Amateur Radio in the next decade are abundant, and the field is open wide enough to allow for contributions in every phase of that development. Take a pioneering step and add your own contribution.

Chapter 14

Shortwave Listening

There's all the time in the world above the 40-meter band — in two languages. Your ham-bands-only transceiver has no WWV band position, but transmissions of Canada's CHU on 7335 kHz, bearing announcements in English and French, lie well within that free 200 kHz of tuning offered by the 7.0-7.5 MHz range on your rig. This particular night, though, something that bids you pause, short of CHU, on 7325 kHz. It's another a-m signal with full carrier — but it sports *two* sidebands, unlike CHU, which transmits only carrier and usb. You listen further. There's music in those sidebands, music 20 seconds before the hour — "Oranges and Lemons" — and time pips. Then you find yourself spending an enjoyable 15 minutes sampling World News and Review of the British Press with the

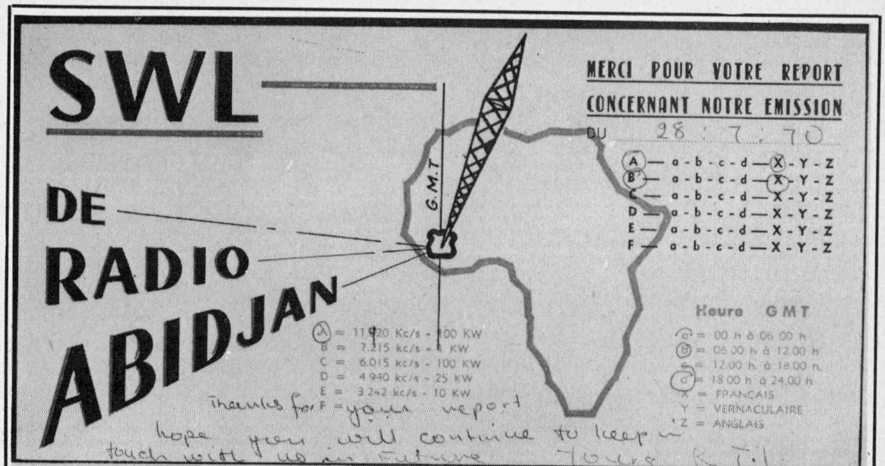

From the Ivory Coast of Africa come the signals of Radio Abidjan broadcasting in French, English and 14 native languages. Listen for them on 4940 and 7215 kHz, and 11.920 MHz.

You're Already There

Most readers can receive signals on frequencies beyond the amateur bands right now — on their amateur-bands-only receivers and transceivers! You can receive signals on the 75- and 41-meter international broadcasting bands (3900-4000 kHz and 7100-7300 kHz, respectively) if your rig covers 80 and 40 meters. Though these amateur bands are exclusive in the U.S. and Canada, elsewhere in the world they are shared with other radio services including broadcasting; the signals of these other services, disregarding national and continental borders as radio is wont to do, can be heard right along with local amateur signals here.

Most gear includes a few tens of kilohertz of tuning beyond the bottom edges of the amateur bands. Below 80 meters you can hear stations in the Aeronautical Mobile Service. Below 40, 20 and 15 lie frequencies allocated to the Fixed Service. Close to the edges of 10 are Fixed and Mobile Service frequencies.

Many rigs have been designed to tune 500-kHz chunks of the radio spectrum. Since only 80/75 and 10 meters use the full tuning range on such rigs, you've been handed free kilohertz carrying fascinating listening. Almost everybody's stepped above 40 to tune the time signals of Canada's CHU on 7335 kHz, but the free 7.3- to 7.5-MHz chunk above 40 is alive with other signals. You'll hear out-of-band broadcasters BBC-7325, Radio Prague-7345, Israeli Broadcasting Authority-7465, and Radio Peking's domestic service on 7504 kHz. Radio Moscow appears at several spots in this range as well. Australia offers time signals via VNG in Lyndhurst on 7500 kHz, audible around UTC noon in North America. Above 20 meters, 14.35 to 14.50 MHz is allocated to the Fixed Service. 21.45 to 21.50 MHz is a better bet for getting your feet wet in SWLing; this is the bottom 50 kHz of the 13-meter SWBC band. The signals of many high-powered broadcasters here may portend of 15-meter amateur DX openings to Europe and Africa.

The WWV band on your rig, if tunable, may offer coverage of one of several bands. If your radio receives WWV at 5 MHz, you may be able to cover a section of the 60-meter tropical broadcasting band. 10 MHz can give you the crowded 31-meter SWBC band. A 15-MHz WWV band may be a boon as well; you may have been given a free ticket to 19-meter shortwave broadcasting.

Try another technique, too: cheat a little. At each band setting, try rocking the preselector tuning through its range. You may find that you have coverage of a few more sections of the high-frequency spectrum by peaking your receiver's front end to image frequencies. If you hear anything new, study your technical manual to figure out what ranges you're tuning. And here's a hint: You may find that you're receiving usb in your rig's lsb position, and vice-versa, because of the different conversion processes necessary to get to image bands.

Antennas for many of these "found" bands are probably already in place. 80/75-meter systems work well on 75-meter SWBC, of course, as well as 40-meter wires handle 41 meters. Your 20-meter beam or dipole will work well on the 19-meter SWBC band. Likewise, your 15-meter antenna will be just what the doctor ordered for that free 50 kHz of the 13-meter SWBC band.

If you're about to protest that you feel left out of such SWBC DX reception because your rig or receiver has no straight a-m detector, we have news for you! Tuning a-m with a BFO and product detector, carrier at zero-beat, is a superior a-m DXing technique on *any* receiver. You already have the radio and the tuning ranges you need to get a running start at some heady listening between and beyond the amateur bands.

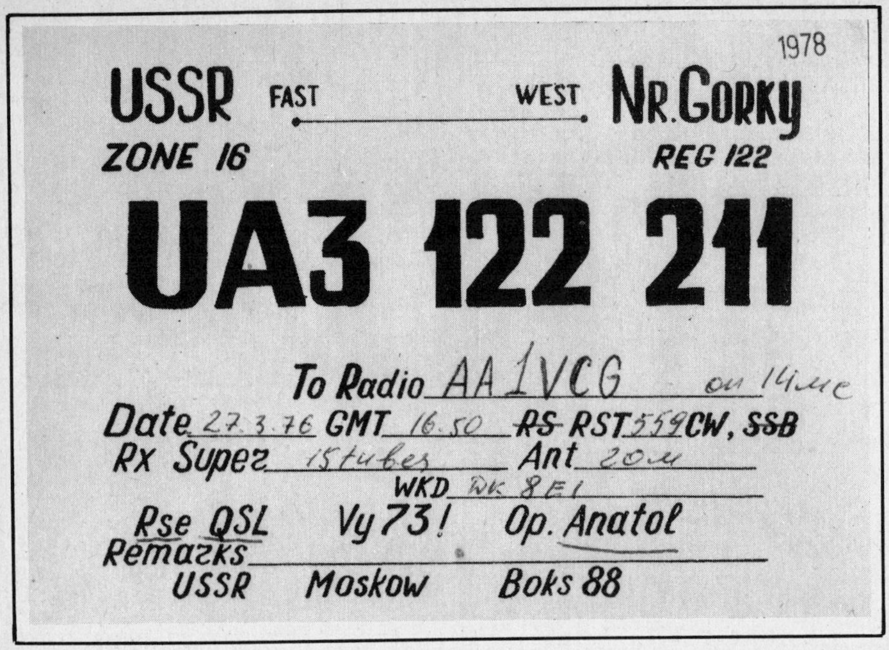

This reception report from SWL Anatol is one of many he'll send to the hams he hears in his study to become a radio amateur. Should you receive one, respond — you'll have helped further a new ham career.

British Broadcasting Corporation.

A surprised but sleepy-eyed searcher finds still more time in the dawn. It comes from far away, this news of time, speaking only once each quarter hour and beeping all seconds in between. Clipped "down under" speech intones the name of VNG, Lyndhurst, Australia, a standard frequency- and time-signal offering of the Australian Telecommunications Commission. This particular band-searcher thought that 7500 kHz would be nothing more than half of 15 MHz this morning, would be radio blackness, would be bare. Now, on an educated hunch, he trundles off to try his luck on 40 sideband. The VKs are loud.

That evening the cycle begins again. *Another* ham ventures to tune past the edge. Ops there are passing traffic for MARS, sure, but this particular listener tunes beyond them, *through* them, beyond what for her is commonplace and into what she thought was radio void. First come RTTY stations seeming scattered randomly as stars. Radio Prague, Czechoslovakia, fires up on 7345 kHz; the Israeli Broadcasting Authority clears its clear throat on 7465. Choppy cw in the Fixed Service surfaces in the noise and sinks again. As the sun wheels around, signals — strangely nonamateur, mysterious — slant in from Asia, then the Americas, then Africa and Europe. The show never ends.

These 200 kHz are gratis; their DX and the DX they imply, exciting. With a bit of effort on your part, there can be even more. There's a pileup tonight for a JY8 on 7002 kHz, maybe, but there's unheard gold in "them thar hills" between and beyond the amateur bands. Here's how to get out there and find it.

A Radio Sky

Every time you press the key or speak into the microphone of your amateur transmitter, you take out another citizenship. You're immediately a citizen not only of a country but of a world, your rig's signal not only on a band but of a spectrum. Direct current through cosmic rays and beyond comprise the electromagnetic spectrum, of which radio is only a part — and of this radio your ham doings are concerned with only a fraction. Imagine it: You will never have transmitted any significant fraction of the ideas, languages and information that can fill your ears in only one half hour's monitoring of the shortwaves. As a listener you can hear it all.

Welcome to the world of shortwave listening. It isn't tame. Here are frequencies where the hiss of Jovian radio storms suffuses through and beyond Megahertz-wide crowds of manmade signals much as the Milky Way straddles the sky of nights — a sky of radio: frequencies where ideas and ideologies, nations and nationalities mix without war and without mutual destruction. All of these signals are yours for the hearing.

"SWL," once the trademark only of those tuning frequencies from 1.6 through 30 MHz, has come to stand for any listener who searches for extraordinary radio on any frequency. You may shrink from the term SWL. Writers of photo captions in amateur journals have for years tacked that label on any non-celebrity in a group of hams luckless enough to be caught without a call. Shortwave listeners, especially those who DX the ham bands, have long been regarded by the amateurs they respect as insincere or halfhearted in their radio interests. Nothing could be further from the truth. Hams are radio enthusiasts who have chosen to be voices as well as ears in the realm of the radio sky; for shortwave listeners, the illimitable variety in active listening is enough. The practical dividing line between the SWL and the amateur is just this: For shortwave listeners, bands are crowded with their quarry and not with their fellows.

What's out there to be heard? You may think at once of broadcasters, and indeed there are many of these. Ships at sea, coast stations, aircraft plying transoceanic routes, point-to-point communications, CB radio — these are all out there in the span 0.5 to 30 MHz, the frequency range we'll explore in detail throughout this chapter. What skills will you need? Those you now enjoy in your amateur work and a few more. The ability to copy cw, even if you prefer phone in the ham bands, will net you a lot of listening pleasure in the forms of aeronautical, fixed and maritime signals. Ssb is heavily used. High-quality a-m is transmitted by broadcasters worldwide and its optimum reception, a dying art in amateur circles, will provide you insights into many new languages, countries and ideological views.

First, some of the sources of signals you may hear as you tune will be enumerated, and they'll be matched with the frequencies on which you'll find them. Following this, how to receive those diverse signals and what to do on the receiving end will be discussed — logging, DX clubs, reporting, QSLing, international call sign prefixes and more. At that point, you'll be excitingly on your own to wander the radio sky.

Regulated Radio

Almost all of the radio signals you will hear are regulated with regard to purpose, frequency, call sign and technical standards. The International Telecommunication Union (ITU) fulfills this function worldwide, dividing the world into three administrative regions to do the job. Region 1 is Europe, Africa, the Middle East and the U.S.S.R.; Region 2, North and South America including Hawaii; Region 3, the rest of the world. Federal governments apply and enforce ITU guidelines at the national level.

Radio signals are placed in the electromagnetic spectrum with respect to their intended purpose or *service*. Each service is allocated groups of frequencies or "bands" on which it can meet its goals under conditions of varying propagation and service growth. Stations in most services operate on strictly assigned frequencies within their allocated bands; nearly worldwide, operators of stations in the

Amateur Service enjoy the privilege of operation on virtually any frequency within the amateur bands at will.

Individual transmitting stations use call signs — alphanumeric words indicative of a signal's service and country of origin — to be transmitted in addition to message or broadcast content as signal identification. The ITU allocates "blocks" of letters (CFA-CFZ, NAA-NZZ, 4UA-4UZ, for example) to each country of the world and suggests call-sign formats distinctive of each service; within their assigned blocks, national governments issue call signs respective to service and consistent with their own regulatory and enforcement objectives.

Stations in the Amateur, Fixed and Maritime Services use their call signs routinely; many low- and medium-frequency navigation beacons transmit only one- to three-letter identifiers often unrelated to their call signs. Domestic broadcasters in Japan, Oceania and the Americas transmit call signs with promotional zeal, while domestics elsewhere and all but a few international broadcasters are lax in this responsibility. You'll find a list of international prefixes in the appendix of this book. We'll detail how to use it as you read about listening techniques further on.

Regulated radio signals must meet internationally recognized technical standards as well as those of service, frequency allocation and call sign. These standards are meant to ensure that radio emissions are confined to their allocated frequencies and that bandwidth and power are no greater than necessary to achieve service objectives.

Here are some of the services you'll hear as you tune from 0.5 to 30 MHz, with the addition of longwave and North American fm and television broadcasting. Several frequency allocations are given for each, but these lists are not all-inclusive because allocations vary according to ITU region — some are shared in parts of the world, some are exclusive. To complicate things further, some of these bands will change in the 1980s as the result of the 1979 World Administrative Radio Conference. At the end of this chapter we'll outline changes affecting any of the bands mentioned for the services to follow. But the world picture as far as frequency allocations are concerned is best reflected in publications of the ITU; these are available by mail from Geneva; we'll mention the address later on as well.

Aeronautical Mobile

Vhf can't fill the bill for traffic control on transoceanic and transcontinental air routes, so shortwave radio is the thing. Audible are ground-to-air and air-to-ground exchanges, both civil and military. Operation is mostly in the A3J (suppressed-carrier-ssb) mode, but some stations use A3H (full-carrier ssb) to reduce bandwidth while retaining compatibility with receivers using envelope detection. China, Czechoslovakia, the U.S.S.R. and others may be heard using cw in the air. Some ground stations broadcast aviation weather around the clock on unchanging frequencies, with several such stations sharing one frequency and alternating their transmission — New York City and Gander, Newfoundland, for example, share 3001, 5652 and 8868 kHz and 13.272 MHz.

Table 1

Aeronautical Mobile Allocations

2850-3155 kHz	10.005-10.100 MHz
5450-5730 kHz	11.175-11.400 MHz
6525-6765 kHz	13.200-13.360 MHz
8815-9040 kHz	15.010-15.100 MHz
17.900-18.030 MHz	

Ground stations generally identify by place name. Aircraft identify with airline identifier and flight number. Military stations may use tactical identifiers such as "Sky King" and "Hackberry."

The Aeronautical Mobile Service is one of what SWLs call the "utilities," along with all other nonbroadcast services except amateur. You must not divulge the content of any addressed message traffic you hear over any of the utilities; international radio regulations ensure the secrecy of these transmissions and absolutely forbid reportage of traffic particulars. Some governments prohibit their citizens from listening to any nonamateur or nonbroadcast services at all. As it is, we may only publish what information is already a matter of public record. Some of the aeronautical bands are listed in Table 1.

Amateur

Personal involvement in this service necessitates the passing of government examinations but precludes the passing of money. Amateurs combine emergency service to the public with personal enjoyment and enrichment along entirely voluntary and noncommercial lines. They utilize a-m, fm and pulse modes — RTTY, SSTV, ssb, cw and more — with most of the countries of the world represented on the amateur bands. It's the only radio service in which fascination with radio itself is as officially valid a reason for participation in its activities as any. Amateur frequency allocations are listed in chapter 3.

Broadcasting

The business of broadcasting is "to make known over a wide area" — to cast out information that it might be heard by citizens, regions, nations, the world. Broadcasts intended for target areas beyond their country of origin are "international" and those intended for reception at home are termed "domestic." You may hear these terms as "external" and "internal" as well. Modes used include a-m, fm, ssb and television. Broadcasting stations in North America are largely operated for profit; both here and elsewhere, broadcasting stations are also run by non-profit government, religious and educational interests.

The 150-285 kHz ("long wave" or LW) band is used for a-m broadcasting only in ITU Region 1. Tremendous powers into the megawatt range are employed. Some of these stations are audible in North America, especially as cold dry air kills local thunderstorm activity in winter months. Even then, North American listeners must dig through signals of other services for such DX.

The 535-1605 kHz ("medium wave" or MW) band is known to just about everybody everywhere. North Americans know it as a domestic band, but worldwide it serves for international purposes as well. A-m is the mode; channel spacing is 10 kHz in the U.S. and Canada but differs in the rest of the world. DX possibilities are excellent. With the excitement offered by medium-wave signals both domestic and international, many SWLs tune into this band and never tune out again. In this chapter you'll find an indepth feature on the organization of medium-wave broadcasting in the U.S. and Canada.

Broadcasting allocations between 2 and 30 MHz — the "short waves" — are used for international and tropical broadcasting. What's the difference? In equatorial regions, the coverage of medium-wave broadcasting stations is severely limited by interference from thunderstorm static. Many such broadcasters in these countries feed their medium-wave programming to transmitters operating in the 2-, 3- and 4-MHz tropical bands to achieve nationwide coverage. In actuality they achieve far more than national coverage — veteran DXers everywhere consider these bands to be among the most exciting sources of exotic loggings from Africa, Asia, Latin America and the Pacific.

International broadcasting bands are just what their name signifies. Here you'll find the supergiants exhaling news, features and propaganda for consumption in target areas far overseas with transmitter powers up to half a million watts. In between you'll still find domestic services trying hard to serve their native audiences without getting stepped on by the big boys. There has been in shortwave radio over the past few years a power struggle — with more powerful transmitters being commissioned yearly as interference levels rise and emerging nations clamor to be heard. It's been said that governments struggling to their feet first install two institutions of national pride and solidarity: a national airline and a broadcasting service.

There are limits to power, but power and position make an unbeatable team.

SWLs of the past decade have seen the rise of a new SWBC institution — the relay station. Nightly, for instance, the 49-meter band is torn to shreds by huge signals dumped into North America by relay stations in the Caribbean. The BBC and the Voice of America, among others, operate relay sites worldwide; the Canadian Broadcasting Corporation relays the BBC and more.

Table 2 shows the frequencies allocated to shortwave broadcasting. Channels are spaced every 5 kHz beginning 5 kHz above the bottom edge of each band. Some of these bands will seem to collide with a few of the allocations we list for other services. They do. We've listed all frequencies allocated to broadcasting because several of the bands utilized in North America by the Aeronautical Mobile, Amateur and Fixed Services offer some of the choicest shortwave broadcast DX going.

Table 2

Shortwave Broadcast Allocations

120 meters:	2300-2495 kHz*
90 meters:	3200-3400 kHz*
75 meters:	3900-4000 kHz**
60 meters:	4750-4995 kHz, 4995-5060 kHz*
49 meters:	5950-6200 kHz
41 meters:	7100-7300 kHz***
31 meters:	9500-9775 kHz
25 meters:	11.700-11.975 MHz
19 meters:	15.100-15.450 MHz
16 meters:	17.700-17.900 MHz
13 meters:	21.450-21.750 MHz
11 meters:	25.600-26.100 MHz

*Tropical broadcasting
**Not in Region 2; 3950-4000 kHz in Region 1.
***Not in Region 2.

Above the shortwaves are fm sound broadcasting and television frequencies. In most of North America, fm frequencies span 88 to 108 MHz, with stations placed on 100 channels 200 kHz wide beginning with 88.1 MHz. Channels from 88.1 to 91.9 MHz (the "college band" to many SWLs) are reserved for noncommercial educational fm broadcasting. Frequencies for television channels 2 to 6 are 54 to 88 MHz with a break at 72 to 76 MHz for remote control of models and activities of several other services. Channels 7 to 13 occupy 174 to 216 MHz, and uhf channels 14 to 83 use 470 to 890 MHz.

CB Radio

Originally conceived as a personal radio service emphasizing ease of licensing and base-to-mobile and mobile-to-mobile communication, this service is allocated 40 channels 10 kHz wide in the range 26.960-27.410 MHz. Modes used are a-m and ssb.

Fixed

Stations in the Fixed Service do no business with mobiles. Their traffic is exchanged from one point to another on the earth's surface, and for this reason they're also known as point-to-point or PTP stations. Modes used include cw, a-m, ssb, RTTY, FAX and exotic modes such as multiplex telegraphy — wide whirring clusters of multichannel RTTY making maximum use of allocated spectrum space. Activities of stations in this service are some of those referred to as "public telecommunications" in questions relating to the desirability of international third-party traffic via Amateur Radio. Amateurs have worked stations in the Fixed Service if they have contacted military stations under the special cross-band conditions of U.S. Armed Forces Day.

The Fixed Service is another of the services grouped together by SWLs under the term "utilities." We do not encourage listening to the traffic content of transmissions in this service because, as mentioned

Medium-wave Broadcasting in the U.S. and Canada

If the frequencies from 1.6 through 30 MHz are likened to "the wide open spaces" — a radio landscape through which static and extraterrestrial noise blow like wind, a landscape now and again populated by "towns" of shortwave broadcasters and amateurs snug in their assigned bands, here and there thicketed with the signals of ships or dotted with lone RTTY fixed stations as trees punctuate a steppe — then the 535- to 1605-kHz medium-wave broadcasting band is Metropolis, replete with its selfish rush-hour subway crowds, its frenzied shoppers and hawkers, its cathedrals calming a busy mind's moment with their offerings of a message of peace.

What's all the hubbub? In the United States and Canada, a scan of the medium-wave band (also called "the standard broadcast band," "the a-m broadcast band," or simply "BCB") from any location reveals signals stacked five to 10 deep on many of the 107 channels spaced every 10 kHz from 540 to 1600 kHz, with 5 kHz above and below this band allowed for sideband energy. Despite what seems to be heavy mutual interference, all's not chaos — medium-wave broadcasting stations in both countries are licensed within an orderly regulatory structure. They can be classified with the application of three criteria: channel/service area, authorized operating time and antenna radiation pattern.

In the channel/service area category, you'll hear stations operating on *clear channels* — frequencies on which a few giants with powers up to 50 kilowatts serve many states or provinces during hours of darkness. On *regional channels*, tens of stations of intermediate power (up to 5 kW) serve areas of intermediate size. And on *local channels* (or, as they are known to the DXing fraternity, "graveyard" channels), low-powered stations numbering in the hundreds compete for local listenerships.

Under the category of authorized operating time, stations may be licensed to operate daytime only, nighttime only, limited time, specified hours, shared time or unlimited time. Nighttime is defined as the hours from sunset to sunrise. Limited time stations must close down to "protect" other stations on frequency from interference. Stations operating with specified hours seek to operate for less than unlimited time. Stations sharing time share a common frequency and "pass it around." Unlimited time stations are licensed to operate 24 hours a day.

A station's antenna radiation pattern may be directional or nondirectional. The times of day a station might operate with a given pattern include daytime (sunrise to sunset), nighttime (sunset to sunrise) and critical hours (the two hours after sunrise and the two hours before sunset).

In Canada, an additional class of station altogether is the low-power relay transmitter. LPRT's are 20- or 40-watt unattended unlimited-time transmitters fed by telephone lines and feeding nondirectional antennas. They fill gaps in and extend the coverage areas of stations they relay.

When authorized operating time is juxtaposed with required radiation patterns, simplicity becomes history. For instance, daytime stations may be either directional or nondirectional. Beyond this, a nondirectional daytimer may have to "go directional" during critical hours, or a directional daytimer may have to become *non*directional during critical hours. Unlimited time stations may be required to perform even more complex feats of legerdemain. Perhaps an unlimited-time station can get by with the same nondirectional pattern day and night. If not, it will probably have to become directional at night. (Some stations reverse this, running directional antennas during the day and nondirectional arrays at night.) Or, an unlimited-time station may have to be directional at all times but with different patterns day and night. It may even use three different directional patterns — one for daytime, another for nighttime and a third for critical hours! The object of these pattern changes (often accompanied with changes in transmitter power to produce the desired effect) is to allow as many stations per channel as possible while keeping mutual interference to a minimum.

Two other concepts affect what the DXer may hear: PSAs and NSP. PSA stands for Pre-Sunrise Authorization. Most daytimers prefer to begin their broadcast days at 0600 local time, but in winter months sunrise may be much later than 0600 — and daytimers could not sign on then without violating the terms of their licenses. To get around this, they may request PSAs — authorizations allowing them to sign on at 0600 with powor reduced until sunrise allows them to operate at full power. Some unlimited-time stations may make use of PSAs as well.

NSP. Most stations have intentional silent periods for transmitter maintenance even if they otherwise operate 24 hours a day. Some close down weekly; some close for the few hours of their silent periods only one or two days each month. The spectre of NSP (No Silent Period) stations is a cause of increasing dismay to MW DXers. Alternate main transmitters keep NSP stations perpetually on the air, one transmitter pumping out the watts while the other undergoes maintenance. Such stations never clear their channels to let prime DX through and more broadcasters jump on the NSP bandwagon each year.

The BCB band requires a challenge demanding some of the most specialized skills in SWLing. Whether you become an aficionado of the regionals, prefer the easier long-distance listening of the clears, or rise to meet the morass of a graveyard channel head on, you'll find yourself among the excellent company of other DXers who prefer their radio action on the medium wave band.

in the description of the Aeronautical Mobile Service, international regulations prohibit listening to, telling about, or benefiting from private traffic on public correspondence frequencies.

What you *can* listen to, as far as the Fixed Service is concerned, are test transmissions, and these are of great value in gauging propagation conditions to and from various areas at a given time. Radiotelephone test transmissions usually consist of identification as a test transmission coupled with transmitter location, sometimes with test counts and declaration of station ownership as well. These repetitive transmissions are sometimes termed "voice mirrors." If they include a melodic sequence of tones you'll find them referred to as "melody mirrors" or "melody markers" in the SWL press.

Cw test transmissions take the forms of V and CQ markers. V markers are series of Vs followed by DE and station call sign; QRA markers are similar except QRA ("the name of my station is") is substituted for the Vs. Other Fixed cw stations merely call CQ. RTTY PTP stations may rely on the familiar RY signal for testing purposes; some call CQ or QRA and some add the Q-signal QJH1 ("transmit test tape by RTTY") as well.

A number of national governments and naval observatories use frequencies allocated to the Fixed Service for transmissions of time signals — Australia's VNG on 4.5, 7.5 and 12 MHz and Ecuador's HD2IOA on 7.6 MHz are lesser-known examples. CHU operates on Fixed frequencies as well.

The list of international prefixes will help you identify the origin of many Fixed stations. Where do they operate? Most amateur and SWBC bands are flanked by Fixed Service allocations. By scanning our listings of allocations for other services in the 3 to 30 MHz range, you can determine chunks of spectrum left unassigned between them — there, you'll find transmissions of the Fixed Service.

Maritime Mobile

Here are transmissions of coastal stations "and all the ships at sea." Inland river traffic is audible as well, especially in the bands below 8 MHz.

Both military and public interests are represented; modes used include cw, ssb and RTTY. Frequency allocations are divided roughly in half, with ships transmitting in the lower part of each band and coastal stations above them. V, CQ and QRA transmissions are common in the Maritime Mobile allocations. Additional signals often heard are QSX ("I am listening on . . .") and QSW ("I am going to send on . . .").

Ship stations are issued call signs consisting of four letters (KEAL, GUQL) or two or three letters followed by four numbers (WZN3907, WU5151). Coastal stations sport call signs of three letters

Table 3

Maritime Mobile Allocations

4063-4438 kHz	12.330-13.200 MHz
6200-6525 kHz	16.460-17.360 MHz
8195-8815 kHz	22.000-22.720 MHz

(KLB, HLG) sometimes followed by a numeral or two (EAD44, HSA2). Coastal stations sometimes transmit traffic lists — strings of call signs of ship stations for which they have traffic. The list of international prefixes is of great value in matching these call signs with their countries of origin.

In addition to routine traffic work, some commercial coastal stations broadcast machine-sent cw press to vessels at sea. This is excellent code practice, especially where names of foreign cities and dignitaries are concerned. Press text carries full punctuation as well — be prepared! One source of press transmissions is WCC, Chatham, Massachusetts, at 0300Z on 4331, 6376 and 8586 kHz and 13.033 MHz.

Observatories of a number of navies of the world also transmit time signals of atomic-clock accuracy over frequencies in the Maritime Mobile, Fixed and Broadcasting allocations.

The Maritime Mobile Service is another of the "utilities," and secrecy of its traffic is guaranteed by international law. An important exception to this are distress messages concerning ships at sea. International calling and distress frequencies include 500 and 2182 kHz. Maritime Mobile frequency allocations include one just below 500 kHz (cw only) and several in the 2-MHz range. Other allocations are shown in Table 3.

The Voice of the German Democratic Republic is RBI — Radio Berlin International. You can hear it using transmitters from 5 to 500 kilowatts on the bands 49 to 13 meters.

Radiodetermination

Not the name of a service itself, *radiodetermination* is a general term for a grouping of various radionavigation and radiolocation services. The Marine Radionavigation and Radionavigation Service signals audible between 0.5 and 30 MHz provide little listening attractive to humans because conversations between machines tend to be rather cut-and-dried. The Radionavigation Service, for instance, is responsible for the rippling phased signals of LORAN, bane of amateurs using the 160-meter band.

The Aeronautical Radionavigation Service is a different story. A-m tone-modulated cw (A2) is the mode used by beacons between 510 and 535 kHz as they transmit long dahs broken by one- to three-letter identifications; these beacons make for excellent DX. An old friend of utility and medium-wave broadcasting DXers alike is beacon "NB" of North Bay, Ontario, on 525 kHz. Below the Maritime Mobile allocation near 500 kHz, more such beacons are arrayed in force.

Standard Frequency and Time Signals

The familiar signals of CHU, WWV, WWVH and others belong to this service. Most operate on internationally allocated frequencies of 2.5, 5, 10, 15, 20 and 25 MHz, although a few use frequencies assigned to the Fixed (CHU, VNG), Broadcast (YVTO, Venezuela, on 6100 kHz) and other services. Hints on their use are contained in "On the Receiving End" later in this chapter and in the section on amateur DX techniques elsewhere in this book.

Radio Beyond Reach

A number of exotic transmission modes are used by the family of radio services, most of them utilizing machine languages and digital techniques. You can't identify these signals merely because you can't translate the machine codes into a recognizable language or alphabet. Short of such cases where they're "just not speaking your language," if you somehow cannot seem to pinpoint meaning, origin or purpose of certain transmissions *it's probably because pains have been taken to keep it that way.* Codes and ciphers, forbidden in the Amateur Service, are routine in others. Tactical identifiers are just that: means employed to thwart unwanted monitoring of transmissions relating to matters of security.

Popular with SWLs are "numbers" transmissions consisting of numeric sequences repeated for several minutes, usually by female voices. The majority of such stations speak Spanish, although a few speak German and even fewer broadcast in English. Though the SWL press terms them "spy" stations, editors

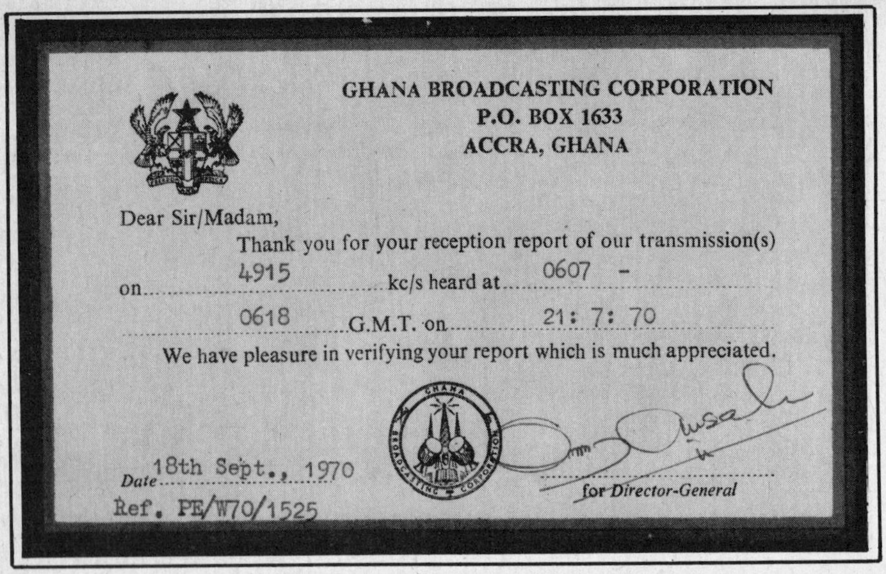
GHANA BROADCASTING CORPORATION
P.O. BOX 1633
ACCRA, GHANA

Dear Sir/Madam,
Thank you for your reception report of our transmission(s) on 4915 kc/s heard at 0607 - 0618 G.M.T. on 21: 7: 70
We have pleasure in verifying your report which is much appreciated.

Date 18th Sept., 1970
Ref. PE/W70/1525
for Director-General

Domestic shortwave stations QSL too. This card verifies reception of the well-heard Ghana Broadcasting Corporation transmitter on 4915 kHz in the 60-meter tropical band.

striving for accuracy can only chalk them up as radio beyond reach.

Transmission methods known as *spread spectrum* techniques are another source of signals meant to escape coherent interception. By these means, message content, already encrypted, is spread out or compressed in both frequency and time to avoid detection.

For the most part, the above uses of radio more or less adhere to the tenets of services on whose frequencies they operate. Three classes of radio do not. They are out-of-band (OOB) stations, pirates and jammers.

A number of powerful shortwave broadcasters operate on frequencies allocated exclusively to Amateur, Fixed and other services. Some of these transmitters have been there since before World War II, their operators refusing to move them even after their frequencies were allocated to other services. Others operate outside of allocated bands as a matter of routine if their regulating governments choose to ignore pressure to desist from the radio community. The SWL can take advantage of such OOB transmissions if they emanate from countries hard to hear within the broadcasting bands. Radio amateurs should lodge official protests with their national governments if stations of other services make unauthorized use of amateur frequencies. In the U.S. and Canada, amateurs may participate in CRRL/ARRL's Intruder Watch to keep amateur bands free of such undesirables.

Pirate stations are most often operated by hobbyists who dislike the strictures of regulated radio. Unlicensed, they generally operate on or near the frequencies of the broadcasting services they mimic. In areas of political unrest, radical factions sometimes take to the air in hopes of marshalling support for their views and undermining popular faith in the opposition. In this capacity, pirates are more likely to be termed "clandestine." A few clandestine stations are operated by national governments with intent to deceive — they claim to represent the revolutionary views of insurgents in their target areas by masquerading as domestic broadcasters. In reality they intend to foment the unrest they claim to mirror.

Almost every radio signal you hear is a symbol of human victory over silence, distance and ignorance. Not so the jammers. They bespeak a government's disbelief in its own dogma, for they take to the air to prevent external doctrines from clashing with carefully nurtured internal "truths." Jamming transmitters are most often heard as 10 to 20 kHz wide masses of rumble, overmodulation and white noise dumped onto the frequencies of "undesirable" broadcasts in *domestic* languages to prevent their reception in domestic target areas; an English transmission to a Lithuanian-speaking populace, for instance, will likely get through unscathed. Oddly, the transmissions of jammers are usually accompanied by Morse identifiers transmitted at intervals so that the administrations generating them can ascertain the results of their handiwork. (We do not spotlight jammers as exemplary of the human spirit; we report their existence because they are, between and beyond the amateur bands, yet another class of citizens, however undesirable, of the radio sky.)

Why do some stations continue to operate on frequencies not allocated to the services they represent? The regulatory efforts of the International Telecommunication Union have been mentioned often enough so that you can tell that its task as the organizational body for world telecommunication is a massive one. The power of the ITU lies only in governments' agreements to be bound by its decisions; over 150 nations have agreed to be so bound. Unlike the United Nations, however, there are no ITU "security troops." Any national administration may assign any station to any frequency, provided that no harmful interference is caused to any station of another country operating in accordance with the allocations table. *It is the responsibility of users of disrupted services to complain.* Success in having an offending station removed depends upon the willingness of its regulating government to comply. Administrations ostensibly signatory to ITU treaties appear oddly deaf-mute on some issues and critically sensitive to others. An offending administration cannot be expected to acknowledge complaints it is actively striving to ignore.

Radio Out of Control

Radio out of control is a realm of unintent and incident. At some point, you'll hear radio signals beyond — or temporarily eluding — the control of Man. Some of these signals have inspired one of the greatest searches in the history of humanity, while others have only had us banging on telephone poles in hopes that they'd disappear.

Every electric spark generates electromagnetic waves from direct current up through visible light and beyond. We call the radio output of lightning static, atmospherics or QRM, and it's pretty well known. Static crashes in conjunction with skip may render a band useless even though local skies are not cloudy all day.

Electric sparks also exist in many of the electromechanical devices contrived by Man. Commutator brushes in motors, arcing thermostats, leaky power-line insulators, spark-igniters in gas stoves . . . it's a busy world. These all add their punctuation to the story of radio. Long ago they were grouped together under the term QRNN. Now they go by many names, they are everywhere and they are totally out of control.

Telecommunication administrations try to have a hand in this sort of stuff by enacting legislation concerning what they term "incidental radiation." Radio itself, of course, cannot be ultimately regulated — only the actions of the people who use it can. People who will never speak over a transmitter in their lives own radio receivers, and if those receivers are superheterodynes, their oscillators are going to leak new radio in exchange for that received. Their pocket calculators will buzz nearby medium-wave broadcast sets through unruly multiplexed digital displays.

Every radio transmitter offers surprises in the form of integral multiples of its output frequency, or harmonics. All a government can do to regulate these and other spurious radiations is set limits on how strong they may be relative to their fundamentals. Some SWLs have made a hobby entirely out of reception of harmonics. If you hear a broadcast harmonic, inform the operators of the station concerned. The electromagnetic spectrum is crowded enough.

Radio out of control also illuminates the minds of dreamers. Interstellar matter shot through with energy gives off other energies. Atomic hydrogen and molecular carbon monoxide expound on the wavelengths of 21 centimeters and 2.6 millimeters, respectively. We might have said "are audible" because the majority of us are accustomed to having our radio served up as sound; old hands can even tune the machine codes of RTTY by ear. Most of the impulses received at the hands of radio astronomers, though, never make it to sound — they're fed to the hungry "intellects" of computers or used to drive chart recorders. One thing those computers keep searching for in the scratchings and hisses of the radio sky is sense — the kind of sense many of us dream might mean that another crew stands similarly hunched over *their* chart recorders a few hundred light-years hence. The sense that we are not alone.

(If you were to stamp and remark pointedly that neither 21 cm nor 2.6 mm can be tuned between 0.5 and 30 MHz, we'd have to admit that you have us there. We've stretched a point a bit to mention this particular form of radio out of control because we wanted to inject mention of a property of radio that has nothing at all to do with physics: mystery. To this end, consider the radio storms of Jupiter.)

It wasn't enough that the planet Jupiter radiates more heat into space than it receives from the sun, for in 1955 Jovian radio emissions similar to earthly static were detected. Subsequent years had them audible — with appropriate gain antennas, of course — at 18, 22 and 27 MHz. Now, the fact that electrical storms rage continuously in Jupiter's weather-torn atmosphere is almost household knowledge, and the origin of Jovian static seems clear. Nonetheless, no QSL has ever been received. We have had to be content merely with having heard.

Radio out of control is everywhere. Sometimes you'll want to destroy it or avoid it; at other times you may find it providing heady challenges worthy of the DX game. Regulated radio is straightforward; you'll likely do most of your listening there. Radio beyond reach is for someone else to know and for you to find out. It edges every signal you'll ever hear with mystery. *That's* the part of radio they don't include in the contest scores.

On the Receiving End

Now that you have an idea of what sorts of signals populate the frequencies from 0.5 to 30 MHz, you'll need to bring them in so that you can hear them. Once

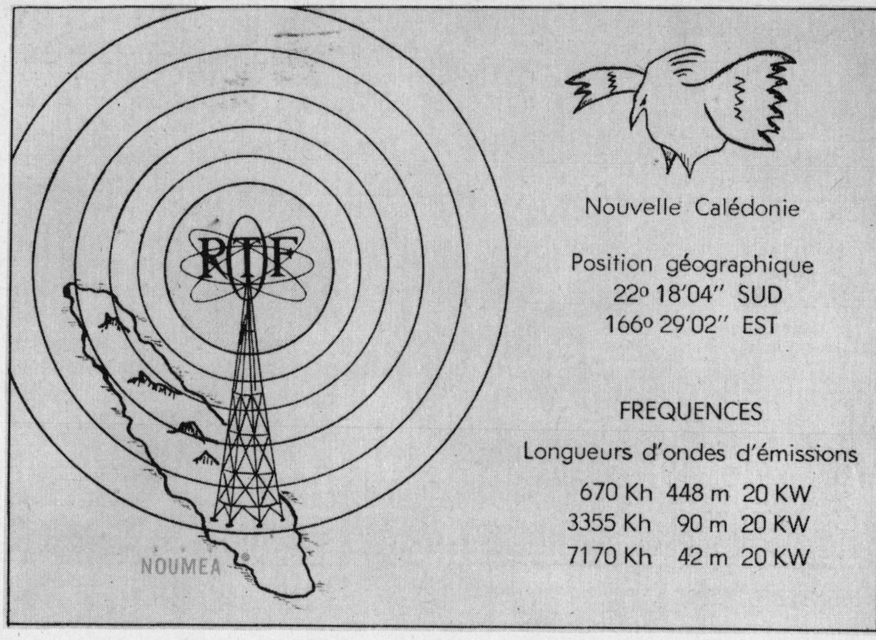

The 20-kilowatt voice of Radio Noumea, New Caledonia, is clear around 0800 UTC in North America on the 41-meter frequency of 7170 kHz. You can hear them tomorrow morning on your ham-band receiver.

you're hearing them, you'll want to know what they are and where they're coming from. Then you may want to write those stations for QSLs confirming your hard-won loggings. Means and methods to these ends are what you'll explore as we look at SWLing on the receiving end.

Hearing

Your receiver is your workhorse in short-wave listening. Its quality, and the quality of its antenna system, will make the difference between your being an aloof stranger or a knowledgeable listener familiar with doings between and beyond the amateur bands. Your receiver should be sensitive — 0.5 microvolts for a 10-dB signal-plus-noise-to-noise ratio for ssb and twice that for a-m are standard for the frequencies between 3 and 30 MHz. Less sensitivity is acceptable and even desirable on the medium-wave and long-wave bands, where atmospherics make high sensitivities unnecessary — and where your receiver's dynamic range may be the only thing between a positive i-d and "not a chance," when you're trying for a weak medium-wave broadcast signal 10 kHz away from one of your 50-kilowatt locals.

Sharp selectivity based on 6-dB bandwidths of 4, 2.4, 1.8 and 0.5 kHz or so are musts if you're going to be listening to all modes. Selectivity options wider than 4 kHz on a-m are not of much utility in serious SWLing unless you want to have them for the times you sit back to enjoy the programming. Naturalness of audio reproduction and optimum interference rejection can't be had at the same time. Some of the stock general-coverage receivers on the market come with 6-kHz ceramic filters as selectivity elements — if you buy one and start digging for weakies, you're in for a lot of grief. A few dealers offer these receivers modified. With selectivity increased and a few other circuit changes made, DXing with them becomes a challenge of propagation and listener skill as opposed to a struggle with insufficient circuit design behind the front panel.

A stable beat-frequency oscillator (BFO) and provision for selectable sideband reception are requisite for cw, ssb and even short-wave and medium-wave a-m reception. Some of the weak ones will get away if you can't resort to product detection.

Finally, dial calibration resettable to at most 5 kHz is a must; utility DXers will want even more precision. A crystal-controlled marker generator can provide dial references that will be enough to turn even a 1930s general-coverage monster into an acceptable SWLing performer.

Receivers embodying these attributes are legion. Recent developments in phase-locked-loop technology have put frequency-synthesized constant-tuning-rate receivers within economic reach. "Two-dial" (main tuning and bandspread) general-coverage receivers and the well-known "crystal-conversion front end" models serve well. Crystal-controlled converters heterodyning LW, MW and SW bands to a range covered by your amateur transceiver or receiver may be just what you need to get started in short-wave listening. One practical circuit can be found on page 51 of January 1972 *QST* and in the 1972 edition of *The Radio Amateur's Handbook*. Take care to ensure safety of a converter if you use it with a transceiver; connecting it to an "alternate receiving antenna" input is one way to avoid combustion. And don't forget to

Join a DX Club!

This book exists because a small group of dedicated radio amateurs got together in the early 1900s to form the American Radio Relay League. SWLs have joined together as well to pool loggings, knowledge and experience under the banners of DX clubs. Many clubs publish these shared experiences in detailed journals, and some have published books on their specialties with the goal of assisting you in *hearing*.

Any list of North American SWL groups wouldn't be complete without *the* club of SWL clubs. ANARC, the Association of North American Radio Clubs, is a federation of SWL clubs dedicated to encouraging contact between clubs and between broadcasters and listeners; in short, the betterment of the listening hobby. All North American clubs aren't listed below by any means; ANARC can supply you with additional information on these and more. Write ANARC, 557 N. Madison Ave., Pasadena, CA 91101. When requesting information from ANARC or any other radio club, be sure to include a self-addressed stamped envelope for reply, just as you do when you write to your League — it'll be appreciated!

Here's the key to the club listing to follow. After each club's name and address we show *C*: coverage, *J*: name of journal and publication frequency, and *P*: additional publications valuable to the DXer. Drop a line to any groups that interest you — tell them you heard about them in this book!

Canadian International DX Club (CIDXC)
169 Grandview Ave.,
Winnipeg, Manitoba
CANADA R2G 0L4
C: all phases
J: *Messenger* (monthly)

International Radio Club of America (IRCA)
P. O. Box 26254,
San Francisco, CA 94126
C: medium waves
J: *DX Monitor* (34 issues per year)
P: *IRCA Almanac, IRCA Foreign Log*

Longwave Club of America (LWCA)
Box 33188,
Granada Hills, CA 91344
C: below 550 kHz
J: *The Lowdown* (monthly)

North American Short Wave Association (NASWA)
P. O. Box 13,
Liberty, IN 47353
C: shortwave broadcasting
J: *FRENDX* (monthly), *Update* (semi-monthly)
P: *DXing According to NASWA*

Newark News Radio Club (NNRC)
P. O. Box 539,
Newark, NJ 07101
C: all phases
J: *Bulletin* (monthly)

National Radio Club (NRC)
P. O. Box 32125,
Louisville, KY 40232
J: *DX News* (30 issues per year)
P: *Domestic Log, Night Antenna Pattern Book, Antenna Reference Manual*, more

The Society to Preserve the Engrossing Enjoyment of DXing (SPEEDX)
P. O. Box E,
Elsinore, CA 92330
C: shortwave broadcast and utilities
J: *SPEEDX* (monthly); *SPEEDX-Gram* (monthly on the half-month)
P: *SPEEDX Utility Guide, A Guide to Soviet Broadcasting,* more

Worldwide TV-FM DX Association (WTFDA)
P. O. Box 97,
Calumet City, IL 60409
C: TV, fm, vhf/uhf utilities
J: *VHF/UHF Digest* (monthly)

try the nonamateur coverage afforded by image tuning and beyond-the-band-edge coverage as detailed in "You're Already There," on the first page of this chapter.

Antennas are as important for SWLing as they are in Amateur Radio. We'll not tread on thin technical ice here when *The Radio Amateur's Handbook* and the *ARRL Antenna Book* give this hobby-within-a-hobby subject the in-depth treatment you need. If you become a confirmed SWL in addition to your amateur ways, you'll want to concentrate on getting wide frequency coverage out of whatever real estate you have to play with. Multiband systems, of which the time-honored "random wire" and matching network is one, will let you achieve good performance inside and outside of the amateur bands.

Here is a fact: *If there is anything worth hearing on the radio tonight, you are going to hear it better with high-fidelity headphones than with a speaker.* A great deal of sonic detail is lost between speaker and ear, and aural distractions from the rest of the house will contribute a layer of QRM that no adjustment of the receiver can alter. Listen with headphones and you will hear things you didn't know were there — a test run with your favorite stereo music recording will bear this out. Choose a comfortable lightweight pair of headphones with as much fidelity as you can afford. Avoid the old "magnetics"; that kind of selectivity belongs in the receiver. Distortionless audio reproduction is pivotal in weak-signal work, especially in the area of intermodulation distortion (IMD). Although the human ear can detect signals several dB weaker than random noise, an audio system suffering from high IMD will cause that noise to modulate the signal and render it useless. Remember at all times that the goal is *hearing*.

References

The guidebook to international and domestic long, medium and shortwave broadcasting is the *World Radio-TV Handbook* or WRTVH. (Available from Gilfer Shortwave, P. O. Box 239, Park Ridge NJ 07656.) It provides the schedules of virtually all short-wave and long-wave, and all major medium-wave, broadcasting stations of the world. Fm and television listings are included for many countries, as are details concerning many time and standard-frequency stations of the world. At the back of the book, all reported LW, MW and SW broadcasting stations are listed by frequency. Once you know what frequency you're tuned to, quick research of the schedules of transmitters listed on that channel can whisk you on your way to that all-important i-d. The WRTVH is a must for amateurs, SWLs and travelers alike. What you hear on your radio may not make much sense without it.

You can tell a station's administering country if you can copy its call. In the utility (Aeronautical Mobile, Fixed and Maritime Mobile) bands, the International Prefixes list will be your key to the call signs. You'll learn how to use it in the upcoming section on utility DXing techniques. In the amateur bands, the latest ARRL DXCC List used to be all you needed to match an amateur prefix to its country, but bizarre commemorative call signs have become all the rage, making the list of international prefixes invaluable in amateur DXing as well. There's a copy of this list in the ARRL *Logbook* and in the appendix of this book.

Next in importance on the list of references are the periodicals of clubs specializing in your particular interests. Just as the ARRL was formed soon after the onset of this century, SWL clubs formed as listeners coalesced into groups strengthened by common interests. All are nonprofit; many publish detailed, high-quality newsletters and bulletins. Many clubs devote themselves entirely to listening to one or two services. They may even specialize in coverage of a given service in specific frequency ranges. Their coverage of their specialties amounts to the constant updating of any related information published in the yearly WRTVH — information important to every SWL but highly volatile just the same. Those club bulletins come together out of information supplied mainly by *hearing* — and hearing is what you are reading this chapter to do. In this chapter you'll find a listing of the larger clubs and what they're set up to hear. Some publish books of special interest; we list these books club by club and include them, where they apply, as reference sources in the listening-techniques sections.

The ultimate in station-frequency references is published by the ITU. If you become the most fanatic of radio fanatics, you may splurge for one of more volumes of the *International Frequency List* — covering all stations known to the ITU with the exceptions of amateurs, ships and experimental stations — in five volumes. Problem is, the set costs as much or more than some of the better short-wave receivers available. There are other ITU publications, though, some of them within reasonable reach of the SWL whose special interests justify them. For a list of them, write the International Telecommunication Union, General Secretariat — Sales Section, Place des Nations, CH-1211 Geneva 20, Switzerland.

A good library listing stations by frequency and location can be likened to a detailed collection of maps — they'll get you where you want to go. But once you get to a certain place, you have to contend with the weather. Just as weather can be predicted with reasonable accuracy, so can propagation conditions. The WRTVH carries a long-term propagation forecast emphasizing conditions on the

Fifty-kilowatt medium wave "clears" operated by the Canadian Broadcasting Corporation provide all who hear them with topnotch programming in addition to showing up in many newly begun logs as "first Canada"!

SWBC bands; these are a good general guide when listening to broadcast DX. *QST* offers something better. Each month in "How's DX?" you'll find propagation charts detailing predictions for the next month for all frequencies from 6 through 42 MHz. Whatever area of the world you seek to hear, you can find the time-of-day transmitters there will likely be audible on a given band.

Propagation bulletins featuring data even more timely than the *QST* charts are transmitted by ARRL's Maxim Memorial Station, W1AW. But the freshest information arrives in the propagation forecast transmitted at 18 minutes after the hour by WWV; see the amateur DXing chapter of this book and *The Radio Amateur's Handbook* for details. With study and practice, the solar and geomagnetic indices given by WWV can tell you the propagation conditions just past, present and expected. If conditions are so bad that you can't copy WWV, you'll still have the story!

With an extensive collection of references, memberships in clubs aligned with your interests and propagation charts staring you in the eye, you may think you've got it made — you haven't even warmed up the receiver and the whole business looks like it's going to be "shooting fish in a barrel." There is one more critical ingredient in DX success: luck. That's the factor that puts the fun into just about anything, reportage of a limitless source of which is truly "beyond the scope of this book."

QSLing and Logging

An electromagnetic impulse breezing through you and your listening post at 186,000 miles per second is about as transient as transient can be. The programs and conversations carried by such waves tend to crash into memory and fade in a matter of hours. One way you can snare radio is through logging. Another radio trap is the QSL. Hence, the QSL card: a verification from a transmitter operator to a listener (or listener/transmitter operator, as in Amateur Radio) that a signal from the transmitter of one was indeed the signal detected in the receiver of the other.

Modern QSLs take many forms. A few examples of SWL QSLs are included as illustrations in this chapter. To fully verify that the received event was the transmitted event, a QSL or verification letter should include frequency, date and time of transmission. It should be addressed to whoever is trying to prove to him or herself, or others, that reception was real. It should state the location of the transmitter. In the cases of amateur and utility stations, verification of transmitter mode is also a must. As important as any of this, a QSL must confirm or verify reception using exactly those words or forms thereof. SWLs of decades have torn out their hair over the "we acknowledge receipt of your report" brush-offs perpetrated by some of the largest and most respected broadcasting organizations in the world.

Some say that the importance of QSLing has been overemphasized. But not enough has been said of the *fun*. They're a "blast" to receive and collect, for one thing. Several of the larger broadcasters issue series of QSLs, each singular and stirringly beautiful, for which SWLs can qualify by submitting correct and detailed reception reports. SWLs aren't just getting "freebies" out of this business, either — far from it. Highly detailed reports, especially with respect to propagation, interference and signal quality, are of pivotal importance to international broadcasters. They can use what they receive from you to ascertain the best frequencies usable to your area. They can tell if anyone's listening at all: recently, Radio Australia requested letters from listeners in China and received over 5000 replies *per day*. Someone was listening, and someone cared enough to write. *Thousands* of someones.

Reports from foreign lands are of dubious value to domestic broadcasters, so assume that what response you receive from them is given in the spirit of international goodwill and fraternity. This is especially true of utility stations that exist only to contact still other utility stations — a "listening audience" may be precisely what they wish to avoid!

In any case, go easy with your requests for verification. See to it that they are nothing more than polite and dignified *requests*.

You QSL a station by submitting a reception report to its staff. A reception report consists of the name/call of the station you heard, its frequency and the date and time of your reception. Details of programming heard — about a half-hour's worth, if possible, in the case of broadcasters — come next. They can include song titles or description of music, content of news items and features, station i-ds and IS (read on for explanation of these terms), and sign-on or sign-off, with times and durations noted for each. Follow this with a verbal report of signal and propagation quality. (There are "code" systems of reporting SWBC signal quality similar to Amateur Radio's RST and subject to similar inconsistency; you can say whatever you want to say better with words.) Mention the receiver that made all of this possible and the antenna used with it; mention how many tubes or transistors are used in that receiver in addition to or in place of its manufacturer and model number. Your geographical coordinates will be of value to a station engineer.

Your request for verification, be it a request of an amateur, broadcast or utility station, ends with precisely that — a polite *request* for verification. Make it all the more polite, if you can, by writing the whole thing in one of the languages the WRTVH lists as being spoken by the station staff. Make it "politer" still by including return postage with reports to domestic broadcasters both MW and SW, to utilities and to amateurs. If you're requesting a QSL via an amateur's QSL manager, include return postage or the customary self-addressed stamped envelope. *Do not* send dollar bills indiscriminately to foreign countries; this practice can land some of our overseas friends in jail in countries where importation of foreign currency is forbidden.

Even in Amateur Radio, logging re-

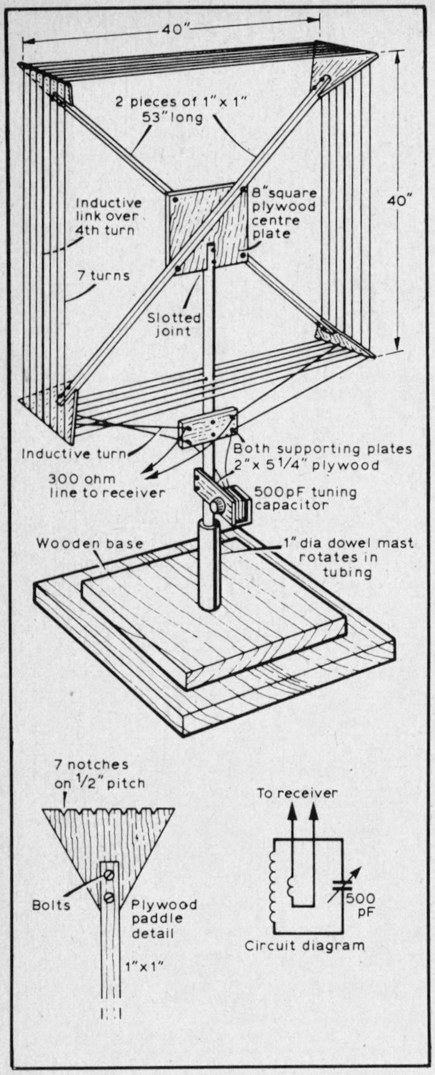

Fig. 1 — Constructional details of loop antenna for operation on medium waves and/or 1.8 MHz and capable of providing deep null on interfering signals (*Electronics Australia*).

quirements have been eased compared to the "old days." A radio signal can be given longer life only in the memory of a mind, the magnetism of a tape recording or the longer-lived journalization of its existence, circumstances and import on some sort of printed page. A well-kept log consisting of most of the details you'd include in a reception report is a keepsake and reference that grows more and more irreplaceable with the passage of years. What you see with your mind's eye from the sounds of your radio is given substance in log pages. Pull the program details you need from that log if you decide to gun for QSLs. The goal, though, remains hearing. Any awards and QSLs you do receive are the gift of the pressure of peers who care enough about listening to want to be shown that you indeed have heard what you're hearing.

Listening on the Medium Waves

Many began a fascination with radio in that place known as "the standard broadcast band." Sooner or later, a tuning knob on even a table clock-radio was jarred, by accident or design, free of the grip of local programming, and another hapless listener fell under the spell of distance.

Monaural a-m is the only transmission mode in broadcast use between 535 and 1605 kHz; a variety of stereo systems have been tested but remain unapplied. Channel spacing in the Americas is nominally 10 kHz, although many Latin American stations either move intentionally or drift to frequencies between assigned channels. These are known as split channels, or simply "splits." Many stations in the rest of the world operate with 9-kHz channel spacing, so when they're breaking through they'll show up as splits as well.

High receiver dynamic range is critical on a band where only a few kilohertz may separate your neighborhood 50-kilowatt station from the irresolute skywave signals of transoceanic DX. Antennas are of prime importance as well. In domestic DXing, a bi- or unidirectional "loop" antenna is unsurpassed in its ability to differentiate between signals arriving from different directions. The Beverage antenna, long a favorite with 160-meter amateur DXers, may be the only way anyone east of the Rockies can hope to hear Alaska or Hawaii. The traditional longwire is of value in early-morning hours when interference on a channel lessens to the point where the DXer can afford to let remaining stations fight it out. The loop-longwire combination is suggested for flexibility. Whichever offers best copy of the signal of your choice, go with it. But keep a hand on that antenna switch.

Elsewhere in this chapter we take a look at the organization of stations in North America by channel, service area and operating time. You should begin your MW DX endeavors with U.S. and Canadian domestics. They're everywhere and they're easy to i d because they spout their call signs incessantly. Start by logging as many stations audible during daylight hours as possible. Some of these will be gone after dark; you'll have to hear some of your nighttime long-haul catches through others, pattern-change or no.

You'll have a lot of fun with the next assignment: Identify and log as many of each night's "clears" and regionals as you can stand. This is a pretty inexhaustible source of loggings compared with daytime listening — you'll have noted the "garden-variety" stuff before long and the toughies will begin to straggle in. You'll find one day that the 5-kilowatt regional four states over that you logged last year as an exciting "first" is now an importunate NSP pest standing between you and North Dakota.

The bewitching ebb and flow of sunrise and sunset DXing await you in addition to what daytimers, nighttime clears and easy regionals have to offer. MW propagation does not take a long day's journey into night. Conditions change quickly; at sunset, and for the similar period at sunrise, the band is alive with signals of daytimers who must close down by night and whose groundwave signals are annihilated by absorption during the day. Many unlimited-time regionals change their radiation patterns at sunrise and sunset for minimum mutual interference as darkness arrives and propagation goes long; they switch back to their daytime patterns at sunrise. Here, then, is a strange radio world in which all of one state's daytimers may disappear in minutes without a trace.

Exactly how does this work? You can hear stations east of you as they sign on at their local dawns because there is darkness between the two of you and because your local daytimers slumber. You can hear daytimers signing on roughly to your north and south — depending on your location and the season — for the scant time propagation strikes a slim balance between radio night and day along the great circle path connecting you. And at your dawn you can hear stations far to the west until the spinning earth breaks the path between you and those western stations with the light of the arriving day.

At sunset, best heard are signoffs north and south and eastern unlimited-time stations normally smothered by local and western interference. Night to the east, sweeping west, will bring the new ones to you; daylight to the west will kill other comers through absorption. Sunrise or sunset, though, the code word must be "darkness."

The local or "graveyard" channels at 1230, 1240, 1340, 1400 and 1450 kHz offer a DX challenge all their own. DXers in metropolitan areas may find some or all of these channels occupied by locals around the clock. Chicago, for instance, has three stations sharing 1240. But if you live in a rural area, you may find the maelstrom presented nightly by the graveyard channels to be your DX cup of tea.

DXing from 0000 to 0600 or so, local time, can be particularly rewarding because longer-haul night-owl domestics and splits may be heard on channels choked with signals a few hours earlier. Many stations still close down in the "wee hours," especially on Monday morning, for transmitter maintenance or because of low listenership or lack of revenue. Some can be heard testing; some of these stations may announce that they are testing on behalf of a particular SWL club. These tests are known as "courtesy programs" and are arranged by MW listening clubs to provide possibilities of logging otherwise-hard-to-hear stations during hours when interference is minimal and propagation is solid. Try one or two Monday-morning DX sessions and you'll be hooked!

Some of the signals you're going to come across in MW and SW broadcast DXing are going to be too weak for your receiver's diode detector to use without assistance. Signal preamplification is *not* the answer because you'll increase noise along with signal and severely compromise your receiver's dynamic range in the process. An excellent technique is the use of product detection. Use your receiver in its selectable sideband mode — demodulate only one of the signal's sidebands with your BFO on and your tuning adjusted for carrier zero beat. This has several advantages. By tuning in one sideband, you've narrowed your receiver bandwidth, thus reducing received noise. Since your BFO provides the signal necessary for distortionless demodulation of the audio, the effects of selective fading (an effect where changes in propagation may attenuate the carrier or one of the sidebands without affecting the rest of the signal) are minimized. When listening to a weak signal adjacent to a strong local station, you can choose to demodulate the sideband farther from the local, thus reducing ear-splitting interference from the local's sideband "splash" (crashes and crackles resulting from high-frequency audio peaks in percussion and speech formants). And the use of your BFO can tell you if there's any signal there at all, even if you happen to tune in at a moment when there's no modulation. You may see this technique referred to as ECSSB, or "exalted carrier single sideband" reception; it isn't. ECSSB is a complex phase-locking technique in which the carrier is filtered from the received signal, amplified separately and reinserted for demodulation. So, while we cannot exalt the carrier, we can exalt the use of your receiver's selectable sideband and BFO for product detection with the carrier tuned at zero beat.

A few notes to wrap up this look at "the broadcast band." If you should seek QSLs from domestics, report in their local time; no reason a station in Herrin, Illinois, for instance, should have to fiddle with UTC. Music titles are nearly worthless for verification purposes because they don't appear in the station log. Copy from local advertisements is excellent verifiable material, as are items from local (not network) news. Return postage is suggested in such domestic QSL endeavors because many stations operate with budgets and personnel stretched to their limits.

References are many. That copy of the WRTVH is first. The National Radio Club's *Domestic Log* lists — by channel, call signs, addresses, transmitter powers, network affiliations, authorized operating times and actual operating hours — all MW broadcasters in the U.S. and Canada. The *North American Radio TV Station Guide,* by Vane A. Jones (Howard W. Sams & Co., Inc.) lists frequency, call sign, location, power and network affiliation for a-m, fm and television stations in North America. These will get you started. Other club publications are included with this chapter's club listings, as are the addresses of the clubs dealing in whole or in part with MW DXing: IRCA, NNRC, NRC and CIDXC.

DXing the SWBC Bands

Shortwave broadcasting is not just medium-wave broadcasting on higher frequencies. Here, broadcasters are labeled by their intent. Is the station you hear transmitting internal programming to domestic audiences or is it hurling ideologies over borders and overseas? This differentiation between international and domestic broadcasting is one great difference in the usage of the MW and SW broadcasting bands.

Channel spacing is another. SW broadcasters utilize 5-kHz channel spacing as opposed to MW's 9- or 10-kHz slots. Although you won't usually find the MW DXers' term "split" used in conjunction with such scofflaws here, SW stations, particularly OOB and tropical broadcasting stations, drift to or choose carrier frequencies between channels.

They may even choose frequencies between bands. On the medium waves, you contend with only one frequency range; SWBC stations choose operating frequencies in a number of allocations throughout the medium- and high-frequency spectrum, and you must match wits with changing propagation *and* station managements if you're to hear the rare ones.

A-m is the predominant mode in shortwave broadcasting. A few long-established broadcasters are testing ssb on a few frequencies in hopes of getting more out of their transmitters and assigned bands. You'll also find a few SWBC "feeder" stations on Fixed Service frequencies transmitting programming to remote transmitter sites via isb (independent sideband) a-m. If you have a receiver that selects sidebands about an unchanging carrier frequency, you can enjoy the novel effect of switching back and forth between upper and lower sidebands carrying separate programming, often in two completely unrelated languages. Usually, a reduced "pilot" carrier is transmitted with the isb feeder to facilitate phase-locked demodulation at the remote transmitter site — necessary because faithful music and voice reproduction in any sort of heterodyne reception tolerates no more tuning error than a number of Hertz countable on the fingers of your hands.

In DXing the SWBC bands you'll come up against the barrier of unintelligibility thrown up by foreign languages. A few MW domestics in the U.S. and Canada cater to ethnic communities in their native tongues, but almost any living language can erupt from the speaker of your SWBC receiver at any time. This brings up a crucial SWBC DXing point: *You are not going to hear many places other than those represented by super-giant broadcasters (VOA, BBC, Radio Moscow and so on) if you expect to hear all of them in English, French or Spanish.* A glance at the broadcasting schedules in the WRTVH will tell you that a station broadcasting in Swahili is not necessarily in Africa; a station crooning in Hakka is not necessarily in China — Radio Moscow alone broadcasts in over 60 languages, many transmissions of which are audible here. The WRTVH staff makes every effort to provide you with transcriptions of foreign-language station slogans and i-d phrases as aids to signal identification, but they're useful mainly after you've made an educated guess as to which station you've tuned and not in picking i-ds out of the air. Find out what you're likely tuned to and then brush up on what you

Use a Tape Recorder

You cannot hear everything going on at once on a crowded channel because you must concentrate on one voice, one song, one advertisement, at a time. A tape recorder can be a great help in listening because it takes in everything applied to its input terminals with equal facility — much as a camera does. Its magnetic memory is one of a *repeatable* everything — the tape can be played and replayed so that the selectivity of your mind's attention may be applied to every bit of audio present for the duration of your vigil on a frequency.

Choose a tape machine with a defeatable automatic record-level feature (or none at all) if possible; the rise and fall of background noise introduced by record-level compression will tire you — and strong static crashes, sideband splash and drumbeats can punch holes in your recording for as long as the compressor takes to recover.

Use a patch cord between recorder and receiver and connect it to a tape output or other constant-level audio tap on the receiver. Adjust record so that the loudest received audio falls well short of the tape recorder's overload point.

At the beginning of taping, use a microphone to give a voice "mark" for the exact time. Upon playback, using a digital 24-hour clock and its "hold" setting feature, restart your clock at the mark to reacquire the actual clock time during the broadcast. Each time you replay the tape you'll have to restart the clock.

Some SWLs use 2- or 4-channel tape machines to record several signals simultaneously — they record several parallel frequencies or WWV/CHU audio contemporaneous with the DX signal's for time reference. Many DXers collect taped i-ds as aural scrapbooks of proof of reception. A number of stations accept tapes of their signals in lieu of transcribed details for verification purposes.

The tape recorder is especially useful in DXing crowded graveyard and regional mediumwave channels, and in cases where foreign language makes repetition desirable for the perception of that all-important i-d. But beware of becoming too dependent on a tape recorder short of cultivating your ability to catch an i-d. A well-trained ear, unlike a machine, enjoys its own hearing as it listens.

think will be the i-d. Listen on the hour, at sign-on and sign-off; try before, after and during news. One day the line noise, sideband splash, interference and fading will abate for that critical five seconds — and you'll have one more for the log.

Domestics have home turf as their service areas. International broadcasters, though, beam their offerings into specific target areas, utilizing languages in use there. Here's where you need to have your propagation charts and forecasts at the ready. Many international shortwave broadcasters — and many domestics as well — utilize several frequencies to a given target area simultaneously. These frequencies may be within one or several bands, and are said to be "in parallel" with each other. This practice allows listeners at different distances from the transmitter to enjoy best reception under varying propagation conditions. Further to this end, many shortwave broadcasters modify their frequency usage seasonally. Information published for December may be nearly worthless as you search for the same station in May. This is where the up-to-date loggings and listenings afforded by the journals of SWL clubs pay off.

Your DXing achievements will dwindle after about 30 loggings if you expect all you hear to be in your language. Likewise, your "heard" totals are going to languish if you break the language barrier but still expect all the transmissions you hear to be beamed to your area. No directional antenna array achieves laserlike dispersion and even the reduced power radiated off-beam can make it around the world if propagation plays into your hand. So even if a station has no broadcasts to your area, all's not lost — it may be clearly audible at another time and frequency even though its transmission is beamed to a different area in another language. Most of the rarer SWBC DX loggings are achieved in this way.

This business of broadcasting to specific target areas in transmissions varying in length from 15 minutes to several hours has caused shortwave broadcasters to adopt programming characteristics peculiar to the short-lived and directional nature of their broadcasts. They know that many of their listeners tune for their signals with receivers of questionable dial calibration and selectivity. They know as well that many of these listeners have no comprehensive sources of schedule and frequency information to help them locate signals they wish to hear. Every season, for instance, listeners who have not heard their favorite station announce its seasonal frequency changes might fail to receive it again until they stumble across it by chance. In this connection, most shortwave broadcasters transmit repetitive recordings of distinctive sounds, birdcalls or musical phrases for several minutes prior to signing on to each target area. Each station's interval signal (IS) remains its listeners' tuning cue when frequency changes or propagation effects make reception difficult. Identification of an interval signal is as definitive as hearing a station i-d because each IS is unlike every other. The WRTVH provides information on the interval signals of many stations verbally or in music notation. You'll learn to recognize the interval signals of many of the larger broadcasters: Radio Peking's mysterious "The East is Red" in chimes, VOA's "Yankee Doodle," Radiotelevisione Italiana's bird chirps, Radio Australia's "Waltzing Matilda" followed at sign-on by the laugh of the Kookaburra bird.

You'll be receiving weak a-m signals when you gun for broadcast DX on the shortwave bands. The heterodyne-reception technique discussed in the medium-wave listening section is even more useful on the shortwaves. Propagation fading is deeper and more rapid here, with selective fading and its annoying distortion being the rule and not the exception. And the ability to choose the more interference-free sideband for demodulation is even more important where 5-kHz channel spacing allows the sidebands of stations on adjacent channels to overlap.

When you tune the SWBC bands, set your 24-hour clock as precisely as you can. International broadcasters will expect accuracy in your reported times if you decide to seek verification of your reception. Make those reports in UTC; use International Reply Coupons to nongovernment and domestic stations. Make every report a report of value to the station staff.

SWBC references include the necessary WRTVH and bulletins from any or all of these clubs: CIDXC, FRENDX, NASWA, NNRC, SPEEDX. The world of SWBC DXing is as changeable as propagation. Station fortunes, frequency and program schedules, interference and jamming prospects — it's a challenge to keep up on all you need to know to jump into the game and keep winning. SWBC is one part of SWLing that amateurs liken to their own DX operations and find wanting. Here, though, is challenge far from "shooting fish in a barrel." A quarter megawatt, you'll find, guarantees no reception at all if you don't know your stuff. You'll find such "stuff" exciting to know and learn as you tune for SWBC DX between and beyond the amateur bands.

Utilities DXing

Listening to the utilities is much like listening on the amateur bands. Modes used are similar (cw, ssb, RTTY, FAX are among the more basic) and transmissions are brief and to the point — with the exceptions of beacons, test transmissions and V, CQ and QRA "wheels" sounding their repetitive messages for hours on end. The first challenge you'll face will be that of determining each signal's origin.

When you hear a call sign, you can ascertain its country of administration with the help of the International Prefixes List included in the appendix of this book. Note that some letters of the alphabet are allocated entirely to one nation and that others are divided among several. All possible K, N and W calls "belong" to the United States, for example, because the U.S. is allotted all calls A through Z after K, N and W. On the other hand, even though the U.S. is allotted AAA-ALZ, other call signs beginning with A may belong to Spain (AMA-AOZ), Pakistan (APA-ASZ), India (ATA-AWZ), the Commonwealth of Australia (AXA-AXZ) and the Argentine Republic (AYA-AZZ). The U.S. can assign "A" call signs only from A through L; that's why we couldn't use "AM" amateur Bicentennial calls in 1976, for example.

The ITU has been under pressure from emerging nations for allocations of call sign blocks when there were no more to be had. Call sign blocks of three letters (such as AAA-ALZ) were all allocated years ago. Blocks consisting of one number followed by three letters were used as well; Nepal is allocated 9NA-9NZ, for instance. Still more blocks were needed.

You may have noticed that countries allocated entire letters of the alphabet (such as the U.S. with KAA-KZZ and the U.K. with GAA-GZZ) may issue call signs bearing only one letter before the insertion of numerals — calls like F5QQ and W9DY. This can be done only when the first letter of a call is sufficient to establish nationality — when a nation is allocated all call signs, A through Z, behind that letter of the alphabet.

Subdivided letters of the alphabet — such as the As — can't be so used because more than the first letter is needed in calls issued from those blocks to establish nationality. Thus, the U.S. cannot assign a call sign such as A1QB — five other nations can lay claim to a call beginning with A; only the second letter of an "A" call sign will tell the tale.

Hard-pressed for more blocks, the ITU took advantage of this by assigning blocks like A2A-A2Z (Republic of Botswana) and S2A-S3Z (Bangladesh) beginning with letters of the alphabet which no single country could use to issue calls bearing one-letter prefixes.

When you copy a call sign, note the letters appearing before any numerals and match them to the country listed for their call sign block in the prefixes list. Some calls will pinpoint a station with no greater accuracy than as being located in "British Overseas Territories" — such generalizations can cover a lot of ground (and water!). That's where your additional references and club bulletins come in.

Anyone who listens on the short waves

should acquire the habit of checking the "ownership" of call signs. Amateurs will want to know if that weird call is the mark of an intruder or just another commemorative mask. Most of the call signs you'll hear on the utility bands will be indecipherable with a list only of amateur prefixes as reference. And for propagation work and all-around fun you'll want to nail every call sign you hear to its country.

Other references you'll find invaluable as you scan the utility bands are ITU publications (in particular, List II: *List of fixed stations operating international circuits;* List IV: *List of coast stations;* List V: *List of ship stations;* and Lists VIIA and VIIB, which are alphabetical lists of call signs), *QST*'s propagation charts, WWV's propagation forecasts and solar/geomagnetic indices broadcast 18 minutes after each hour, and the *Utility Guide* published by club SPEEDX. Speaking of clubs, get in touch with CIDXC, LWCA, NNRC, SPEEDX and WTFDA. Their utilities coverage will provide you right off with station/frequency listings of great use as you tune the bands. Join and give them your support!

FM and Television DXing

Even though these broadcasts are far afield of the shortwave range, we urge you to jump into the swim of DX reception on the fm and television bands. Your amateur work on the vhf/uhf bands will benefit from the propagation know-how you'll acquire when the fm and TV bands come alive with E skip and tropo signals. (Do these terms throw you? Read the vhf/uhf chapter of this book.) If you're already hooked on SWLing of any kind, the new world of loggings offered by the fm and television bands will make your head spin. You can't lose.

We know of hams who have persuaded themselves through lack of experimentation that there's no DX on 80 meters. Meanwhile, dedicated DXers ignorant of this ignorance qualify for Five-Band DXCC every month. Many radio enthusiasts write off frequencies above 30 MHz as DX desert. In the SWL world, a few pioneering listeners have logged over 1000 fm stations from one location.

Rotatable gain antennas are important in both fm and television DXing. Get them up as high as you can. Use an antenna cut for the frequencies you want to tune (not a "vhf/uhf/fm" general-purpose special) when you dig for fm DX. The same holds for vhf and uhf television — don't compromise in antenna performance if you can avoid it.

The potential fm or television DXer has many more receivers to choose from than the MW or SW listener; ironically, none of this glut of merchandise is designed with DX in mind. You may be shopping for more selectivity but find your choices wider in the all-important questions of cabinet finish and amplifier power output. Club publications can help you choose because you can benefit from the experience of longtime DXers.

In an fm receiver, you'll need selectivity and sensitivity, of course, but in signal-saturated metropolitan areas, immunity from cross-modulation is important to keep your locals from appearing at five or 10 spots on the dial. Two other important features are becoming harder and harder to find: defeatable automatic frequency control (afc) and provision for true monaural reception. If you can't defeat your receiver's afc, DX reception will be impossible on frequencies adjacent to those of your locals because each local will grab the afc and lock your tuning there over a wide excursion of the dial. True monaural detection (as opposed to the "mono" switch that leaves the stereo detector in and shorts the left and right channels together) is a must because the wider bandwidth utilized by a stereo detector passes much unwanted noise when its input signal is too weak for "full quieting."

Adequate dial calibration is another must and another rarity. The digital age has arrived in fm land and it's now possible to know where you're tuned to the 10th of a megahertz. Short of this, a high-resolution logging scale can be used to note dial settings for each fm channel to the limits of dial parallax and backlash.

You have even less to choose from as far as television receivers go if you must split hairs in specifications, but you're less likely to buy a television just for DXing if you're starting out in the hobby. Your home television receiver will capture a lot of the good ones if you give it a chance with a gain antenna.

Fm and television DXing is even more dependent on knowing the radio neighborhood than DXing on the medium waves. Through experience you'll come to know when signals on the fringe of reliable reception have risen enough above the noise to portend of an extensive opening. Long-distance propagation on the vhf/uhf bands usually favors relatively restricted areas at a given time; when the bands open you'll rush through your station listings to get an idea of what might be "in" from the skip area. You may have a tape recorder running continuously to catch i-ds your harried senses may miss. Television DXers make use of their cameras to photograph the elusive images of i-d slides broadcast by distant stations. In TV DXing circles, such shots are almost as good as QSLs.

Speaking of QSLs, we'll mention that return postage and use of local time are as important in fm and television reception reports as they are in those to MW domestics. Very little DX reception is expected by these stations because of the local coverage vhf/uhf signals normally afford.

References for your fm/television listening post should include the WRTVH, the *North American Radio TV Station Guide* by Vane A. Jones (Howard W. Sams & Co., Inc.) and the *FM Atlas and Station Directory* (available from Gilfer Shortwave). Tune into these clubs for columns devoted to fm and television DXing: CIDXC, NNRC and WTFDA.

1979 WARC Changes

Changes in frequency allocations for a number of services will be among the results of the World Administrative Radio Conference which was held from September to December of 1979 in Geneva, Switzerland. Some of these changes are highlighted here. They are keyed to affected services as they appear in the Regulated Radio portion of this chapter.

In the Broadcasting Service, the long-wave band shifts to 148.5 to 283.5 kHz, and the medium-wave band will expand to 525 to 1705 kHz in Region 2. As far as the shortwaves go, no changes were made in the tropical bands and the international broadcasting allocations at 5950 and 7100 kHz. But these changes *were* made to high-frequency bands pending the successful and satisfactory relocation of stations now on affected frequencies:

9500-9775 kHz to 9500-9900 kHz
11.700-11.975 MHz to 11.650-12.050 MHz
new allocation at 13.600-13.800 MHz
15.100-15.450 MHz to 15.100-15.600 MHz
17.700-17.900 MHz to 17.550-17.900 MHz
21.450-21.750 MHz to 21.450-21.850 MHz
25.600-26.100 MHz to 25.670-26.100 MHz

The Maritime Mobile Service received slight increases in its allocations at 12 MHz above, and netted two new bands as well: 18.780 to 18.900 MHz and 19.680 to 19.800 MHz.

Standard Frequency and Time Signal allocations remain as they are, except for the addition of standard frequencies of 4 and 8 MHz in Region 3.

None of these changes is yet in effect. Before the conference, a number of administrations moved shortwave broadcasting stations to out-of-band frequencies adjacent to pre-WARC-79 allocations; these operations would have been immediately made legitimate if the WARC-79 stipulation that nonbroadcast stations now on these frequencies be relocated first was not in force. At this writing, these OOB broadcasting stations are not operating in accordance with the frequency allocations table. Successful relocation of any stations dispossessed of their frequencies by WARC-79 is expected to take the better part of a decade.

Chapter 15

References

ARRL Numbered Radiograms

The letters ARL are inserted in the preamble in the check and in the text before spelled out numbers, which represent texts from this list. Note that some ARL texts include insertion of numerals. *Example:* NR 1 R W1AW ARL 5 NEWINGTON CONN JUNE 1 DONALD R SMITH $\overline{\text{AA}}$ 164 EAST SIXTH AVE $\overline{\text{AA}}$ NORTH RIVER CITY MO $\overline{\text{AA}}$ PHONE 773 3968 $\overline{\text{BT}}$ ARL FIFTY ARL SIXTY ONE $\overline{\text{BT}}$ DIANA $\overline{\text{AR}}$.

Group One — For Possible "Relief Emergency" Use

ONE	Everyone safe here. Please don't worry.
TWO	Coming home as soon as possible.
THREE	Am in ____ hospital. Receiving excellent care and recovering fine.
FOUR	Only slight property damage here. Do not be concerned about disaster reports.
FIVE	Am moving to new location. Send no further mail or communication. Will inform you of new address when relocated.
SIX	Will contact you as soon as possible.
SEVEN	Please reply by Amateur Radio through the amateur delivering this message. This is a free public service.
EIGHT	Need additional ____ mobile or portable equipment for immediate emergency use.
NINE	Additional ____ radio operators needed to assist with emergency at this location.
TEN	Please contact ____. Advise to standby and provide further emergency information, instructions or assistance.
ELEVEN	Establish Amateur Radio emergency communications with ____ on ____MHz.
TWELVE	Anxious to hear from you. No word in some time. Please contact me as soon as possible.
THIRTEEN	Medical emergency situation exists here.
FOURTEEN	Situation here becoming critical. Losses and damage from ____ increasing.
FIFTEEN	Please advise your condition and what help is needed.
SIXTEEN	Property damage very severe in this area.
SEVENTEEN	REACT communications services also available. Establish REACT communications with ____on channel ____.
EIGHTEEN	Please contact me as soon as possible at ____.
NINETEEN	Request health and welfare report on ____. (State name, address and telephone number.)
TWENTY	Temporarily stranded. Will need some assistance. Please contact me at ____.
TWENTY ONE	Search and Rescue assistance is needed by local authorities here. Advise availability.
TWENTY TWO	Need accurate information on the extent and type of conditions now existing at your location. Please furnish this information and reply without delay.
TWENTY THREE	Report at once the accessibility and best way to reach your location.
TWENTY FOUR	Evacuation of residents from this area urgently needed. Advise plans for help.
TWENTY FIVE	Furnish as soon as possible the weather conditions at your location.
TWENTY SIX	Help and care for evacuation of sick and injured from this location needed at once.

Emergency/priority messages originating from official sources must carry the signature of the originating official.

Group Two — Routine Messages

FIFTY	Greetings by Amateur Radio.
FIFTY ONE	Greetings by Amateur Radio. This message is sent as a free public service by ham radio operators here at ____. Am having a wonderful time.
FIFTY TWO	Really enjoyed being with you. Looking forward to getting together again.
FIFTY THREE	Received your ____. It's appreciated; many thanks.
FIFTY FOUR	Many thanks for your good wishes.
FIFTY FIVE	Good news is always welcome. Very delighted to hear about yours.
FIFTY SIX	Congratulations on your ____, a most worthy and deserved achievement.
FIFTY SEVEN	Wish we could be together.
FIFTY EIGHT	Have a wonderful time. Let us know when you return.
FIFTY NINE	Congratulations on the new arrival. Hope mother and child are well.
SIXTY*	Wishing you the best of everything on ____.
SIXTY ONE	Wishing you a very merry Christmas and a happy New Year.
SIXTY TWO*	Greetings and best wishes to you for a pleasant ____ holiday season.
SIXTY THREE	Victory or defeat, our best wishes are with you. Hope you win.
SIXTY FOUR	Arrived safely at ____.
SIXTY FIVE	Arriving ____on ____. Please arrange to meet me there.
SIXTY SIX	DX QSLs are on hand for you at the ____QSL Bureau. Send ____self-addressed envelopes.
SIXTY SEVEN	Your message number ____undeliverable because of ____. Please advise.
SIXTY EIGHT	Sorry to hear you are ill. Best wishes for a speedy recovery.

*Can be used for all holidays.

ARL numbers should be spelled out at all times.

Prefix	Country	Continent	Zone		Worked/ Confirmed		5BDXCC				
			ITU	CQ	W	C	80	40	20	15	10
A2	Botswana	Af	57	38							
A3	Rep. of Tonga	Oc	62	32							
A4X	Oman	As	39	21							
A5	Bhutan	As	41	22							
A6X	United Arab Emirates	As	39	21							
A7X	Qatar	As	39	21							
A9X	Bahrain	As	39	21							
AP	Pakistan	As	41	21							
BV	Taiwan	As	44	24							
BY	China	As	[1]	23							
C2	Rep. of Nauru	Oc	65	31							
C3	Andorra	Eu	27	14							
C5	The Gambia	Af	46	35							
C6	Bahamas	NA	11	08							
C9	Mozambique	Af	53	37							
CE9AA/KC4	Antarctica		[2]	[3]							
CE	Chile	SA	14,16	12							
CEØA	Easter Is.	SA	63	12							
CEØX	San Felix	SA	14	12							
CEØZ	Juan Fernandez	SA	14	12							
CO	Cuba	NA	11	08							
CN	Morocco	Af	37	33							
CP	Bolivia	SA	12,14	10							
CR3/J5	Guinea-Bissau	Af	46	35							
CR9	Macao	As	44	24							
CT	Portugal	Eu	37	14							
CT2	Azores	Eu	36	14							
CT3	Madeira Is.	Af	36	14							
CX	Uruguay	SA	14	13							
D2	Angola	Af	52	36							
D4	Rep. of Cape Verde	Af	46	35							
D6	Comoros	Af	53	39							
DJ	Fed. Rep. of Germany	Eu	28	14							
DM	German Dem. Rep.	Eu	28	14							
DU	Rep. of Philippines	Oc	50	27							
EA	Spain	Eu	37	14							
EA6	Balearic Is.	Eu	37	14							
EA8	Canary Is.	Af	36	33							
EA9	Ceuta & Melilla	Af	37	33							
EI	Rep. of Ireland	Eu	27	14							
EL	Liberia	Af	46	35							
EP	Iran	As	40	21							
ET	Ethiopia	Af	48	37							
F	France	Eu	27	14							
FB8W	Crozet	Af	68	39							
FB8X	Kerguelen Is.	Af	68	39							
FB8Z	Amsterdam-St. Paul	Af	68	39							
FC	Corsica	Eu	28	15							
FG	Guadeloupe	NA	11	08							
FS	Saint Martin	NA	11	08							
FH	Mayotte	Af	53	39							
FK	New Caledonia	Oc	56	32							
FM	Martinique	NA	11	08							
FO	Clipperton I.	NA	10	07							
FO	Fr. Polynesia	Oc	63	32							
FP	St. Pierre	NA	09	05							
FR	Glorioso Is.	Af	53	39							
FR	Juan de Nova	Af	53	39							
FR	Reunion	Af	53	39							
FR	Tromelin	Af	53	39							
FW	Wallis & Futuna	Oc	62	32							
FY	Fr. Guiana	SA	12	09							
G	England	Eu	27	14							
GD	Isle of Man	Eu	27	14							
GI	N. Ireland	Eu	27	14							
GJ	Jersey	Eu	27	14							
GM	Scotland	Eu	27	14							

Prefix	Country	Continent	Zone		Worked/Confirmed		5BDXCC				
			ITU	CQ	W	C	80	40	20	15	10
GU	Guernsey	Eu	27	14							
GW	Wales	Eu	27	14							
H4	Solomon Is.	Oc	51	28							
HA	Hungary	Eu	28	15							
HB	Switzerland	Eu	28	14							
HBØ	Liechtenstein	Eu	28	14							
HC	Ecuador	SA	12	10							
HC8	Galapagos Is.	SA	12	10							
HH	Haiti	NA	11	08							
HI	Dominican Rep.	NA	11	08							
HK	Colombia	SA	12	09							
HKØ	Bajo Nuevo	NA	11	08							
HKØ	Malpelo I.	NA	12	10							
HKØ	San Andres	NA	11	07							
HL	Korea	As	44	25							
HP	Panama	NA	11	07							
HR	Honduras	NA	11	07							
HS	Thailand	As	49	26							
HV	Vatican	Eu	28	15							
HZ/7Z	Saudi Arabia	As	39	21							
I	Italy	Eu	28	15							
IS	Sardinia	Eu	28	15							
J2	Djibouti	Af	48	37							
J3/VP2G	Grenada	NA	11	08							
J6/VP2L	St. Lucia	NA	11	08							
J7/VP2D	Dominica	NA	11	08							
JA	Japan	As	45	25							
JD	Minami Torishima	As	45	27							
JD	Ogasawara	As	45	27							
JD	Okino Tori-shima	As	45	27							
JT	Mongolia	As	32,33	23							
JW	Svalbard	Eu	18	40							
JX	Jan Mayen	Eu	18	40							
JY	Jordan	As	39	20							
K/W/N/A	United States	NA	6,7,8	3,4,5							
KB/KH1	Am. Phoenix	Oc	61,62	31							
KC4/KP1	Navassa I.	NA	11	08							
KC6	E. Caroline Is.	Oc	65	27							
KC6	W. Caroline Is.	Oc	64	27							
KG4	Guantanamo Bay	NA	11	08							
KG6/KH2	Guam	Oc	64	27							
KG6R/S/T	Mariana Is.	Oc	64	27							
KH6	Hawaiian Is.	Oc	61	31							
KH7	Kure I.	Oc	61	31							
KJ/KH3	Johnston Is.	Oc	61	31							
KL7	Alaska	NA	1,2	1							
KM/KH4	Midway Is.	Oc	61	31							
KP4	Puerto Rico	NA	11	08							
KP4	Desecheo Is.	NA	11	08							
KP6/KH5K	Kingman Reef	Oc	61	31							
KP6/KH5	Palmyra Is.	Oc	61	31							
KS4/KP3/HKØ	Serrana Bank	NA	11	07							
KS6/KH8	Am. Samoa	Oc	62	32							
KV/KP2	Virgin Is.	NA	11	08							
KW/KH9	Wake Is.	Oc	65	31							
KX	Marshall Is.	Oc	65	31							
LA	Norway	Eu	18	14							
LU	Argentina	SA	14,16	13							
LX	Luxembourg	Eu	27	14							
LZ	Bulgaria	Eu	28	20							
OA	Peru	SA	12	10							
OD	Lebanon	AS	39	20							
OE	Austria	Eu	28	15							
OH	Finland	Eu	19	15							
OHØ	Aland Is.	Eu	18	15							
OJØ	Market	Eu	18	15							
OK	Czechoslovakia	Eu	28	15							

Prefix	Country	Continent	Zone		Worked/Confirmed		5BDXCC				
			ITU	CQ	W	C	80	40	20	15	10
ON	Belgium	Eu	27	14							
OX	Greenland	NA	5,75	40							
OY	Faroe Is.	Eu	18	14							
OZ	Denmark	Eu	18	14							
P2	Papua New Guinea	Oc	51	28							
PA	Netherlands	Eu	27	14							
PJ	Neth. Antilles	SA	11	09							
PJ	St. Maarten	NA	11	08							
PY	Brazil	SA	[4]	11							
PYØ	Fernando	SA	13	11							
PYØ	St. Peter & Paul's Rocks	SA	13	11							
PYØ	Trindade	SA	15	11							
PZ	Suriname	SA	12	09							
S2	Bangladesh	As	41	22							
S7	Seychelles	Af	53	39							
S9/CR5	Sao Tome	Af	47	36							
SM	Sweden	Eu	18	14							
SP	Poland	Eu	28	15							
ST	Sudan	Af	48	34							
STØ	Southern Sudan	Af	48	34							
SU	Egypt	Af	38	34							
SV	Greece	Eu	28	20							
SV	Crete	Eu	28	20							
SV	Dodecanese	Eu	28	20							
SV	Mt. Athos	Eu	28	20							
TA	Turkey	As	39	20							
TF	Iceland	Eu	17	40							
TG	Guatemala	NA	11	07							
TI	Costa Rica	NA	11	07							
TI9	Cocos I.	NA	11	07							
TJ	Cameroon	Af	47	38							
TL	Cen. Af. Empire	Af	47	36							
TN	Congo	Af	52	36							
TR	Gabon	Af	52	36							
TT	Chad	Af	47	36							
TU	Ivory Coast	Af	46	35							
TY	Benin	Af	46	35							
TZ	Mali	Af	46	35							
UA	European RFSFR	Eu	[5]	16							
UA1	Franz Josef Land	Eu	75	40							
UA2	Kaliningradsk	Eu	29	15							
UA9-Ø	Asiatic RFSFR	As	[6]	[7]							
UB	Ukraine	Eu	29	16							
UC2	White RFSFR	Eu	29	16							
UD6	Azerbaijan	As	29	21							
UF6	Georgia	As	29	21							
UG6	Armenia	As	29	21							
UH8	Turkoman	As	30	17							
UI8	Uzbek	As	30	17							
UJ8	Tadzhik	As	30	17							
UL7	Kazakh	As	30	17							
UM8	Kirghiz	As	31	17							
UO5	Moldavia	Eu	29	16							
UP2	Lithuania	Eu	29	15							
UQ2	Latvia	Eu	29	15							
UR2	Estonia	Eu	29	15							
VE/VO	Canada	NA	[8]	1-5							
VE1	Sable Is.	NA	09	05							
VE1	St. Paul Is.	NA	09	05							
VK	Australia	Oc	[9]	29,35							
VK	Lord Howe Is.	Oc	60	30							
VK9	Willis I.	Oc	60	30							
VK9C	Christmas I.	Oc	54	29							
VK9	Cocos I.	Oc	54	29							
VK9	Mellish Reef	Oc	55	30							
VK9	Norfolk I.	Oc	60	32							
VKØ	Heard I.	Af	68	39							

Prefix	Country	Continent	Zone		Worked/Confirmed		5BDXCC				
			ITU	CQ	W	C	80	40	20	15	10
VKØ	Macquarie I.	Oc	60	30							
VP1	Belize	NA	11	07							
VP2A	Antigua, Barbuda	NA	11	08							
VP2E	Anguilla	NA	11	08							
VP2K	St. Kitts, Nevis	NA	11	08							
VP2M	Montserrat	NA	11	08							
VP2S	St. Vincent	NA	11	08							
VP2V	Brit. Virgin Is.	NA	11	08							
VP5	Turks & Caicos Is.	NA	11	08							
VP8	Falkland Is.	SA	16	13							
VP8/LU	S. Georgia Is.	SA	73	13							
VP8/LU	S. Orkney Is.	SA	73	13							
VP8/LU	S. Sandwich Is.	SA	73	13							
VP8/LU/CE9	S. Shetland Is.	SA	73	13							
VP9	Bermuda	NA	11	05							
VQ9	Chagos	Af	53	39							
VR1/T3P	Cent. Kiribati	Oc	62	31							
VR1/T3K	W. Kiribati	Oc	65	31							
VR3/T3L	E. Kiribati	Oc	61	31							
VR6	Pitcairn I.	Oc	63	32							
VR8/T2	Tuvalu	Oc	65	31							
VS5	Brunei	Oc	54	28							
VS6	Hong Kong	As	44	24							
VS9K	Kamaran I.	As	39	21							
VU	India	As	41	22							
VU7	Andamar & Nicobar I.	As	49	26							
VU7	Laccadive Is.	As	41	22							
XE	Mexico	NA	10	06							
XF4	Revilla Gigedo	NA	10	06							
XT	Upper Volta	Af	46	35							
XU	Khmer Rep.	As	49	26							
XV	Vietnam	As	49	26							
XW	Lao People's Dem. Rep.	As	49	26							
XZ	Burma	As	49	26							
YA	Afghanistan	As	40	21							
YB	Indonesia	Oc	54	28							
YI	Iraq	As	39	21							
YJ	New Hebrides	Oc	56	32							
YK	Syria	As	39	20							
YN	Nicaragua	NA	11	07							
YO	Romania	Eu	28	20							
YS	Salvador	NA	11	07							
YU	Yugoslavia	Eu	28	15							
YV	Venezuela	SA	12	09							
YVØ	Aves I.	NA	11	08							
ZA	Albania	Eu	28	15							
ZB	Gibraltar	Eu	37	14							
ZD7	St. Helena	Af	66	36							
ZD8	Ascension I.	Af	66	36							
ZD9	Tristan & Gough I.	Af	66	38							
ZE	Rhodesia	Af	53	38							
ZF	Cayman Is.	NA	11	08							
ZK1	S. Cook Is.	Oc	62	32							
ZK1	N. Cook Is.	Oc	62	32							
ZK2	Niue	Oc	62	32							
ZL	New Zealand	Oc	60	32							
ZL	Auckland & Campbell	Oc	60	32							
ZL	Chatham I.	Oc	60	32							
ZL	Kermadec Is.	Oc	60	32							
ZM7	Tokelaus	Oc	62	31							
ZP	Paraguay	SA	14	11							
ZS1-6	South Africa	Af	57	38							
ZS2	Marion Is.	Af	57	38							
ZS3	Southwest Africa	Af	57	38							
1S	Spratly Is.	As	50	26							
3A	Monaco	Eu	27	14							
3B6	Agalega & St. Brandon	Af	53	39							

Prefix	Country	Continent	Zone		Worked/Confirmed		5BDXCC				
			ITU	CQ	W	C	80	40	20	15	10
3B8	Mauritius	Af	53	39							
3B9	Rodriguez I.	Af	53	39							
3C	Equatorial Guinea	Af	47	36							
3CØ	Annobon	Af	52	36							
3D2	Fiji Is.	Oc	56	32							
3D6	Swaziland	Af	57	38							
3V	Tunisia	Af	37	33							
3X	Rep. of Guinea	Af	46	35							
3Y	Bouvet	Af	67	38							
4S	Sri Lanka	As	41	22							
4U1ITU	ITU Geneva	Eu	28	14							
4U1UN	U.N. hq., New York	NA	08	05							
4W	Yemen	As	39	21							
4X	Israel	As	39	20							
5A	Libya	Af	38	34							
5B	Cyprus	As	39	20							
5H	Tanzania	Af	53	37							
5N	Nigeria	Af	46	35							
5R	Malagasy Rep.	Af	53	39							
5T	Mauritania	Af	46	35							
5U	Niger	Af	46	35							
5V	Togo	Af	46	35							
5W	Western Samoa	Oc	62	32							
5X	Uganda	Af	48	37							
5Z	Kenya	Af	48	37							
6O	Somali	Af	48	37							
6W	Senegal	Af	46	35							
6Y	Jamaica	NA	11	08							
7O	People's D.R. of Yemen	As	39	21							
7P	Lesotho	Af	57	38							
7Q	Malawi	Af	53	37							
7X	Algeria	Af	37	33							
8P	Barbados	NA	11	08							
8Q	Maldive Is.	As	41	22							
8R	Guyana	SA	12	09							
8Z4	Saudi Arabia/Iraq Neut. Zone	As	39	21							
9A/M1	San Marino	Eu	28	15							
9G	Ghana	Af	46	35							
9H	Malta	Eu	28	15							
9J	Zambia	Af	53	36							
9K	Kuwait	As	30	21							
9L	Sierra Leone	Af	46	35							
9M2	W. Malaysia	As	54	28							
9M8	E. Malaysia	As	54	28							
9N	Nepal	As	42	22							
9Q	Rep. of Zaire	Af	52	36							
9U	Burundi	Af	52	36							
9V	Singapore	As	54	28							
9X	Rwanda	Af	52	36							
9Y	Trinidad & Tobago	SA	11	09							
	Abu Ail, Jabal at Tair	As	39	21							

Continent

Af = Africa
As = Asia
Eu = Europe
NA = North America
Oc = Oceania
SA = South America

Notes

[1]33,42,43,44
[2]67,69-74
[3]12,13,29,30,32,38,39
[4]12,13,15
[5]19,20,29,30
[6]20-26,30,35,75
[7]17,18,19,23
[8]2,3,4,9,75
[9]55,58,59

Table of Allocation of International Call Signs[1]

Call Sign Series	Allocated to
AAA-ALZ	United States of America
AMA-AOZ	Spain
APA-ASZ	Pakistan (Islamic Republic of)
ATA-AWZ	India (Republic of)
AXA-AXZ	Australia
AYA-AZZ	Argentine Republic
A2A-A2Z	Botswana (Republic of)
A3A-A3Z	Tonga (Kingdom of)
A4A-A4Z	Oman (Sultanate of)
A5A-A5Z	Bhutan (Kingdom of)
A6A-A6Z	United Arab Emirates
A7A-A7Z	Qatar (State of)
A8A-A8Z	Liberia (Republic of)
A9A-A9Z	Bahrain (State of)
BAA-BZZ	China (People's Republic of)
CAA-CEZ	Chile
CFA-CKZ	Canada
CLA-CMZ	Cuba
CNA-CNZ	Morocco (Kingdom of)
COA-COZ	Cuba
CPA-CPZ	Bolivia (Republic of)
CQA-CUZ	Portugal
CVA-CXZ	Uruguay (Oriental Republic of)
CYA-CZZ	Canada
C2A-C2Z	Nauru (Republic of)
C3A-C3Z	Andorra (Principality of)
C4A-C4Z	Cyprus (Republic of)
C5A-C5Z	Gambia (Republic of the)
C6A-C6Z	Bahamas (Commonwealth of the)
C7A-C7Z*	World Meteorological Organization
C8A-C9Z	Mozambique (People's Republic of)
DAA-DRZ	Germany (Federal Republic of)
DSA-DTZ	Republic of Korea
DUA-DZZ	Philippines (Republic of the)
D2A-D3Z	Angola (People's Republic of)
D4A-D4Z	Cape Verde (Republic of)
D5A-D5Z	Liberia (Republic of)
D6A-D6Z	Comoros (Federal and Islamic Republic of the)
D7A-D9Z	Republic of Korea
EAA-EHZ	Spain
EIA-EJZ	Ireland
EKA-EKZ	Union of Soviet Socialist Republics
ELA-ELZ	Liberia (Republic of)
EMA-EOZ	Union of Soviet Socialist Republics
EPA-EQZ	Iran (Islamic Republic of)
ERA-ESZ	Union of Soviet Socialist Republics
ETA-ETZ	Ethiopia
EUA-EWZ	Byelorussian Soviet Socialist Republic
EXA-EZZ	Union of Soviet Socialist Republics
FAA-FZZ	France
GAA-GZZ	United Kingdom of Great Britain and Northern Ireland
HAA-HAZ	Hungarian People's Republic
HBA-HBZ	Switzerland (Confederation of)
HCA-HDZ	Ecuador
HEA-HEZ	Switzerland (Confederation of)
HFA-HFZ	Poland (People's Republic of)
HGA-HGZ	Hungarian People's Republic
HHA-HHZ	Haiti (Republic of)
HIA-HIZ	Dominican Republic
HJA-HKZ	Colombia (Republic of)
HLA-HLZ	Republic of Korea[2]
HMA-HMZ	Democratic People's Republic of Korea[2]
HNA-HNZ	Iraq (Republic of)
HOA-HPZ	Panama (Republic of)
HQA-HRZ	Honduras (Republic of)
HSA-HSZ	Thailand
HTA-HTZ	Nicaragua
HUA-HUZ	El Salvador (Republic of)
HVA-HVZ	Vatican City State
HWA-HYZ	France
HZA-HZZ	Saudi Arabia (Kingdom of)
H2A-H2Z	Cyprus (Republic of)
H3A-H3Z	Panama (Republic of)
H4A-H4Z	Solomon Islands
H6A-H7Z	Nicaragua
H8A-H9Z	Panama (Republic of)
IAA-IZZ	Italy
JAA-JSZ	Japan
JTA-JVZ	Mongolian People's Republic
JWA-JXZ	Norway
JYA-JYZ	Jordan (Hashemite Kingdom of)
JZA-JZZ	Indonesia (Republic of)
J2A-J2Z	Djibouti (Republic of)
J3A-J3Z	Grenada
J4A-J4Z	Greece
J5A-J5Z	Guinea-Bissau (Republic of)
J6A-J6Z	Saint Lucia
J7A-J7Z	Dominica
KAA-KZZ	United States of America
LAA-LNZ	Norway
LOA-LWZ	Argentine Republic
LXA-LXZ	Luxembourg
LYA-LYZ	Union of Soviet Socialist Republics
LZA-LZZ	Bulgaria (People's Republic of)
L2A-L9Z	Argentine (Republic of)
MAA-MZZ	United Kingdom of Great Britain and Northern Ireland
NAA-NZZ	United States of America
OAA-OCZ	Peru
ODA-ODZ	Lebanon
OEA-OEZ	Austria
OFA-OJZ	Finland
OKA-OMZ	Czechoslovak Socialist Republic
ONA-OTZ	Belgium
OUA-OZZ	Denmark
PAA-PIZ	Netherlands (Kingdom of the)
PJA-PJZ	Netherlands Antilles
PKA-POZ	Indonesia (Republic of)
PPA-PYZ	Brazil (Federative Republic of)
PZA-PZZ	Suriname (Republic of)
P2A-P2Z	Papua New Guinea
P3A-P3Z	Cyprus (Republic of)
P4A-P4Z	Netherlands Antilles
P5A-P9Z	Democratic People's Republic of Korea
QAA-QZZ	(Service abbreviations)
RAA-RZZ	Union of Soviet Socialist Republics
SAA-SMZ	Sweden
SNA-SRZ	Poland (People's Republic of)
SSA-SSM	Egypt (Arab Republic of)
SSN-STZ	Sudan (Democratic Republic of the)
SUA-SUZ	Egypt (Arab Republic of)
SVA-SZZ	Greece
S2A-S3Z	Bangladesh (People's Republic of)
S6A-S6Z	Singapore (Republic of)
S7A-S7Z	Seychelles (Republic of)
S9A-S9Z	Sao Tome and Principe (Democratic Republic of)
TAA-TCZ	Turkey
TDA-TDZ	Guatemala (Republic of)
TEA-TEZ	Costa Rica
TFA-TFZ	Iceland
TGA-TGZ	Guatemala (Republic of)
THA-THZ	France
TIA-TIZ	Costa Rica
TJA-TJZ	Cameroon (United Republic of)
TKA-TKZ	France
TLA-TLZ	Central African Republic
TMA-TMZ	France
TNA-TNZ	Congo (People's Republic of the)
TOA-TQZ	France
TRA-TRZ	Gabon Republic
TSA-TSZ	Tunisia
TTA-TTZ	Chad (Republic of the)
TUA-TUZ	Ivory Coast (Republic of the)
TVA-TXZ	France
TYA-TYZ	Benin (People's Republic of)
TZA-TZZ	Mali (Republic of)
T2A-T2Z	Tuvalu
T3A-T3Z	Kiribati Republic
T4A-T4Z	Cuba
T5A-T5Z	Somali Democratic Republic
T6A-T6Z	Afghanistan (Democratic Republic of)
UAA-UQZ	Union of Soviet Socialist Republics
URA-UTZ	Ukrainian Soviet Socialist Republic
UUA-UZZ	Union of Soviet Socialist Republics
VAA-VGZ	Canada
VHA-VNZ	Australia
VOA-VOZ	Canada
VPA-VSZ	United Kingdom of Great Britain and Northern Ireland
VTA-VWZ	India (Republic of)
VXA-VYZ	Canada
VZA-VZZ	Australia
WAA-WZZ	United States of America
XAA-XIZ	Mexico
XJA-XOZ	Canada
XPA-XPZ	Denmark
XQA-XRZ	Chile
XSA-XSZ	China (People's Republic of)
XTA-XTZ	Upper Volta (Republic of)
XUA-XUZ	Democratic Kampuchea
XVA-XVZ	Viet Nam (Socialist Republic of)

XWA-XWZ	Lao People's Democratic Republic
XXA-XXZ	Portugal
XYA-XZZ	Burma (Socialist Republic of the Union of)
YAA-YAZ	Afghanistan (Democratic Republic of)
YBA-YHZ	Indonesia (Republic of)
YIA-YIZ	Iraq (Republic of)
YJA-YJZ	New Hebrides
YKA-YKZ	Syrian Arab Republic
YLA-YLZ	Union of Soviet Socialist Republics
YMA-YMZ	Turkey
YNA-YNZ	Nicaragua
YOA-YRZ	Romania (Socialist Republic of)
YSA-YSZ	El Salvador (Republic of)
YTA-YUZ	Yugoslavia (Socialist Federal Republic of)
YVA-YYZ	Venezuela (Republic of)
YZA-YZZ	Yugoslavia (Socialist Federal Republic of)
Y2A-Y9Z	German Democratic Republic
ZAA-ZAZ	Albania (Socialist People's Republic of)
ZBA-ZJZ	United Kingdom of Great Britain and Northern Ireland
ZKA-ZMZ	New Zealand
ZNA-ZOZ	United Kingdom of Great Britain and Northern Ireland
ZPA-ZPZ	Paraguay (Republic of)
ZQA-ZQZ	United Kingdom of Great Britain and Northern Ireland
ZRA-ZUZ	South Africa (Republic of)
ZVA-ZZZ	Brazil (Federative Republic of)
2AA-2ZZ	United Kingdom of Great Britain and Northern Ireland
3AA-3AZ	Monaco
3BA-3BZ	Mauritius
3CA-3CZ	Equatorial Guinea (Republic of)
3DA-3DM	Swaziland (Kingdom of)
3DN-3DZ	Fiji
3EA-3FZ	Panama (Republic of)
3GA-3GZ	Chile
3HA-3UZ	China (People's Republic of)
3VA-3VZ	Tunisia
3WA-3WZ	Viet Nam (Socialist Republic of)
3XA-3XZ	Guinea (People's Revolutionary Republic of)
3YA-3YZ	Norway
3ZA-3ZZ	Poland (People's Republic of)
4AA-4CA	Mexico
4DA-4IZ	Philippines (Republic of the)
4JA-4LZ	Union of Soviet Socialist Republics
4MA-4MZ	Venezuela (Republic of)
4NA-4OZ	Yugoslavia (Socialist Federal Republic of)
4PA-4SZ	Sri Lanka (Democratic Socialist Republic of)
4TA-4TZ	Peru
4UA-4UZ*	United Nations Organization
4VA-4VZ	Haiti (Republic of)
4WA-4WZ	Yemen Arab Republic
4XA-4XZ	Israel (State of)
4YA-4YZ*	International Civil Aviation Organization
4ZA-4ZZ	Israel (State of)
5AA-5AZ	Libya (Socialist People's Libyan Arab Jamahiriya)
5BA-5BZ	Cyprus (Republic of)
5CA-5GZ	Morocco (Kingdom of)
5HA-5IZ	Tanzania (United Republic of)
5JA-5KZ	Colombia (Republic of)
5LA-5MZ	Liberia (Republic of)
5NA-5OZ	Nigeria (Federal Republic of)
5PA-5QZ	Denmark
5RA-5SZ	Madagascar (Democratic Republic of)
5TA-5TZ	Mauritania (Islamic Republic of)
5UA-5UZ	Niger (Republic of the)
5VA-5VZ	Togolese Republic
5WA-5WZ	Western Samoa
5XA-5XZ	Uganda (Republic of)
5YA-5ZZ	Kenya (Republic of)
6AA-6BZ	Egypt (Arab Republic of)
6CA-6CZ	Syrian Arab Republic
6DA-6JZ	Mexico
6KA-6NZ	Republic of Korea
6OA-6OZ	Somali Democratic Republic
6PA-6SZ	Pakistan (Islamic Republic of)
6TA-6UZ	Sudan (Democratic Republic of the)
6VA-6WZ	Senegal (Republic of the)
6XA-6XZ	Madagascar (Democratic Republic of)
6YA-6YZ	Jamaica
6ZA-6ZZ	Liberia (Republic of)
7AA-7IZ	Indonesia (Republic of)
7JA-JNZ	Japan
7OA-7OZ	Yemen (People's Democratic Republic of)
7PA-7PZ	Lesotho (Kingdom of)
7QA-7QZ	Malawi (Republic of)
7RA-7RZ	Algeria (Algerian Democratic and Popular Republic)
7SA-7SZ	Sweden
7TA-7YZ	Algeria (Algerian Democratic and Popular Republic)
7ZA-7ZZ	Saudi Arabia (Kingdom of)
8AA-8IZ	Indonesia (Republic of)
8JA-8NZ	Japan
8OA-8OZ	Botswana (Republic of)
8PA-8PZ	Barbados
8QA-8QZ	Maldives (Republic of)
8RA-8RZ	Guyana
8SA-8SZ	Sweden
8TA-8YZ	India (Republic of)
8ZA-8ZZ	Saudi Arabia (Kingdom of)
9AA-9AZ	San Marino (Republic of)
9BA-9DZ	Iran (Islamic Republic of)
9EA-9FZ	Ethiopia
9GA-9GZ	Ghana
9HA-9HZ	Malta (Republic of)
9IA-9JZ	Zambia (Republic of)
9KA-9KZ	Kuwait (State of)
9LA-9LZ	Sierra Leone
9MA-9MZ	Malaysia
9NA-9NZ	Nepal
9OA-9TZ	Zaire (Republic of)
9UA-9UZ	Burundi (Republic of)
9VA-9VZ	Singapore (Republic of)
9WA-9WZ	Malaysia
9XA-9XZ	Rwanda (Republic of)
9YA-9ZZ	Trinidad and Tobago

1) The series of call signs with as asterisk indicate the international organization to which they are allocated.

2) The two administrations concerned undertake to change their existing use of HLA-HLZ and HMA-HMZ Call Sign series to conform with the 1979 Table of Allocations as soon as practicable, in order to clarify their operational arrangements for other administrations. In this regard, the Administration of the Republic of Korea will take action to change the existing call signs registered with the ITU in the HMA-HMZ series as changes occur in the use of call signs in this series. The above-mentioned actions shall, in any case, be completed by 1 January 1984.

Spanish Phonetics

America	ah-MAIR-ika	Norvega	nor-WAY-gah	W		DOE-bleh-vay
Brasil	brah-SIL	Ontario	on-TAR-eeoh	Ø	cero	SEH-roe
Canada	cana-DAH	Portugal	portu-GAL	1	uno	OO-no
Dinamarca	dina-MAR-ka	Quito	KEY-toe	2	dos	DOS
Espana	es-PAHN-yah	Roma	ROW-mah	3	tres	TRAYCE
Francia	FRAHN-seeah	Santiago	santee-AH-go	4	cuatro	KWAT-roe
Grenada	gre-NAH-dah	Toronto	tor-ON-toe	5	cinco	SINK-oh
Holanda	oh-LONN-dah	Uniforme	oonee-FORM-eh	6	seis	SAYCE
Italia	i-TAL-eeah	Victoria	vic-TOR-eeah	7	siete	see-AY-teh
Japon	hop-OWN	Washington, Wisky	washingtone, wisky	8	ocho	OCH-oh
Kilowatio	kilo-WAT-eeoh	Xilofono	see-LOW-phono	9	nueve	new-AY-veh
Lima	LIMA	Yucatan	yuca-TAN			
Mejico	MEH-heeco	Zelandai	see-LAND-eeah			

— John Mason Jr., TG9ML

Q Signals

Given below are a number of Q signals whose meanings most often need to be expressed with brevity and clearness in amateur work. (Q abbreviations take the form of questions only when each is sent followed by a question mark.)

QRG	Will you tell me my exact frequency (or that of . . .)? Your exact frequency (or that of . . .) is . . . kHz.
QRH	Does my frequency vary? Your frequency varies.
QRI	How is the tone of my transmission? The tone of your transmission is . . . (1. Good; 2. Variable; 3. Bad).
QRK	What is the intelligibility of my signals (or those of . . .)? The intelligibility of your signals (or those of . . .) is . . . (1. Bad; 2. Poor; 3. Fair; 4. Good; 5. Excellent).
QRL	Are you busy? I am busy (or I am busy with . . .). Please do not interfere.
QRM	Is my transmission being interfered with? Your transmission is being interfered with . . . (1. Nil; 2. Slightly; 3. Moderately; 4. Severely; 5. Extremely.)
QRN	Are you troubled by static? I am troubled by static . . . (1-5 as under QRM).
QRO	Shall I increase power? Increase power.
QRP	Shall I decrease power? Decrease power.
QRQ	Shall I send faster? Send faster (. . . wpm).
QRS	Shall I send more slowly? Send more slowly (. . . wpm).
QRT	Shall I stop sending? Stop sending.
QRU	Have you anything for me? I have nothing for you.
QRV	Are you ready? I am ready.
QRW	Shall I inform . . . that you are calling him on . . . kHz? Please inform . . . that I am calling . . . kHz.
QRX	When will you call me again? I will call you again at . . . hours (on . . . kHz).
QRY	What is my turn? Your turn is numbered . . .
QRZ	Who is calling me? You are being called by . . . (on . . . kHz).
QSA	What is the strength of my signals (or those of . . .)? The strength of your signals (or those of . . .) is . . . (1. Scarcely perceptible; 2. Weak; 3. Fairly good; 4. Good; 5. Very good).
QSB	Are my signals fading? Your signals are fading.
QSD	Are my signals mutilated? Your signals are mutilated.
QSG	Shall I send . . . messages at a time? Send . . . messages at a time.
QSK	Can you hear me between your signals and if so can I break in on your transmission? I can hear you between my signals; break in on my transmission.
QSL	Can you acknowledge receipt? I am acknowledging receipt.
QSM	Shall I repeat the last message which I sent you, or some previous message? Repeat the last message which you sent me [or message(s) number(s) . . .].
QSN	Did you hear me (or . . .) on . . . kHz? I did hear you (or . . .) on . . . kHz.
QSO	Can you communicate with . . . direct or by relay? I can communicate with . . . direct (or by relay through . . .).
QSP	Will you relay to . . .? I will relay to . . .
QST	General call preceding a message addressed to all amateurs and ARRL members. This is in effect "CQ ARRL."
QSU	Shall I send or reply on this frequency (or on . . . kHz)? Send or reply on this frequency (or on . . . kHz).
QSV	Shall I send a series of Vs on this frequency (or . . . kHz)? Send a series of Vs on this frequency (or . . . kHz).
QSW	Will you send on this frequency (or on . . . kHz)? I am going to send on this frequency (or on . . . kHz).
QSX	Will you listen to . . . on . . . kHz? I am listening to . . . on . . . kHz.
QSY	Shall I change to transmission on another frequency? Change to transmission on another frequency (or on . . . kHz).
QSZ	Shall I send each word or group more than once? Send each word or group twice (or . . . times).
QTA	Shall I cancel message number . . .? Cancel message number . . .
QTB	Do you agree with my counting of words? I do not agree with your counting of words. I will repeat the first letter or digit of each word or group.
QTC	How many messages have you to send? I have . . . messages for you (or for . . .).
QTH	What is your location? My location is . . .
QTR	What is the correct time? The time is . . .

Some Abbreviations for CW Work

Although abbreviations help to cut down unnecessary transmission, make it a rule not to abbreviate unnecessarily when working an operator of unknown experience.

AA	All after
AB	All before
ABT	About
ADR	Address
AGN	Again
ANT	Antenna
BCI	Broadcast interference
BCL	Broadcast listener
BK	Break; break me; break in
BN	All between; been
BUG	Semi-automatic key
B4	Before
C	Yes
CFM	Confirm; I confirm
CK	Check
CL	I am closing my station; call
CLD-CLG	Called; calling
CQ	Calling any station
CUD	Could
CUL	See you later
CW	Continuous wave (i.e., radiotelegraph)
DLD-DLVD	Delivered
DR	Dear
DX	Distance, foreign countries
ES	And, &
FB	Fine business, excellent
FM	Frequency modulation
GA	Go ahead (or resume sending)
GB	Good-by
GBA	Give better address
GE	Good evening
GG	Going
GM	Good morning
GN	Good night
GND	Ground
GUD	Good
HI	The telegraphic laugh; high
HR	Here, hear
HV	Have
HW	How
LID	A poor operator
MA, MILS	Milliamperes
MSG	Message; prefix to radiogram
N	No
NCS	Net control station
ND	Nothing doing
NIL	Nothing; I have nothing for you
NM	No more
NR	Number
NW	Now; I resume transmission
OB	Old boy
OC	Old chap
OM	Old man
OP-OPR	Operator
OT	Old timer; old top
PBL	Preamble
PSE	Please
PWR	Power
PX	Press
R	Received as transmitted; are
RCD	Received
RCVR (RX)	Receiver
REF	Refer to; referring to; reference
RFI	Radio frequency interference
RIG	Station equipment
RPT	Repeat; I repeat
RTTY	Radioteletype
RX	Receiver
SASE	Self-addressed, stamped envelope
SED	Said
SIG	Signature; signal
SINE	Operator's personal initials or nickname
SKED	Schedule
SRI	Sorry
SSB	Single sideband
SVC	Service; prefix to service message
T	Zero
TFC	Traffic
TMW	Tomorrow
TNX-TKS	Thanks
TT	That
TU	Thank you
TVI	Television interference
TX	Transmitter
TXT	Text
UR-URS	Your; you're; yours
VFO	Variable-frequency oscillator
VY	Very
WA	Word after
WB	Word before
WD-WDS	Word; words
WKD-WKG	Worked; working
WL	Well; will
WUD	Would
WX	Weather
XCVR	Transceiver
XMTR (TX)	Transmitter
XTAL	Crystal
XYL (YF)	Wife
YL	Young lady
73	Best regards
88	Love and kisses

ARRL Station Appointments

Amateur licensees who are full ARRL members and hold the required grade of license are invited to contact their SCMs to take on "station" appointments along the lines of their special interests. At the present time, such two-year appointments are open to amateurs interested in traffic handling, emergency operating, vhf operating, observer-monitoring, and transmitting bulletins to all amateurs. A special Junior OTS appointment is available to Novices, details to follow. All appointees receive handsome certificates from the SCM and are entitled to wear ARRL membership pins with the distinctive blue background. All SCM appointees are required to submit monthly reports to the SCM in order to hold their appointments, to keep their stations active on the air, to follow standard ARRL operating practices and to take active part in ARRL on-the-air activities to the extent practical.

The monthly report is the criterion of activity. An appointee who misses three consecutive monthly reports is subject to cancellation by the SCM, who cannot know what or how much you are doing if you do not tell him. An appointee whose appointment is cancelled for this or other reasons must earn reinstatement by demonstrating his activity and adherence to the requirements. Reinstatement of cancelled appointments, and indeed judgment of whether or not a candidate meets the requirements, is at the discretion of the SCM.

The detailed qualifications for the five "station" grade appointments are given below. If you are interested, your SCM will be glad to receive your application.

Official Traffic Station

This is a traffic-handling appointment requiring a Technician Class license or higher. It replaces both the time-honored ORS and OPS appointments which have existed separately in the cw and phone traffic-handling fields for decades and have always been marks of distinction in the traffic-handling field. This new appointment applies equally to all modes and all parts of the spectrum. It is for traffic-handlers, regardless of how they do it or in what part of the spectrum.

The potential value of the operator who has traffic know-how to his country and community is enhanced by his ability and the readiness of his station to function in the community interest in case of emergency. Traffic awareness and experience are often the signs by which mature amateurs may be distinguished.

Traditionally, there have been considerable differences between procedures for traffic handling by cw, phone, RTTY and other modes. Appointment requirements for OTS do not deal with these, but with factors equally applicable to all modes. The appointed OTS may confine his activities to one mode or one part of the spectrum if he wishes. There is no versatility requirement, although versatility does indeed make it possible for anyone to perform a more complete public service. There is, however, the expectation that the OTS will set the example in traffic-handling *however* he does it. To the extent that he is deficient in performing traffic functions by any mode, to that extent he does not meet the qualifications for the appointment. Here are the basic requirements:

1) Full ARRL membership and Technician Class license or higher.

2) Code and voice transmission.

3) Transmission quality, by whatever mode, must be of the highest quality, both technically and operationally. For example, cw signals must be pure, chirpless, clickless, code sending must be well spaced and properly formed. Voice transmission must be of proper modulation percentage or deviation, precisely enunciated with minimum distortion. RTTY must be clickless, proper shift, etc.

4) All OTS are expected to follow standard ARRL operating practices (message form, ending signals, abbreviations or prowords, courtesy, etc.).

5) Regular participation in traffic activities, either free-lance or ARRL-sponsored. The latter is encouraged, but not required.

6) Handle all record communications speedily and reliably and set the example in efficient operating procedures. All traffic is relayed or delivered promptly after receipt.

7) Report monthly to the SCM, including a breakdown of traffic handled during the past calendar month.

Official Traffic Station-II

In order to bring Novice operators into the traffic-handling field, an appointment as OTS-II has also been created. This appointment has the same qualifications as OTS except that it may be held by a Novice Class licensee. OTS-II is *not* available to amateurs of General Class or higher. A holder of OTS-II who acquires his general class license must qualify as OTS or lose the appointment altogether.

Official Emergency Station

Amateur operators of Technician class and above may be appointed OES by their SCMs at the recommendation of their ECs or SECs (if no EC) if they are ARRL members and interested in setting high standards of emergency preparedness and operating. In addition to candidates for this appointment operating within their own local EC jurisdictional areas, we want to recruit OESs who can set similar examples in "off shore" emergencies, such as those which frequently occur in foreign countries (other than Canada, which is not considered foreign for field organizational purposes). Here are the standard qualifications and functions of this appointment.

1) Possession of full ARRL membership and a Technician Class license or higher.

2) Regular participation in the local ARES, if any, including all drills and tests, emergency nets and, or course, real emergency situations.

3) Ability to operate independent of commercial mains on at least two amateur bands; this includes mobile capability.

4) Must be fully acquainted with standard ARRL message form and capable of using it in handling any third party messages.

5) Report monthly to the SCM on form provided, or on CD-210.

Official VHF Station

This appointment is open to any licensed amateur who is a full member of the League. Its basis is vhf experimental on-the-air activity such as repeaters, space communications, moonbounce, propagation, etc. If you are an amateur who is interested in vhf work of any kind you probably can be eligible for OVS appointment. Here are the requirements and functions.

1) Available to any ARRL full member who operates above 50 MHz.

2) Report contacts, observations, on-air experiments, etc., monthly to SCM, on standard form provided. After extracting any data for his report, the SCM passes these reports along to Headquarters where they are noted by the communications manager, the vhf editor and then passed along to the editor of *QST*'s vhf column for possible use therein.

3) The OVS appointee is expected to set the example of excellence in all fields of vhf endeavor, including both signal quality and operating ability. Participation may be in vhf repeater activities, satellite or space communications, propagation, contests, or in any or all of these and other vhf activities.

4) Traffic and emergency preparedness activities are no longer considered a part of the function of OVS appointees. Such work can now qualify for appointment as OTS or OES (see above).

Official Bulletin Station

Once each week an official Bulletin is issued by ARRL headquarters and transmitted several times daily on a regular schedule over Headquarters station W1AW. Official Bulletin Station appointees have the function of copying these transmissions and retransmitting them for the benefit of amateurs in their areas who may have difficulty receiving them direct from W1AW. This is particularly applicable to vhf transmissions, and amateurs who can copy the transmissions and retransmit them on vhf are much in demand and of maximum usefulness. However, OBS appointments are also conferred on hf stations where the need is apparent. OBS appointees are expected to copy the bulletins on the air and proceed to transmit them on regular schedule — copying direct from W1AW or another OBS.

1) OBS appointees must be full ARRL members and have a Technician license or higher.

2) Retransmission of the official bulletin must be made a minimum of three times per week in order to retain appointment.

3) Appointees must adhere to a regular schedule of transmissions agreed upon with the SCM, and notify the SCM when changes are made.

4) Send a monthly activity report to the SCM, indicating transmissions made and/or missed and apparent extent of reception.

Applicants for OBS should submit their qualifications to the SCM with the proposed days, times and frequencies for transmission of the bulletins. If your other qualifications are in order and the SCM feels that your transmissions will be useful, chances are good that the appointment will be made.

Official Observer

The Official Observer program has been sponsored by the League for over 50 years to help amateurs help others. Official Observer appointees have aided thousands of amateurs to maintain their transmitting equipment and operating procedures in compliance with the regulations. The object of the OO program is to notify amateurs by mail of operating/technical irregularities before they come to the attention of the FCC or DOC (Canada).

The OO performs his functions by *listening* rather than transmitting, keeping a watchful ear out for such things as frequency instability, harmonics, hum, key clicks, broad signals, distorted audio, overdeviation, out-of-band operation, etc. The program is neither a vigilante nor a law-enforcement function. The OO completes his task once the notification card is sent. Reimbursement for postage expenses are provided for through the SCM. Requirements follow:

1) Must be an ARRL full member and have been a licensee of Technician Class or higher for at least four years.

2) To qualify for "precise frequency measurement," the Observer must qualify annually in the Frequency Measuring Test held quarterly to a tolerance of 100 Hz. Observers who have not so qualified must report all frequencies as "approximate," when reporting discrepancies.

3) Must report to the SCM each month on CD-23.

4) Maintain regular activity in sending out notices as observed.

(Continued on next page)

ARRL Station Appointments (cont.)

Applications for this most serious endeavor are available from your SCM or directly from Headquarters. Once the appointment is made by the SCM, the new appointee will receive a complete set of detailed operating instructions (CD-100) and an initial supply of notification report cards.

Periodically, the OO will receive *The Observation Post*, a publication distributed to all OO appointees giving latest information to assist in daily observations.

The OO program is one of the most important functions of the League. A sincere dedication to helping our brother and sister amateurs is required for appointment. Only the "very best" are sought.

The Origin of "73"

The traditional expression "73" goes right back to the beginning of the landline telegraph days. It is found in some of the earliest editions of the numerical codes, each with a different definition, but each with the same idea in mind — it indicated that the end, or signature, was coming up. But there are no data to prove that any of these were used.

The first authentic use of 73 is in the publication *The National Telegraphic Review and Operators' Guide*, first published in April 1857. At that time, 73 meant "My love to you!" Succeeding issues of this publication continued to use this definition of the term. Curiously enough, some of the other numerals used then had the same definition as they have now, but within a short time, the use of 73 began to change.

In the National Telegraph Convention, the numeral was changed from the Valentine-type sentiment to a vague sign of fraternalism. Here, 73 was a greeting, a friendly "word" between operators and it was so used on all wires.

In 1859, the Western Union Company set up the standard "92 Code." A list of numerals from one to 92 was compiled to indicate a series of prepared phrases for use by the operators on the wires. Here, in the 92 Code, 73 changes from a fraternal sign to a very flowery "accept my compliments," which was in keeping with the florid language of that era.

Over the years from 1859 to 1900, the many manuals of telegraphy show variations of this meaning. Dodge's *The Telegraph Instructor* shows it merely as "compliments." *The Twentieth Century Manual of Railway and Commercial Telegraphy* defines it two ways, one listing as "my compliments to you;" but in the glossary of abbreviations it is merely "compliments." Theodore A. Edison's *Telegraphy Self-Taught* shows a return to "accept my compliments." By 1908, however, a later edition of the Dodge Manual gives us today's definition of "best regards" with a backward look at the older meaning in another part of the work where it also lists it as "compliments."

"Best regards" has remained ever since as the "put-it-down-in-black-and-white" meaning of 73 but it has acquired overtones of much warmer meaning. Today, amateurs use it more in the manner that James Reid had intended that it be used — a "friendly word between operators." — *Louise Ramsey Moreau, W3WRE*

The RST System

READABILITY

1 — Unreadable.
2 — Barely readable, occasional words distinguishable.
3 — Readable with considerable difficulty.
4 — Readable with practically no difficulty.
5 — Perfectly readable.

SIGNAL STRENGTH

1 — Faint signals barely perceptible.
2 — Very weak signals.
3 — Weak signals.
4 — Fair signals.
5 — Fairly good signals.
6 — Good signals.
7 — Moderately strong signals.
8 — Strong signals.
9 — Extremely strong signals.

TONE

1 — Sixty-cycle ac or less, very rough and broad.
2 — Very rough ac, very harsh and broad.
3 — Rough ac tone, rectified but not filtered.
4 — Rough note, some trace of filtering.
5 — Filtered rectified ac but strongly ripple-modulated.
6 — Filtered tone, definite trace of ripple modulation.
7 — Near pure tone, trace of ripple modulation.
8 — Near perfect tone, slight trace of modulation.
9 — Perfect tone, no trace of ripple or modulation of any kind.

The "tone" report refers only to the purity of the signal, and has no connection with its stability or freedom from clicks or chirps. If the signal has the characteristic steadiness of crystal control, add X to the report (e.g., RST 469X). If it has a chirp or "tail" (either on "make" or "break") add C (e.g., 469C). If it has clicks or noticeable other keying transients, add K (e.g., 469K). Of course a signal could have both chirps and clicks, in which case both C and K could be used (e.g., RST 469CK).

W1AW

ARRL operates a headquarters station using W1AW, the original call of its founding father, Hiram Percy Maxim; thus, it is often known as the Maxim Memorial Station. Many services are performed for the operating amateur over W1AW every day, including bulletins of information and latest news, code practice at speeds from 5 through 35 wpm, and certificate-qualifying runs, frequency-measuring transmissions, etc.

IN A COMMUNICATIONS EMERGENCY monitor W1AW for bulletins as follows: *Phone*, on the hour. *RTTY*, 15 minutes past the hour. *Cw*, on the half hour.

The complete W1AW operating schedule appears in April and October *QST*. The schedule is also available in sheet form for anyone asking. (An s.a.s.e. is always appreciated with such requests.)

DX Operating Code (For W/VE Amateurs)

Some amateur DXers have caused considerable confusion and QRM in their efforts to work DX stations. The points below, if observed by all W/VE amateurs, will help make DX more enjoyable for all.

1) *Call* DX only after he calls CQ, QRZ? or signs $\overline{SK}$, or voice equivalents thereof. Make your calls *short*.

2) Do *not* call a DX station:

a) On the frequency of the station he is calling until you are *sure* the QSO is over ($\overline{SK}$).

b) Because you hear someone else calling him.

c) When he signs $\overline{KN}$, $\overline{AR}$, or CL.

d) Exactly on his frequency.

e) After he calls a directional CQ, unless of course you are in the right direction or area.

3) Keep within frequency band limits. Some DX stations can get away with working outside, but you cannot.

4) Observe calling instructions given by DX stations. *Example:* 15U means "call 15 kHz *up* from my frequency." 15D means *down*, etc.

5) Give honest reports. Many DX stations *depend* on W/VE reports for adjustment of station and equipment.

6) Keep your signal clean. Key clicks, ripple, feedback or splatter give you a bad reputation and may get you a citation from FCC.

7) *Listen* and call the station you want. Calling CQ DX is not the best assurance that the *rare* DX will reply.

8) When there are several W or VE stations waiting, avoid asking DX to "listen for a friend." Also avoid engaging him in a ragchew against his wishes.

(For Overseas Amateurs)

To all overseas amateur stations:

In their eagerness to work you, many W and VE amateurs resort to practices that cause confusion and QRM. Most of this is good-intentional but ill-advised; some of it is intentional and selfish. ***The key to the cessation of unethical DX operating practices is in your hands.*** We believe that your adoption of certain operating habits will increase your enjoyment of Amateur Radio and that of amateurs on this side who are eager to work you. We recommend your adoption of the following principles:

1) Do not answer calls on your own frequency.

2) Answer calls from W/VE stations only when their signals are of good quality.

3) Refuse to answer calls from other stations when you are already in contact with someone, and do not acknowledge calls from amateurs who indicate they wish to be "next."

4) Give *everybody* a break. When many W/VE amateurs are patiently and quietly waiting to work you, avoid complying with requests to "listen for a friend."

5) Tell listeners where to call you by indicating how many kilocycles *up* (U) or *down* (D) from your frequency you are listening. *Examples:* Cw — CQ DX CQ DX CQ DX 15U DE AC4YN AC4YN AC4YN 15U K; Voice — "Answer 15 kilohertz up from my frequency."

6) Use the ARRL-recommended ending signals, especially $\overline{KN}$ to indicate to impatient listeners the status of the QSO. $\overline{KN}$ means "Go ahead (specific station), all others keep out."

7) Let it be known that you avoid working amateurs who are constant violators of the above principles.

ARRL Field Organization

The United States and Canada are divided into 16 ARRL Divisions. Every two years the full ARRL members in each of these divisions elect a director and a vice director to represent them on the League's Board of Directors. The Board determines the policies of the League, which are carried out by the Headquarters staff. A director's function is principally policymaking at the highest level, but the Board of Directors is all-powerful in the conduct of League affairs.

The 16 divisions are further broken down into 73 "sections," and the ARRL full members in each section elect a Section Communications Manager. Just as a director seldom concerns himself with operating-administration, except in an overall-policy way, the SCM has no function regarding overall ARRL policy. The SCM as such is not a subordinate of the director. They serve two entirely different, although allied, functions.

The number of geographical divisions of ARRL sections can change from time to time, but this seldom occurs and the operating field organization remains fairly stable from year to year. Following is a breakdown of sections within each division and counties within each split-state section:

Atlantic Division: *Delaware; Eastern Pennsylvania* (Adams, Berks, Bradford, Bucks, Carbon, Chester, Columbia, Cumberland, Dauphin, Delaware, Juniata, Lackawanna, Lancaster, Lebanon, Lehigh, Luzerne, Lycoming, Monroe, Montgomery, Montour, Northhampton, Northumberland, Perry, Philadelphia, Pike, Schuylkill, Snyder, Sullivan, Susquehanna, Tioga, Union, Wayne, Wyoming, York); *Maryland-D.C.; Southern New Jersey* (Atlantic, Burlington, Camden, Cape May, Cumberland, Gloucester, Mercer, Ocean, Salem); *Western New York* (Allegany, Broome, Cattaraugus, Cayuga, Chautauqua, Chemung, Chenango, Clinton, Cortland, Delaware, Erie, Essex, Franklin, Fulton, Genesee, Hamilton, Herkimer, Jefferson, Lewis, Livingston, Madison, Monroe, Montgomery, Niagara, Oneida, Onondaga, Ontario, Orleans, Oswego, Otsego, St. Lawrence, Schoharie, Schuyler, Seneca, Steuben, Sullivan, Tioga, Tompkins, Wayne, Wyoming, Yates); *Western Pennsylvania* (those counties not listed under Eastern Pennsylvania).

Central Division: *Illinois, Indiana, Wisconsin.*

Dakota Division: *Minnesota, North Dakota, South Dakota.*

Delta Division: *Arkansas, Louisiana, Mississippi, Tennessee.*

Great Lakes Division: *Kentucky, Michigan, Ohio.*

Hudson Division: *Eastern New York* (Albany, Columbia, Dutchess, Greene, Orange, Putnam, Rensselaer, Rockland, Saratoga, Schenectady, Ulster, Warren, Washington, Westchester); *N.Y.C.-L.I.* (Bronx, Kings, Nassau, New York, Queens, Richmond, Suffolk); *Northern New Jersey* (Bergen, Essex, Hudson, Hunterdon, Middlesex, Monmouth, Morris, Passaic, Somerset, Sussex, Union, Warren).

Midwest Division: *Iowa, Kansas, Missouri, Nebraska.*

New England Division: *Connecticut; Maine; Eastern Massachusetts* (Barnstable, Bristol, Dukes, Essex, Middlesex, Nantucket, Norfolk, Plymouth, Suffolk); *New Hampshire; Rhode Island; Vermont; Western Massachusetts* (those counties not listed under Eastern Massachusetts).

Northwestern Division: *Alaska, Idaho, Montana, Oregon, Washington.*

Pacific Division: *East Bay* (Alameda, Contra Costa, Lake, Napa, Solano); *Nevada; Pacific* (Hawaii and U.S. possessions in the Pacific); *Sacramento Valley* (Alpine, Butte, Colusa, El Dorado, Glenn, Lassen, Modoc, Nevada, Placer, Plumas, Sacramento, Shasta, Sierra, Siskiyou, Sutter, Tehama, Trinity, Yolo, Yuba); *San Francisco* (Del Norte, Humboldt, Marin, Mendocino, San Francisco, Sonoma); *San Joaquin Valley* (Amador, Calaveras, Fresno, Kern, Kings, Madera, Mariposa, Merced, Mono, San Joaquin, Stanislaus, Tulare, Tuolumne); *Santa Clara Valley* (Monterey, San Benito, San Mateo, Santa Clara, Santa Cruz).

Roanoke Division: *North Carolina, South Carolina, Virginia, West Virginia.*

Rocky Mountain Division: *Colorado, Utah, New Mexico, Wyoming.*

Southeastern Division: *Alabama; Georgia; Northern Florida* (Alachua, Baker, Bay, Bradford, Calhoun, Citrus, Clay, Columbia, Dixie, Duval, Escambia, Flagler, Franklin, Gadsden, Gilchrist, Gulf, Hamilton, Hernando, Holmes, Jackson, Jefferson, Lafayette, Lake, Leon, Levy, Liberty, Madison, Marion, Nassau, Okaloosa, Orange, Pasco, Putnam, Santa Rosa, Seminole, St. Johns, Sumter, Suwanee, Taylor, Union, Volusia, Wakulla, Walton, Washington); *Southern Florida* (those counties not listed under Northern Florida); *West Indies* (Puerto Rico and U.S. possessions in the Caribbean.

Southwestern Division: *Arizona; Los Angeles; Orange* (Inyo, Orange, Riverside, San Bernardino); *San Diego* (Imperial, San Diego); *Santa Barbara* (San Luis Obispo, Santa Barbara, Ventura).

West Gulf Division: *Northern Texas* (Anderson, Andrews, Archer, Armstrong, Bailey, Baylor, Bell, Borden, Bosque, Bowie, Briscoe, Brown, Callahan, Camp, Carson, Cass, Castro, Cherokee, Childress, Clay, Cochran, Coke, Coleman, Collin, Collingsworth, Comanche, Cooke, Coryell, Cottle, Crane, Crosby, Dallam, Dallas, Dawson, Deaf Smith, Delta, Denton, Dickens, Donley, Eastland, Ector, Ellis, Erath, Falls, Fannin, Fisher, Floyd, Foard, Franklin, Freestone, Gaines, Garza, Glasscock, Gray, Grayson, Gregg, Hale, Hall, Hamilton, Hansford, Hardeman, Harrison, Hartley, Haskell, Hempshill, Henderson, Hill, Hockley, Hood, Hopkins, Howard, Hunt, Hutchinson, Irion, Jack, Johnson, Jones, Kaufman, Kent, King, Knox, Lamar, Lamb, Lampasas, Limestone, Lipscomb, Loving, Lubbock, Lynn, McLennan, Marion, Martin, Midland, Mills, Mitchell, Montague, Moore, Morris, Motley, Nacogdoches, Navarro, Nolan, Ochiltree, Oldham, Palo Pinto, Panola, Parker, Parmer, Potter, Rains, Randall, Regan, Red River, Roberts, Rockwall, Runnels, Rusk, Scurry, Shackelford, Shelby, Sherman, Smith, Somervell, Stephens, Sterling, Stonewall, Swisher, Tarrant, Taylor, Terry, Throckmorton, Titus, Tom Green, Upshur, Upton, Van Zandt, Ward, Wheeler, Wichita, Wilbarger, Winkler, Wise, Wood, Yoakum, Young); *Oklahoma; Southern Texas* (those counties not listed under Northern Texas).

Canadian Division: *Alberta, British Columbia; Manitoba; Maritime* (Nova Scotia, New Brunswick, Prince Edward Island, Labrador, Newfoundland); *Ontario, Quebec, Saskatchewan.*

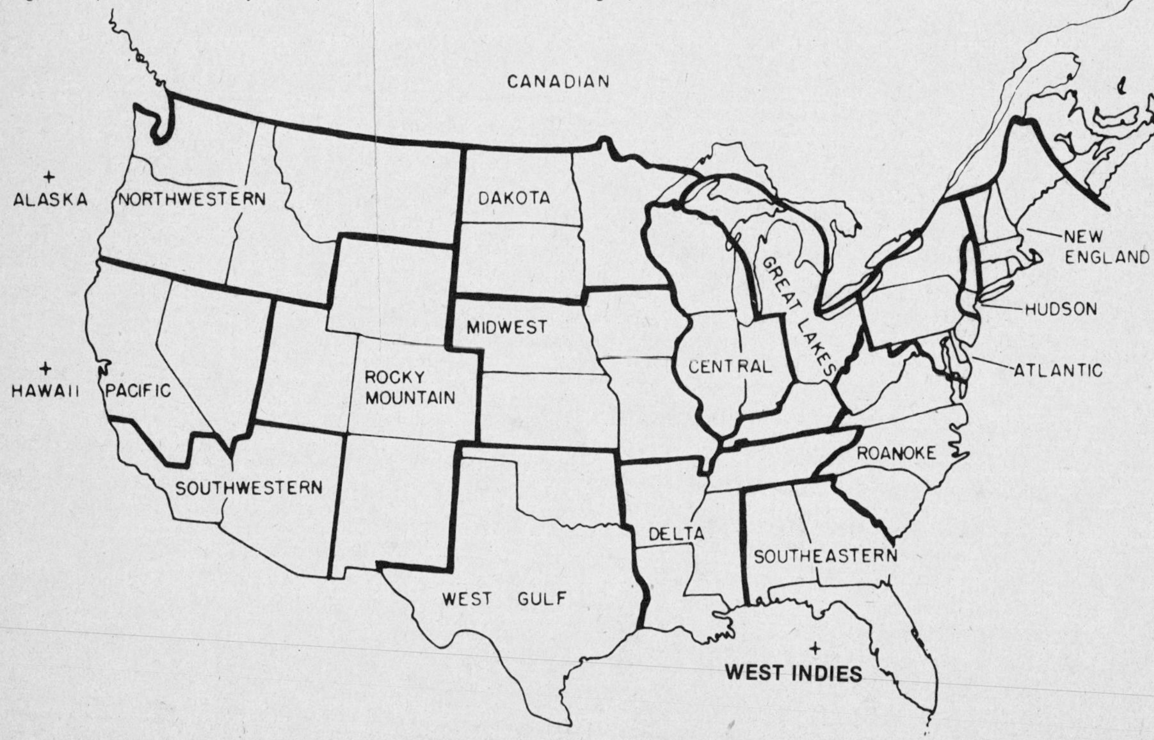

THE AMERICAN RADIO RELAY LEAGUE, INC.

ADMINISTRATIVE HEADQUARTERS NEWINGTON, CONNECTICUT, U S A 06111

TO SCM IN THE .... SECTION OF ARRL. (Addresses on page 8, *QST*).

After reading the reverse side of this sheet, I feel that I can qualify and wish to apply for the following ARRL station appointment(s):

☐ OTS ☐ OTS-II ☐ OES ☐ OVS ☐ OBS ☐ OO

Full name ______________________________ Call ______________

The following address should be the same as your ARRL/QST mailing address.

Number and Street ______________________________

City, State/Province, ZIP/Postal Code ______________________________

ARRL membership expiration date __________ Current age ________ Home Telephone ____________

Class of present FCC/DOC license ______________ Date of FIRST license ______________

BANDS WORKED AND EQUIPMENT USED IN MY PRESENT STATION
(use additional sheets, as necessary)

BAND	TRANSMITTER	RECEIVER	ANTENNAS
MHz			
MHz			
MHz			
MHz			
MHz			

Current code speed __________ wpm Net participation ______________

Modes used ______________ Experimental work on ______________

Measuring equipment ______________________________

____

Achievements and certificates ______________________________

Current appointments __________. I'm a member of ____ ARES ____ NTS ____ MARS ____ RACES.

OFFICIAL BULLETIN STATION SCHEDULE

Day	Frequency	Mode	Time

Day	Frequency	Mode	Time

I understand that my appointment requires continuing activity, regular reports and maintenance of my ARRL membership as prequisite to endorsement, and that it may be suspended or cancelled at the discretion of the SCM for violation of any of the bases for appointment.

DATE ______________ SIGNATURE ______________________________ CALL ______________

CD-187 (578)

INDEX

When YOU Want

SERVICE BEYOND SUBSCRIPTION

You Need the American Radio Relay League

Intruder Watch
Hq. for the International Amateur Radio Union
Representation before government agencies
Inexpensive Equipment Insurance
Operating Awards and on the air activities
Low cost outgoing QSL Bureau and free incoming bureau
QST, the monthly Journal for Amateur Radio
Field Organization for public service
W1AW Code Practice and Bulletins
TVI/BCI and Tower Ordinance assistance
Reciprocal Operating guidance
Licensing Training Aids
OSCAR Education Program
Blind and Handicapped help
Operating Aids
Club liaison
Amateur Radio Publicity through mass media
Technical Information Service
ARRL Lab research and development
Testing of Commercial Products

...AND MORE

"Since 1914, Of, by and for the Amateur"

HELP US SERVE YOU

MEMBERSHIP APPLICATION

Name ______________________ Call ____________

Street ______________________________________

City ____________ State/Prov. ____________ Zip/PC ____________

1 year
$18.00 U.S. ____________ Enclosed Visa No. ____________
$20.00 Canada Master Charge No. ____________
$21.00 Elsewhere Bank Code ____________

ARRL 225 Main Street Newington, CT 06111

OP

Membership and QST cannot be separated. Fifty percent of dues is allocated to QST, the balance for membership.